T0220066

Problems of
Instrumental Analytical Chemistry
A Hands-On Guide

Essential Textbooks in Chemistry   ISSN: 2059-7738

Essential Textbooks in Chemistry

Problems of
Instrumental
Analytical Chemistry
A Hands-On Guide

**JM Andrade-Garda • A Carlosena-Zubieta
MP Gómez-Carracedo • MA Maestro-Saavedra
MC Prieto-Blanco • RM Soto-Ferreiro**

University of A Coruña, Spain

World Scientific

NEW JERSEY • LONDON • SINGAPORE • BEIJING • SHANGHAI • HONG KONG • TAIPEI • CHENNAI • TOKYO

Published by

World Scientific Publishing Europe Ltd.
57 Shelton Street, Covent Garden, London WC2H 9HE
Head office: 5 Toh Tuck Link, Singapore 596224
USA office: 27 Warren Street, Suite 401-402, Hackensack, NJ 07601

Library of Congress Cataloging-in-Publication Data
Names: Andrade-Garda, José Manuel.
Title: Problems of instrumental analytical chemistry : a hands-on guide / by J.M. Andrade-Garda
    (University of A Coruña, Spain) [and five others].
Description: New Jersey : World Scientific, 2016. | Series: Essential textbooks in chemistry
Identifiers: LCCN 2016036786| ISBN 9781786341792 (hc : alk. paper) |
    ISBN 9781786341808 (pbk : alk. paper)
Subjects: LCSH: Chemistry, Analytic. | Spectrum analysis. | Chromatographic analysis.
Classification: LCC QD75.22 .P76 2016 | DDC 543--dc23
LC record available at https://lccn.loc.gov/2016036786

British Library Cataloguing-in-Publication Data
A catalogue record for this book is available from the British Library.

Desk Editors: Herbert Moses/Mary Simpson

Typeset by Stallion Press
Email: enquiries@stallionpress.com

Printed in Singapore

PREFACE

That the world is changing very fast is an understatement in the digital age. How this should translate into higher education and teaching, however, is a subject which is up for much debate. Universities are responsible for training students in the development of their professional careers, however without a doubt it is impossible to teach every concept within the short amount of time allocated to each course. The increasing complexity of the sciences means that students need to understand deeper and greater fundamental concepts in order to have the confidence to face different working environments in laboratories, hospitals and offices. Their backgrounds and basic knowledge must be sound and deeply rooted to enable the ability of applying concepts and ideas to new situations.

The strongest recent effort undertaken by Europe to integrate higher education has been the so-called European Credit Transfer and Accumulation System (ECTS). This is a tool of the European Higher Education Area (EHEA) to make courses more transparent and thus help to enhance the quality of higher education. From a pragmatic viewpoint (which, unfortunately, we feel has been the main underlying reason), it is a way to homogenize university degrees throughout the European Union in order to get a free exchange of professionals. In many countries (including Spain, Italy, France, and Portugal), the ECTS system implied a (dramatic) reduction in the time available for teaching and increased the requirement for the student to implement individual training and study.

This sounds reasonable in theoretical terms but it is very hard to achieve in practice. Students need to attend laboratory and theoretical classes, develop oral presentation skills and find solutions for numerical exercises for several subjects within the semester — a difficult task for teachers confronted with less contact time. Teachers are — somehow — forced to reduce content on fundamental concepts in order to include new skills and updated chapters for their subjects within the time allocated. Impetus for learning therefore lies largely with the student, a big ask when looking at the complexity and level of knowledge needed to be able to understand even basic theories.

A constant comment in almost every higher education conference is the reference to the low level of background understanding of the fundamentals of chemistry undergraduates demonstrate daily — this begs the question of at which point and for what reason is the system failing you, the student?

We identified a gap in current literature on analytical chemistry whereby there was a lack of dedicated exercises for students to apply what they had learnt. Feedback from you throughout teaching was that examples and information found in individual study were useful, but did not provide enough opportunity for practice and further learning. We as teachers attempted to provide in-class examples and exercises, however time constraints and different learning needs meant that these were not as useful as they could be outside the classroom. We decided at that point to create a learning resource which you could use during individual study to learn the basics of instrumental analytical study — this book was conceived.

The first problem we faced was to limit its extent — we did not want an overwhelming textbook full of incomprehensible equations. The many instrumental techniques and possible content had to be restricted somehow. We considered two issues: First, we knew about very good books dealing with numerical exercises on the electroanalytical fields. Similarly, classical methods (such as titimetry or gravimetry) are very well treated in many general textbooks.

Accordingly, we could avoid considering these matters within the first edition. We then decided to focus on exercises based on typical instrumental analytical techniques, as they are considered in almost every fundamental undergraduate training throughout Europe. With this in mind, we started off on our task to write a clear, cohesive textbook which is designed to be used in undergraduate study of chemistry and the chemical sciences.

Thanks must be given here to Merlin Fox and Mary Simpson for their kind help, their continuous support and patience. The editorial staff of World Scientific Publishing are also acknowledged for their wonderful and professional work.

The chapters are organized so that a general review of the basic concepts addressed are presented first. Here, general explanations, equations and guidelines to study separate topics are given. Then, a set of example questions are shown and solved in detail and, finally, a set of exercises are presented, designed to test students' learning.

A recommendation is in order just here. We strongly encourage you to try to solve the numerical exercises before reading their solutions. You will learn much more effectively. In some occasions, we know the exercises are particularly difficult, and hints are included to help you solve them. We hope they are useful.

Last, although far from least, we would like to thank all students that have suffered us throughout our years' teaching. We are proud to see many of them in relevant and high profile positions. Sometimes they visit us (or we visit them to ask for collaboration with the university) and reflect on education in the sciences, and the learning of young graduates. Time goes by ... for them as well as us (!) and they hardly remember themselves as undergraduates in our classrooms. They complain about the low background of young graduates and encourage us to be less permissive. Yes, it is true that university teachers should demand a minimum (unchanging) background on the disciplines they teach. However, this is more easily said than done, particularly in this fast-changing environment.

We hope this textbook can help both us and undergraduates to keep an eye on relevant issues that we should teach/learn. Who knows, maybe even postgraduates can take advantage of exercises for enhanced learning and/or on-the-job training of their staff. That would be a marvelous use of this textbook they helped create.

To close this preface, the authors want to encourage readers of this book to contact us if they detect errors or if they have a say on particular sections. If you would like to pose a particular numerical exercise because you feel it is relevant, or an explanation that would help students, we would be delighted to include that contribution in future versions (citing the source of course!)

Thanks to all and good luck!

JM Andrade, A Carlosena, MP Gómez,
MA Maestro, MC Prieto and RM Soto

ABOUT THE AUTHORS

José Manuel Andrade-Garda is a full professor at the University of A Coruña (Galicia, Spain) since 2011. He was in charge of teaching different subjects on analytical chemistry since 1995. His main interests are quality control and chemometrics, in particular, multivariate regression and pattern recognition methods (either, unsupervised and supervised). He works on infrared spectrometry and atomic spectrometry. There, he applied both optimization techniques and multivariate regression tools to cope with spectral and chemical interferences in ETAAS. In the infrared spectrometry arena, he developed analytical methods for the petrochemical field. He published around 100 papers in peer-reviewed journals, co-authored 12 chapters of books and edited two editions of a book with the RSC.

Alatzne Carlosena-Zubieta is an associate professor of analytical chemistry at the University of A Coruña, where she is in charge of several subjects on analytical chemistry since 1994. She currently works on atomic spectrometry, particularly quantitation of trace metals in samples of environmental interest. She published many papers on slurry-sampling ETAAS. There, robust analytical protocols were developed based on multivariate chemometric optimization methods (namely, experimental design and simplex optimization). She published around 60 papers in peer-reviewed journals and co-authored five chapters of books, including two editions of an RSC book on atomic spectrometry.

María Paz Gómez-Carracedo is a post doctoral researcher at the Department of Analytical Chemistry, University of A Coruña. She got her PhD in 2005 with a special award. Her research areas cover characterization of refinery products by infrared spectrometry and multivariate regression modeling, particularly partial least squares and artificial neural networks. She also works on variable selection by genetic algorithms. She taught analytical chemistry issues at the Universities of Vigo and A Coruña. She published around 30 papers in peer-reviewed journals and co-authored 10 chapters of books, including two editions of an RSC book on atomic spectrometry. At present, she is a teacher at a secondary school.

María del Carmen Prieto-Blanco is an assistant professor at the University of A Coruña since 2004. There, she is in charge of teaching different subjects on analytical chemistry. Her research is focused on developing analytical methods for determining organic compounds in the environment (emerging and persistent pollutants and surfactants), using liquid chromatography in combination with different treatment systems. At present, she works in automated sample preparation techniques coupled to miniaturized chromatographic systems (CapLC), and the development of *in situ* analysis devices for environmental analysis and food analysis. She was a quality control laboratory manager in a fine chemistry and surfactants industry for 17 years. She published around 20 papers in peer-reviewed journals and co-authored two chapters or books.

Rosa María Soto-Ferreiro is an associate professor of analytical chemistry at the University of A Coruña, where she is in charge of different subjects on analytical chemistry. She has been working on atomic spectroscopy since 1989. In 1994, she obtained her PhD, where she developed new analytical procedures to determine butyltin compounds in marine waters by HPLC–ETAAS. Her interests focus on sample treatment procedures for metal determination in petroleum products, development of green analytical methods using emulsions and pressurized liquid extraction. She also works with inductively coupled plasma mass spectrometry (ICP-MS). She published around 20 papers and co-authored a chapter of an RSC book on atomic spectrometry.

Miguel Angel Maestro-Saavedra is an associate professor of organic chemistry at the University of A Coruña. There, he is in charge of teaching different subjects on organic chemistry and structural elucidation since 1996. His fields of expertise are organic synthesis and structural elucidation. In the former, his main interests are in preparing vitamin D metabolites and analogs and isotopically labeled compounds. In the second field, he contributed to the foundation of the laboratories for scientific services of the University of A Coruña, being its first Director (1993–2004). He currently applies spectroscopic and spectrometric techniques (NMR, MS, UV, single crystal XRD) to characterize natural products and synthetic intermediates. He published around 140 papers in peer-reviewed journals.

CONTENTS

Chapter 2. Basic Data Analysis   43

José Manuel Andrade-Garda and María Paz Gómez-Carracedo

Chapter 5. Atomic Spectrometry   263

Rosa María Soto-Ferreiro and Alatzne Carlosena-Zubieta

Chapter 6. Chromatographic Techniques   333

María del Carmen Prieto-Blanco

**Chapter 7. Nuclear Magnetic Resonance and
Mass Spectrometries** **379**

Miguel Angel Maestro-Saavedra

CHAPTER 1

FUNDAMENTAL CALCULATIONS IN ANALYTICAL CHEMISTRY

Alatzne Carlosena-Zubieta and José Manuel Andrade-Garda

OBJECTIVES AND SCOPE

This chapter summarizes the units and the expressions of concentrations employed more often in the calculations associated to the analytical techniques presented in this book. How to (inter) convert them is also discussed, with special emphasis on conversion factors and mathematical relations between expressions/definitions. In our experience, this is a point where students present difficulties frequently throughout their B.S. degree. Another main objective here is to present a systematic approach for the student to deliver a final result taking into account every working step of the operational treatment of the samples or standards (drying, dissolution, dilution, concentration, digestion, etc.). This is an issue that will be reinforced in many examples throughout the other chapters contained in this book.

1. INTRODUCTION

It is critical for students to be aware of the importance of using the measurement units properly, to express the concentrations correctly and to consider the adequate number of significant figures. Throughout their professional careers, they will unavoidably have to master every operation related to scientific notation, conversions among units and relationships between concentration forms. The adequate development of any analytical procedure depends on that, from its inception and planning to the preparation of the reagents, the choice of suitable laboratory material and the final result and its interpretation.

The calculations reflect the different stages of the analytical process and, so, the more complex the analytical procedures become, the more intertwined the required calculations will be. In general, calculations are not too complex although it is essential to understand what is going on chemically in each step of the analysis. This would allow students to establish the appropriate mathematical relationships between the various stages of the analytical process.

To achieve satisfactory learning outcomes, you should be able to understand and justify every operation or calculation rather than only performing them mechanically, so that you yourself should be able to detect any gross error in the final solution. There is no single 'magic' formula to solve all exercises. Students have to learn problem-solving strategies and start developing some chemical intuition ('the chemical criterion'). This means that they must develop critical thinking skills to succeed in their professional lives.

A fundamental starting point that falls outside the scope of this book is that the students should know how to formulate (and name) all the compounds involved in the exercises to correctly write and balance the chemical reactions. (S)He has to know how to perform the necessary basic stoichiometric calculations. Without this essential background, (s)he will not achieve satisfactory results in their degrees or, worse, they will fail as chemists. When the compounds mentioned into the numerical exercises are not of common use, their chemical structure will be shown. This will be particularly so in Chapter 6.

1.1. Relevant units and expressions of concentration

Initially, it is worth starting this chapter by presenting some basic recommendations. As a general rule, scientific notation must be used in order to avoid working with very large or very small numbers, being advisable to obtain a number between 1 and 10, and to express its magnitude through an exponent. Thus, large numbers have positive exponents and small numbers have negative exponents.

Another way for chemists to avoid using very large or very small numbers is by selecting the most suitable measuring unit or concentration expression for each number. Furthermore, the units of measurement have a major role in making the numbers meaningful. In effect, a number itself makes little sense. If you read in a laboratory manual: mix 1 of sodium carbonate and 2 of sodium hydrogen carbonate, what would you do? You should ask for the lost units!

1.1.1. Units

From the previous discussions, it turns out that the value of a quantity should be reported as the product of a number and a unit. The number multiplying the unit is the numerical value of the quantity expressed in that unit.

For scientific measurements, the most convenient metric system is the *International System of Units (Système International d'Unités)* with the international abbreviation **SI**, which is used worldwide, although in the United States of America and most countries associated in the Commonwealth, the so-called *English system* is applied frequently. This uses traditional units such as inches, yards and pounds instead of the metric system such as centimeters, meters and kilograms (see Table 1 for some examples).

The SI defines a set of basic units, prefixes and derived units. All related information is exhaustively compiled and updated at the official website of the responsible organization, the *Bureau International des Poids et Mesures*, BIPM [1].

Following International Union for Pure and Applied Chemistry (IUPAC) recommendations [2], the symbol of the unit is placed after the numerical value, separated by a space (i.e., $t = 25$ mL instead of 25mL); it is written in roman (upright) type (km instead of *km*); it should remain unaltered in the plural (i.e., 10 cm, not 10 cms) and

Table 1. Some common units and equivalences between the SI and the English system.

	International system (SI)	English system
Mass	1 kilogram (kg)	2.205 pounds (lb)
	453.59 grams (g)	1 pounds (lb)
	28.35 grams (g)	1 ounce (oz)
Volume	1 liter (L)	1.057 quarts (qt)
	3.785 liters (L)	1 U.S. gallon (gal)
	100 milliliters (mL)	6.10 cubic inch (in^3)

should not be followed by a full stop (but at the end of a sentence). The symbols should be printed in lowercase letters, except for those derived from a personal name. An exception is the symbol for the liter, which can be uppercase (L, more advisable) and lowercase (l), to avoid its confusion with a one.

It is also advisable not to mix information with unit symbols. For example, the form 'the copper content is 5 mg/kg' should be used instead of '5 mg Cu/kg' or '5 mg of copper/kg'. When mathematical operations are performed, it should be made clear to which unit symbol a numerical value belongs to. Thus, 12 cm × 3 cm is used and not 12 × 3 cm; 54 mg ± 1 mg or (54 ± 1) mg, but not 54 ± 1 mg [3].

Commas are not used to separate digits into groups of three but a plain space, counting from both the left and the right of the decimal symbol. For example, 5 425.123 12 instead of 5,425.12312 is preferred.

When scientific measurements or results are reported, the number should be written according to some additional criteria as, for instance, the number of significant figures. The result should only contain digits which are considered 'correct' except for the last digit, which indicates that such digit is not 'certain', but estimated. Students should be familiar with the basic rules for setting significant figures for the results, as well as their rounding. These issues are not considered here due to space restrictions but the student can find them in almost all classical textbooks on analytical chemistry.

1.1.2. Some important terminologies

In this section, we will try to summarize different terminologies intended to indicate the concentration of a mixture, solution, sample, etc. Please bear in mind that not always the recommendations from competent organizations, as NIST or IUPAC, are applied in every academic and scientific text. Nevertheless, we will resume here, as practically as possible, the most widely used terms, as well as some indications on their limitations.

The term *concentration* is related to the composition of a mixture in which we need to describe the amount of one or more substances ('constituents' or 'components' are preferred terms) that are present. According to IUPAC, the term concentration includes four quantities characterizing the composition of a mixture with respect to the volume of the mixture [4]: mass, amount, volume and number concentration. Also, the term concentration is a shortcut for *amount-of-substance concentration* or *amount concentration*.

The SI unit for the amount of substance is the mole (n): defined as the *amount of substance of a system which contains as many elementary entities as there are atoms in* 0.012 *kilogram of carbon-12* [2]. IUPAC recommends that this physical quantity ('amount of substance') is no longer called 'number of moles', just as the physical quantity 'mass' should not be called 'number of kilograms'. This unit is selected so that the *atomic mass* (expressed in the unified atomic mass unit, u) and the *molar mass* (in grams/mol) have the same numerical value (*Note*: the abbreviation amu is not an acceptable unit symbol nowadays for the unified atomic mass unit).

In general, the calculations that involve moles of a substance require the use of the molar mass of that substance. The terms *atomic weight* and *molecular weight* are obsolete and, thus, should be avoided. They have been replaced by the equivalent but preferred terms *relative atomic mass* and *relative molecular mass*, respectively, which are simple dimensionless numbers. When they are expressed as g/mol, they are designated as *molar mass*, represented by M, for both atoms and molecules.

The concentration or *amount-of-substance concentration* (c_A) is the amount of a substance, A, divided by the volume of the mixture in which it is present: $c_A = n_A / V$. The common unit is mol/dm^3 or mol/L. Molarity (M) is commonly used instead of these terms. Thus, the molarity for a 0.5 mol/dm^3 solution is said to be 0.5 M (0.5 molar solution). However, NIST's Guide [3] prefers the original term because it is unambiguous; they recommend that the term molarity and the M symbol should no longer be used because they are obsolete. However, its use is so rooted and broadly extended in chemistry that this change will, likely, be difficult.

Related terms are: (i) *substance content*, defined as the amount-of-substance of a component (n) divided by the mass of the system (m); (ii) *number concentration* (C), the number of entities of a constituent in a mixture divided by the volume of the mixture; (iii) *mass concentration*, mass of a constituent divided by the volume of the mixture; (iv) *volume fraction*, volume of a constituent of a mixture divided by the sum of volumes of all constituents before mixing them.

The relationship between all these terms is summarized in Figure 1. In general, the term level as a synonym for content is obsolete.

1.1.3. Specifying the concentrations of solutions

A *solution* is a liquid or a solid phase containing more than one substance. One of them is called the solvent, whereas the other(s) is(are) called solute(s). When the sum of the mole fractions of solutes is small compared with unity, as it is frequently (but not necessarily) the case, the solution is called a *diluted solution*. That is, the solutes are the minor constituents and the solvent is the major constituent. A *concentrated solution* contains large amounts of solute relative to the solvent.

A *standard solution* is a solution of accurately known concentration, prepared using standard substances in one of several ways.

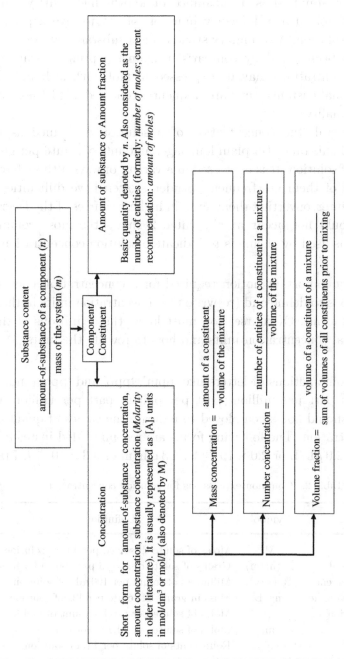

Figure 1. Definitions of the main terms related to the composition of a mixture (adapted from Ref. [4]).

A primary standard is a substance of known high purity which may be dissolved in a known volume of solvent to give a primary standard solution. A secondary standard is a substance whose active agent has been found by comparison against a primary standard. Their concentrations may be expressed in mol/dm^3, although due to traditional customs, they are frequently reported as M (molarity) or N (normality).

In general, the concentration of a solution is defined as the amount-of-substance (in plain language, amount) of solute per given amount of solution. It is expressed in a variety of ways although only a handful of them are frequent. Students often have difficulties in correctly interconverting them, even in highest courses of the Degree, which should not occur at all. Table 2 shows the most common expressions used by chemists to indicate a solution concentration.

The most important notion required for a concentration conversion step is to define and (re)write the concentration units in their simplest forms. Of course, you must know the definitions of the individual concentration units and how to rewrite the units.

The combinations of letters like 'ppm', 'ppb' and 'ppt' (and, so, the terms part per million, part per billion, part per trillion and the like) should not be employed to indicate the values of quantities or concentration. The following forms are recommended instead: for example, $2.0\,\mu L/L$ or 2.0×10^{-6} V, $4.3\,nm/m$ or 4.3×10^{-9} l, $7\,ps/s$

Table 2.   Common expressions for solution concentration.

Expression	Symbol	Definition
Molarity	M	Moles of solute dissolved per 1 L of solution
Mass percent	% (m/m)	Grams of solute dissolved per 100 g of solution
Volume percent	% (v/v)	Milliliters of solute per 100 mL of solution
Mass per volume	mg/L	Mass in grams of solute per 1 L of solution
Mole fraction	χ	Moles of solute per total amount of moles
Molality	m	Moles of solute per 1 kg of solvent
Normality	N	Equivalents of solute per 1 L of solution.

or $7 \times 10^{-12} t$, where V, l and t are the adequate quantity symbols for volume, length, and time respectively [4].

Some old expressions for the concentration are ubiquitous since they are still present in many fields such as regulations, textbooks and scientific reports, to cite but a few. Thus, as it was commented already for the molarity, the term *normality* and its symbol, N, should no longer be used because they are obsolete. The main limitation of this term is that there is no unique expression for its calculation because it is a function of the reaction that takes place, which usually represents a big difficulty for students. However, it is very useful and convenient to carry out many calculations. Two typical examples are the Faraday's statements on electrical deposition and the calculations associated to many electroanalytical measurements.

Normality is an old term in chemistry based on the concept of equivalence between the amounts of reacting substances. It has played a fundamental role in the history of quantitative analysis (all reactions take place on an equivalent-to-equivalent basis). Today, its role in titrimetric analysis and electrochemistry studies is still very important. The equivalent of a species must be specified with respect to a definite reaction. For instance:

(i) For acid–base reactions, 1 equivalent equals 1 mole of hydrogen ions donated or accepted.
(ii) For oxidation–reduction reactions, 1 equivalent equals 1 mole of electrons.
(iii) For determining electrolyte concentration, 1 equivalent equals 1 mole of electrical charge.

A direct relation between molarity and normality is derived easily considering their definitions: $N = M \cdot K$, where K denotes the amount of equivalents (corresponding to moles of hydrogen ions, electrons or charge, depending on the type of reaction). To reach this formula, the term equivalent weight (EW) must be considered as the mass of a compound that contains an equivalent (e.g., the equivalent weight of an acid is the mass of acid required to generate one mole of hydrogen ion — hydronium — in aqueous solution). It is calculated as: EW = M/amount of equivalents.

The use of percentages also leads to confusing situations for students, mainly due to the diversity of terms related to this expression: percentage by weight, percentage by mass, percentage by volume, percentage by amount of substance, percentage solution, etc.; as well as the symbology to denote them: % (by weight), % (by volume), % (mol/mol), etc. Thus, somehow contrary to common practices in laboratories, NIST [3] discourages the use of phrases like 'percentage by weight', 'percentage by mass', and the like. Similarly, one should avoid writing, for instance, '% (m/m)', '% (by weight)', '% (v/v)', '% (by volume)', or '% (mol/mol)'. The preferred forms are 'the mass fraction is 0.10' or 'the mass fraction is 10 %', or 'the amount-of-substance fraction is 0.10', or 'the amount-of-substance fraction is 10 %'.

In the next sections, the term 'percent solution' will be always related to either 'mass percent' (often designed by m/m) or 'volume percent' (v/v), avoiding other ambiguous terms. The resulting number is dimensionless. The term mass/volume percent (m/v) is discouraged because numerator and denominator have different units and, therefore, this concentration unit is not a true relative unit. However, it is often employed as a simple-to-use concentration unit since volumes of solvent and solutions are easier to measure than weights. Moreover, since the density of dilute aqueous solutions is close to $1 \, g/cm^3$, if the volume of a solution is measured in mL (as per definition), then this well approximates the mass of the solution in grams (making a true relative unit) [5].

1.2. Preparing solutions from solids or liquids

Solutions are commonly made in the laboratory from solid materials, from liquids or from other solutions. It is very advisable to use the correct terminology in each instance. Thus, the process of dissolving a solute in a solvent is called *dissolution* (dissolving) and when the solution is prepared from a liquid, the process is called *dilution* (diluting).

Note that when solid materials or reagents are considered, a solvent will only dissolve a limited quantity of solute, at a given temperature, according to the salt solubility (a physical property of the salt). However, the rate at which the solute dissolves can be accelerated by triturating the solid, heating the solvent and/or agitating the mixture.

To calculate the mass of a reagent required to prepare a solution, it is very important to consider whether that reagent has water molecules of hydration. If it does, they must be included in the molar mass because, otherwise, a serious error (by default) is committed.

In many cases, solutions are prepared from a liquid reagent or stock solution by addition of a known amount of solvent, usually deionized water. Thus, a volume of a stock solution is combined with the appropriate volume of solvent to achieve the desired concentration in the diluted solution.

Usually, a very simple, common equation is applied to perform the calculations related to dilution. It combines the concentrations and volumes of both the concentrated and the diluted dissolutions, and considers that the amount of solute that is finally present in the diluted solution proceeds uniquely from the volume of the concentrated stock solution. Figure 2 shows the basic operations implied in this typical procedure.

$$C_{\text{concentrated}} \cdot V_{\text{concentrated}} = C_{\text{diluted}} \cdot V_{\text{diluted}}$$
$$C_1 \cdot V_1 = C_2 \cdot V_2$$

Although it might be ambiguous and confusing for students, the notation ':' is still used to designate dilutions (this option is not recommended by NIST). For example, a 1:10 dilution (verbalize as '1 to 10' dilution) entails combining one unit volume of the solution to be diluted plus nine unit volumes of the solvent.

An important caveat is not to confound the act of preparing a solution by mixing parts or volumes of different solutions (even deionized water) with a simple dilution step. In doubt, ask always to

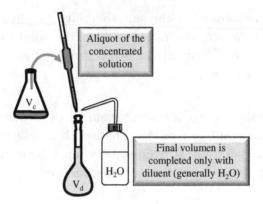

Figure 2. Basic operations for dilution.

avoid misunderstanding. Thus, to prepare a 1:4 acetic acid–ethanol solution, you should mix one unit volume of acetic acid and four unit volumes of ethanol. However, to make a 1:4 dilution of acetic acid in ethanol, you should mix one unit volume of acetic acid with three unit volumes of ethanol (to make four unit volumes of solution).

Another aspect that may mislead students is that many times, the volume to prepare the final solution is not indicated explicitly. They must decide it in accordance to the needs of the assay.

1.3. General operational treatments before the analysis

Another embarrassing question for students is usually associated to the calculations involved in multistage sample (standard) treatment to bring it to a state suitable to perform an experimental measurement. For example, the sample is solid and we have to analyze a solution which is obtained after applying a series of treatments. This will be discussed in this section and many examples will be proposed throughout Chapters 5 and 6, mainly.

When we consider the analysis of a *sample* (consider this term in a chemical sense, not in a statistical one as in Chapter 2), we are thinking on withdrawing a portion of material from a larger quantity of that material and, then, analyzing only that portion. But that term can be ambiguous if we do not specify what type of sample we are considering; namely, bulk sample, representative sample, primary sample, test sample, etc. The existence of a sampling error when obtaining such portion has to be assumed, which means that the results gathered from different portions extracted from the bulk (or lot) are only estimates of the true concentration of the constituent of interest in it. The portion removed from the bulk is called a *test portion* or *aliquot* (or specimen, in some cases).

The term *aliquot*, a known amount of a homogeneous material, is usually applied properly to fluids, although there are some difficulties when solids are considered. Sometimes, it matches with *test portion*, which is the amount or volume of the test sample taken for the analysis, usually of known weight or volume.

The solution prepared from the test portion to perform the analytical measurement is called *test solution* (proportions of the test portion and solvent are normally known). If the test solution has been subjected to reaction or separation procedures prior to measurement, the resulting solution is called *treated solution* [4].

In general, the objective of an analysis is to determine/quantitate an *analyte*. Thus, the analyte is the component of a system we are interested in, and which has to be measured [4]. It is differentiated from the other components of the sample which, in conjunction, is called *matrix*. *Qualitative analysis* involves identifying the analyte: atoms, molecules or functional groups. It indicates the *presence* of the analyte in the matrix. *Quantitative analysis* involves measuring the amount or concentration of the analyte in the sample. It indicates the magnitude of the property being measured and it commonly (but not necessarily) relates to the *concentration* of analyte in the matrix, it could be any other physical, chemical or biological property.

1.3.1. Drying

Many samples are wet originally and they must be dried before they can be analyzed. Sometimes, this step is mandatory because of the nature of the next step of the analytical procedure (grinding, sieving, digestion, etc.) and in other cases, it is made to get a dried sample with a constant mass to which the content of the analyte investigated in the analysis can be easily referred to.

Drying can also be required for chemical reagents prior their use. This is usually the case for hygroscopic salts which should be dried first in order to weight them accurately. Drying is very important whenever standard solutions are prepared from solid primary standards and its correct performance is crucial to achieve reliable results, because drying a substance under improper conditions can result in the loss of target elements (or even in the transformation to a different compound).

Constant weight is obtained by removing the water from the sample, that is, by drying it. Mathematically, the water content (moisture) of the sample can be calculated usually in the form of a mass percent (%). Experimentally, moisture is determined by measuring the mass of a sample before and after water is removed by drying. Then, we apply the mass percent expression, as follows:

$$\text{Moisture } (\%) = \frac{M_{\text{initial}} - M_{\text{dried}}}{M_{\text{initial}}} \cdot 100$$

Here, M_{initial} and M_{dried} are the mass of sample before and after drying, respectively. Their difference represents the water content of the sample.

1.3.2. Dissolution, digestion and extraction

Most methods of analysis necessitate a more or less simple (more often than not, complex) preparation of the sample. Only few direct methods allow the introduction of the sample without any preparation.

When solid samples are analyzed, and even some complex liquid matrices, a partial or total dissolution is needed prior to their instrumental analysis. A variety of 'sample treatment' methods have been developed and reported in literature, depending on the type of matrix, analyte, methodology, instrumentation used, etc. They can involve dissolution, digestion, extraction of the sample, and any combination among them.

In quantitative analysis, it is essential to consider mathematically all treatments of the sample to reach the objective, which is to determine the amount or concentration of analyte in the sample. The process can be schematized briefly as: a given mass of sample is submitted to a series of treatments to finally obtain a solution. This is the test portion which, when analyzed yields a result that must be converted through adequate calculations to the concentration of the constituent we are interested in.

The biggest difficulties that many students face consist often of relating each intermediate dissolution with the corresponding changes in analyte concentration, deciding if a particular treatment modifies the analyte concentration, i.e., whether it increases (decreases) the concentration.

In almost all circumstances, the solution to these problems is very simple: depict graphically the process you are studying, set each step and the numerical information you have, write each number with its correct units and identify somehow (by a letter, a number, etc.) each solution obtained during the analytical procedure and, finally, evaluate whether it is a dilution or a concentration. In this way, you will assure that each unit will be canceled out correctly, and check easily whether the correct units for your numerical result were obtained.

The steps of a basic analytical procedure to analyze a solid sample or complex liquid sample are presented in Figure 3a. Despite how complex it may appear, it can be simplified largely to perform the necessary calculations (Figure 3b).

Finally, remember that the enunciates of the numerical exercises will contain (almost all) data you will need to solve them (tables are contained in the introductory sections and students have to realize when they need them). Therefore, read them carefully and discriminate between queries and supplied information, otherwise mistakes occur frequently. For instance, in some exercises, the questions may be formulated in one direction or in another, that is, you may be asked to calculate the concentration in the sample

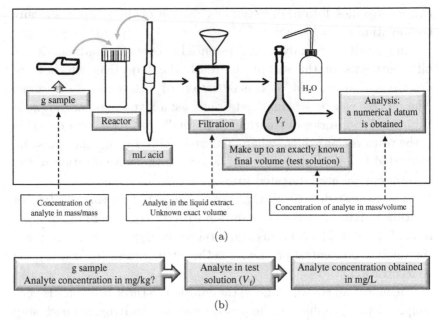

(a)

| g sample Analyte concentration in mg/kg? | Analyte in test solution (V_f) | Analyte concentration obtained in mg/L |

(b)

Figure 3. (a) Graphical representation of a typical sample treatment procedure and (b) its conceptual simplification to perform the final calculations.

or in the test portion. For both situations, the multistep conversion procedure seen above will be valid.

> Units are critical in calculations. You should always include units in your calculations, they will help you determine the correct solving approach. Converting from one unit to another is simply made by multiplying, dividing and canceling units like any other algebraic quantity.

1.3.3. Dilution and concentration of test solutions

Sometimes, when analyzing a test portion, the result gathered from the instrumental system is out of the calibration range where accuracy was validated (this is explained in more detail in Chapter 2). When this occurs, the test portion of the sample must be diluted, if it is above the limits; or concentrated, if it is below them. Next, they have to be measured again and the new result used to calculate the concentration of analyte in the original sample. Dilution and concentration factors, respectively, are considered usually in such situations, although IUPAC discourages their use.

The dilution factor (DF) is equal to the final volume of the solution to be measured (volume of the test portion plus volume of the diluent) divided by the volume of the original test portion.

Dilution factor (DF)
$$= \frac{\text{Final volume of the solution to be measured } (V_f)}{\text{Original test portion } (V_o)} = \frac{V_f}{V_o}$$

The concentration factor (CF) is calculated analogously as the final volume of the solution to be measured divided by the original volume of the test portion.

A simplified way to solve this type of calculation consists of using the same approach as for preparing solutions by diluting a concentrated stock solution:

$$C_1 \cdot V_1 = C_2 \cdot V_2$$
$$C_{\text{initial}} \cdot V_{\text{initial}} = C_{\text{final}} \cdot V_{\text{final}}$$

Here, C_{initial} and V_{initial} refer to the concentration and volume of the concentrated solution, respectively. When a known and exact amount of solvent is added to this solution, a diluted solution is obtained, being C_{final} its concentration and V_{final} its volume.

Note that it does not matter what the units of C and V are; but of course, they have to be the same on each side of the equation. Keep your units coherent!

REFERENCES

[1] Bureau International des Poids et Mesures, BIPM. Available at: http://www. bipm.org/ (Accessed 10 October 2016).

[2] Mills, I.; Cvitas, T.; Homann, K.; Kallay, N.; Kuchitsu, K. (1998). *Quantities, Units and Symbols in Physical Chemistry. International Union of Pure and Applied Chemistry*, 2nd ed. Blackwell Science Ltd, Oxford.

[3] Thompson, A.; Taylor, B. N. (2008). *Guide for the Use of the International System of Units (SI)*. NIST Special Publication 811. Edition. U.S. Department of Commerce. Available at: http://www.nist.gov/pml/pubs/sp811/.

[4] McNaught, A. D.; Wilkinson, A. (1997). *IUPAC. Compendium of Chemical Terminology*, 2nd ed. (the 'Gold Book'). Blackwell Scientific Publications, Oxford. XML on-line corrected version: http://goldbook.iupac.org (2006-) created by M. Nic, J. Jirat, B. Kosata; updates compiled by A. Jenkins. ISBN 0-9678550-9-8. doi:10.1351/goldbook. Last update: 2014-02-24; version: 2.3.3.

[5] Stoker, H. S. (2014). *General, Organic & Biological Chemistry*, 7th ed. Cengage Learning, Boston.

WORKED EXERCISES

Some preliminary notes:

(1) Every number (result) you write down in science should have some units accompanying it. Quite often, undergraduates perform calculations and they report only numbers in their notebooks, without units. This is a serious mistake that may cause them many problems in their professional life.

(2) Identify the solute, solvent and solution (solution = solute + solvent).

(3) Identify each quantity symbol with their corresponding substance. For this, it is generally preferable to place symbols for substances in parentheses immediately after the quantity symbol, for example $n(NaOH)$, $M(H_2SO_4)$; or as a subscript, C_{Hg}.

(4) Whenever you solve any kind of quantitative exercise/calculation, you must check whether it makes sense.

(5) Usually, there are many ways to solve a problem. If the one you apply has no conceptual drawbacks (too frequent in

students!), makes you feel comfortable, you really understand it, you can justify it, and it delivers correct results always, go ahead and follow it!

(6) When the sample to be analyzed has to be submitted to several sample treatment steps, we strongly recommend to depict the analytical process in order to avoid typical missing calculations associated with sample treatment.

(7) In this chapter, the numerical exercises will discriminate with detail the data which are given in the enunciate and those which are requested for you to calculate. In the other chapters, this task is let to the student. Remember that it is always advisable to write down this scheme before starting the exercise.

1. The visible region of the electromagnetic spectrum corresponds to wavelengths from about 0.00000078 m to 0.00000038 m. Express this range in (a) scientific notation, (b) adequate units.

SOLUTION:

Question a:

To convert these numbers to scientific notation, move the decimal point to the right 7 decimal places. For this reason, the exponent will be 7. The sign of the exponent is negative as the decimal point moves to the right.

Strategy: 0.00000078
1234567

Solution: 7.8×10^{-7} m and 3.8×10^{-7} m.

Question b:

To select the most adequate units (SI) for a value, it is advisable to obtain a number higher than 1 and lower than 1000. Moreover, scientific criteria should be applied, according to our previous knowledge of the particular subject, to assess its significance.

Taking into account that the exponent that converts these numbers into an integer is -7, the unit that has a similar relation with the meter is nanometer (10^{-9} m).

$$7.8 \cdot 10^{-7} \, \cancel{m} \cdot \frac{10^9 \, \text{nm}}{1 \, \cancel{m}} = 7.8 \cdot 10^2 \, \text{nm} = 780 \, \text{nm}$$

The results you were asked for are 780 nm and 380 nm.

2. Calculate the concentration of a solution prepared by mixing 24.0 g of potassium chloride with 126.0 g of water. Express it as (a) mass percent and (b) molality.

SOLUTION:

You can follow the next steps: (1) Define clearly the quantities you get in the enunciate and the quantities you are requested to calculate, (2) write down the concentration expression you need, (3) substitute solute and solvent (solution) quantities into the equation and perform the calculations.

Data	Given by enunciate:	Have to find out:
	24.0 g solute	% KCl
	126.0 g H_2O	m

Question a:

$$\%(\text{m/m}) = \frac{\text{mass of solute}}{\text{mass of solution}} \cdot 100$$

$$\%(\text{m/m}) = \frac{24.0}{24.0 + 126.0} \cdot 100 = 16\,\%$$

Question b:

$$m = \frac{n \text{ solute}}{\text{mass (kg) solvent}} = \frac{\text{mass (g) solute/molar mass solute}}{\text{mass (kg) solvent}}$$

$M(\text{KCl})$: $39.1 + 35.5 = 74.6 \, \text{g/mol}$

$$m = \frac{24.0 \, \cancel{\text{g solute}}}{126.0 \, \text{kg solvent}} \cdot \frac{1 \text{ mole solute}}{74.6 \, \cancel{\text{g solute}}} = 0.0025 \, \text{m}$$

The final solution has 16 % of potassium chloride and a molality of 2.5×10^{-3}.

3. The European guideline value for total inorganic mercury in drinking water is $6\,\mu g/L$. Two laboratories reported average results for the analysis of two spring water samples, which were $1\,mM$ (sample A) and $2 \cdot 10^{-5}\,\%$ (in w/v, sample B), can you drink them safely?

SOLUTION:

To compare these concentrations is mandatory to express them in the same units. Thus, taking into account that the guideline value is given in $\mu g/L$, we will consider this unit as the reference one (C_{Hg}).

Data	Given by enunciate:	Have to find out:
	Guideline value: $6\,\mu g/L$ Hg	C_{Hg} in $\mu g/L$
	Sample A: $1\,mM$ Hg	
	Sample B: $2 \cdot 10^{-5}\,\%$ (m/v) Hg	

Now, the next steps for sample A should be done:

— Rewrite $1\,mM$ (millimolar) as a function of M: $1\,mM = 10^{-3}$ M
— Express M in its simplest possible units: 10^{-3} mol/L
— Convert molarity to g/L multiplying by the molar mass of the analyte $(M(Hg) = 206.6\,g/mol)$, and finally convert g to μg:

$$C_{Hg} = 10^{-3}\,\frac{\text{mol}}{L} \cdot \frac{206.6\,\text{g}}{1\,\text{mol}} \cdot \frac{10^3\,\mu g}{1\,\text{g}} = 206.6\,\mu g/L$$

For sample B:

— Rewrite % (w/v) in its simplest possible units: $2 \cdot 10^{-5}\,g/100\,mL$
— Divide to obtain: $2 \cdot 10^{-7}\,g/mL$

— Convert g in μg and mL in L with the appropriate conversion factor:

$$C_{Hg} = 2 \cdot 10^{-7} \frac{\text{g}}{\text{mL}} \cdot \frac{10^6 \, \mu\text{g}}{1 \, \text{g}} \cdot \frac{10^3 \, \text{mL}}{1 \, \text{L}} = 2 \cdot 10^2 \, \mu\text{g/L}$$

Accordingly, the solution to this exercise is: Both samples, A and B, exceeded the guideline value for total inorganic mercury as $206.6 \, \mu$g/L and $200 \, \mu$g/L and are higher than $6 \, \mu$g/L and, so, they should not be drunk.

> Observe that, as it will occur many times in Chapter 2, the solution to the exercise is not a value, but a statement or conclusion. When solving out the exercises, do not forget to keep your units!

4. What is the concentration of the Cu^{2+} ion, expressed as molarity, in a solution prepared by dissolving 1.25 g of $CuSO_4 \cdot 5H_2O$ in enough water to give 50.0 mL of solution?

SOLUTION:

The highlight of this problem is that the salt used to prepare the solution is hydrated. Thus, we have to consider those molecules of hydration water in the calculation of its molar mass: $M(CuSO_4 \cdot 5H_2O)$: $159.5 + 5 \cdot 18 = 249.5$ g/mol.

Data	Given by enunciate:	Have to find out:
	1.25 g $CuSO_4 \cdot 5H_2O$	$M(Cu^{2+})$
	50 mL solution	

$$\frac{1.25 \; \text{g } \cancel{CuSO_4 5H_2O}}{50 \; \cancel{\text{mL solution}}} \cdot \frac{1 \; \cancel{\text{mol } CuSO_4 5H_2O}}{249.5 \; \text{g } \cancel{CuSO_4 5H_2O}}$$

$$\cdot \frac{1 \; \text{mol } Cu^{2+}}{1 \; \cancel{\text{mol } CuSO_4 5H_2O}} \cdot \frac{10^3 \; \cancel{\text{mL}}}{1 \; \text{L}}$$

$$= 0.10 \, \text{mol/L}$$

Molarity concentration of $Cu^{2+} = 0.10$ M

5. Prepare 150 mL of 4 % Na_2CO_3 from a 6.00 M stock solution of this salt.

SOLUTION:

This problem calls for the use of the $V_1 \cdot C_1 = V_2 \cdot C_2$ formula, since a stock concentrated solution is used to prepare a diluted solution of the same solute. But the concentrations given in the enunciate are in different units; therefore, one of them must be converted to match the other. It does not matter which unit of concentration is chosen. Let us convert them to molarity.

Firstly, we must know the type of percent solution. When no description accompanies the unit, a mass percent solution is assumed (m/m).

A practical way to convert % to M, as well as to other expressions, is to use concentrations as *conversion factor*:

Mass percent	Meaning	Conversion factors	
4 %	4 g of Na_2CO_3 in 100 g of solution	$\dfrac{4\,g \text{ of } Na_2CO_3}{100\,g \text{ of solution}}$	$\dfrac{100\,g \text{ of solution}}{4\,g \text{ of } Na_2CO_3}$

Now, the adequate conversion factor is employed to transform the 4 % value, as well as additional conversion factors to pass g of salt to mole (using molar mass) and to convert g of solution to L of solution. This requires a value for the density of the solution but that was not given, so we considered it equal to the water density ($1\,g/cm^3$, at 25 °C). This is true in dilute aqueous solutions.

$$\frac{4 \text{ g } Na_2CO_3}{100 \text{ g solution}} \cdot \frac{1 \text{ mol } Na_2CO_3}{106 \text{ g } Na_2CO_3} \cdot \frac{1 \text{ g solution}}{cm^3 \text{ solution}}$$

$$\cdot \frac{10^3 \text{ cm}^3 \text{ solution}}{1 \text{ L solution}} = 0.38 \text{ mol/L}$$

So, the 4% Na_2CO_3 stock solution has a molarity of 0.38 M.

Now using the dilution formula, we can calculate the volume of 6.00 M Na_2CO_3 stock that is needed to prepare 150 mL of 0.38 M solution.

$$C_1 \cdot V_1 = C_2 \cdot V_2$$

Data are replaced: $6.00 \text{ M} \cdot V_1 = 0.38 \text{ M} \cdot 150 \text{ mL}$

$$V_1 = \frac{0.38 \text{ M} \cdot 150 \text{ mL}}{6.00 \text{ M}}$$

After canceling the units, $V_1 = 9.5 \text{ mL}$.

Finally, to prepare a 4% Na_2CO_3 solution, 9.5 mL of a 6 M solution will be measured and the volume completed with distilled water to 150 mL.

> The use of concentrations as *conversion factor* simplified the calculations, but it is impossible to correctly apply them if the corresponding units are omitted!

> When a percent concentration is given without specifying the type of percent concentration (not a desirable situation, although frequent), you have to assume that it is mass percent (m/m).

6. A 0.10 M solution of oxalic acid will be used as titrant in an acid–base titration. To facilitate further volumetric calculations, express its concentration as normality.

SOLUTION:

The relation between both units is: $N = M \cdot K$, where K is an integer constant ≥ 1 and for a particular species is defined by the reaction type and the balanced chemical reaction. In this case, the enunciate indicates that the type of reaction is acid–base. Hence, K is the number of moles of H^+ produced or neutralized per mole of acid or base supplied. In this case, we consider the ionization equilibrium of the oxalic acid:

$$C_2H_2O_4 \Leftrightarrow C_2O_4^{2-} + 2H^+$$

— This acid produces 2 moles of H^+
— Calculate: $N = 0.10 \cdot 2 = 0.20$ N.

The normality of the oxalic acid is twice its molarity, as it can offer two protons per mole of compound.
The solution we are looking for is: 0.20 N.

> It is common (mainly in exams and laboratories) that students get doubts on whether the correct equation is $N = M \cdot K$ or $M = N \cdot K$. In doubt, first write their definitions and see where K is needed to make them equal.

7. Although HCl is not a primary standard, their solutions are widely used as titrants. How would you prepare 100 mL of a 0.10 M HCl solution knowing that the density and richness of a commercial acid are 1.37 g/mL and 37 %, respectively?

SOLUTION:

We start the calculations by considering the volume and concentration of the solution that we must prepare. Then, using the density and richness of the commercial solution (commercial solution is denoted as c.s. in the calculations), it is possible to convert the moles of solute required to prepare the solution into the volume of commercial solution.

Data	Given by enunciate:	Have to find out:
	100 mL solution	V(HCl c.s.)
	0.10 M HCl	
	$d = 1.37$ g/mL	
	37 %	

Taking into account that the units of the density refer always to the solution and not to the solute/solution ratio, the exercise can be solved.

> For instance, 1.37 g/mL verbalizes 1.37 g of solution per mL of solution, instead of 1.37 g of solute per mL of solution. Be careful, this is a common confusion!

Write the necessary conversion factors, operate and check that the units cancel out properly.

$$100 \; \text{mL} \cdot \frac{1 \; \text{L}}{10^3 \; \text{mL}} \cdot \frac{0.1 \; \text{mol HCl}}{1 \; \text{L}} \cdot \frac{36.45 \; \text{g HCl}}{1 \; \text{mol HCl}} \cdot \frac{100 \; \text{g c.s.}}{37 \; \text{g HCl}}$$

$$\cdot \frac{1 \; \text{mL c.s.}}{1.37 \; \text{g c.s.}} = 0.719 \; \text{mL}$$

Therefore, 0.72 mL of the commercial acid must be pipetted, introduced into a volumetric flask (which already has some pure water) and finally, the volume made up to 100 mL.

8. From a 100 mg/L sodium stock standard solution, five calibration solutions (2–10 mg/L) have to be prepared by dilution. Present the adequate calculations to get: (a) 500 mL of the stock standard solution from commercial solid sodium chloride, and (b) 50 mL of each calibration solution.

   SOLUTION:

Data	Given by enunciate:	Have to find out:
	500 mL of 100 mg/L	Mass NaCl (g)
	Na^+, stock solution	stock solution volumes
	from NaCl	

Question a:

Many solutions used in the laboratory are prepared from solid chemical reagents, according to the general steps depicted in Figure 4.

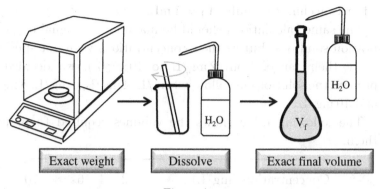

Figure 4.

We have to calculate the mass of salt to be weighed to obtain the $100\,mg/L$ Na^+ solution. For this, we should consider the stoichiometric ratio: $NaCl \rightarrow Na^+ + Cl^-$. Thus, 1 mole of sodium ion is obtained per mole of salt. The richness of the salt is not specified, so we consider it pure (100%).

$$500\;\cancel{mL} \cdot \frac{1\;\cancel{L}}{10^3\;\cancel{mL}} \cdot \frac{100\;\cancel{mg\text{-}Na^+}}{1\;\cancel{L}} \cdot \frac{1\;\cancel{g}}{10^3\;\cancel{mg}} \cdot \frac{1\;\cancel{mol\text{-}Na^+}}{23\;\cancel{g\text{-}Na^+}}$$
$$\cdot \frac{1\;\cancel{mol\text{-}NaCl}}{1\;\cancel{mol\text{-}Na^+}} \cdot \frac{58.5\,g\;NaCl}{1\;\cancel{mol\text{-}NaCl}}$$
$$= 0.127\,g$$

Accordingly, $0.127\,g$ of NaCl should be weighed, dissolved in deionized water and then made up to 100 mL in a volumetric flask with water.

Question b:

The calibration solutions are prepared by dilution of the sodium stock standard solution. Hence, we must calculate the required volume of the concentrated solution (V_1) for each one. This volume will be brought to final volume of $50\,mL$ (V_2) with distilled water.

$$100\,mg/L \cdot V_1 = 2\,mg/L \cdot 50\,mL$$

$$V_1 = \frac{2\;\cancel{mg/L} \cdot 50\,mL}{100\;\cancel{mg/L}}$$

After canceling the units, $V_1 = 1\,\text{mL}$.

The same calculations should be made for the remaining calibration solutions, but as their concentrations were not specified (only their range, from $2\,\text{mg/L}$ to $10\,\text{mg/L}$), we decided to prepare the solutions as follows: $2\,\text{mg/L}$, $4\,\text{mg/L}$, $6\,\text{mg/L}$, $8\,\text{mg/L}$ and $10\,\text{mg/L}$.

The following table shows the volumes required to prepare them.

Concentration (mg/L)	2	4	6	8	10
Stock solution volume (mL)	1	2	3	4	5

9. According to the World Health Organization (WHO), household bleach is one of the recommended standard disinfectants to use against the Ebola virus. Household bleach is a solution of sodium hypochlorite which generally contains 5 % available chlorine. Diluted solutions 1:10 are used as strong disinfectants for bodies, spills of blood/body fluids; whereas 1:100 bleach solutions are recommended to disinfect surfaces, medical equipment, etc. How would you prepare 2 L of each solution? Express the concentration of both solutions in percent and in g/L.

SOLUTION:

These solutions are prepared directly by dilution of the concentrated solution, 5 %.

Data	Given by enunciate:	Have to find out:
	5 % available Cl_2	Final concentration in
	2 L solutions	% and g/L
	1:10	
	1:100	

Before applying the formula that relates the concentrated and diluted solutions, it is necessary to convert the 1:10 and 1:100

ratios to percentages. They are verbalized as a part of solute (in this case, the 5 % available Cl_2) per 10 parts of solution, and a part of solute per 100 parts of solution, respectively.

$$(5\,\%)\frac{1}{10} = 0.5\,\%$$

$$(5\,\%)\frac{1}{100} = 0.05\,\%$$

Consequently, the final solutions will exhibit a 10-fold lower concentration and a 100-fold lower concentration than the original, respectively.

Now, the dilution formula can be applied.

$$C_1 \cdot V_1 = C_2 \cdot V_2$$

Data are replaced to get the first solution: $5\,\% \cdot V_1 = 0.5\,\% \cdot 2\,L$

$$V_1 = \frac{0.5\,\% \cdot 2\,L}{5\,\%} = 0.2\,L$$

$$V_1 = 0.20\,\cancel{L} \cdot \frac{10^3\,mL}{1\,\cancel{L}} = 200\,mL$$

In the same way, the second solution is obtained:

$$V_1 = 0.02\,\cancel{L} \cdot \frac{10^3\,mL}{1\,\cancel{L}} = 20\,mL$$

In both cases, the volumes obtained for household bleach (200 mL and 20 mL) are diluted to a 2 L final volume with water.

Finally, the concentrations must also be reported in g/L. By definition, a 0.5 % solution contains 0.5 g of solute per 100 g of solution. Considering a diluted aqueous solution, its density is taken as that for pure water (1 g/mL, 25 °C).

$$C = \frac{0.5\,g\ solute}{100\,\cancel{g\ solution}} \cdot \frac{1\,\cancel{g\ solution}}{1\,\cancel{mL\ solution}} \cdot \frac{10^3\,\cancel{mL}}{1\,L} = 5\,g/L$$

For the 1:100 solution, calculations are performed in the same manner, obtaining 0.5 g/L.

The following table summarizes the results obtained.

Data	Given by enunciate:	Calculated:
	1:10	5 g/L, 0.5 %
	1:100	0.5 g/L, 0.05 %

10. A method based on the complexation of metals with ammonium pyrrolidine dithiocarbamate (APDC) and extraction into methyl isobutyl ketone (MIBK) is employed to determine Cd, Co, Cu and Pb in a seawater sample [1]. Thus, 100 mL of seawater were placed in a separating funnel and pH was adjusted to 4.1 with an acetic buffer. Around 1 mL of 1 % APDC and 5 mL of MIBK were added. The funnel was shaken for 5 min and allowed to stand for 20–30 min for phase separation. The lower seawater phase was drained and the organic phase was transferred to a separating funnel for back-extraction with 5 mL of 4 M HNO_3, shaking for 10 min. After phase separation was obtained, the lower aqueous phase was transferred to a test tube for its final analysis by inductively coupled plasma mass spectrometry (ICP-MS) (this technique will be studied in Chapter 5). The extract showed the following metal concentrations (value ± confidence interval, in $\mu g/L$): 13.62 ± 0.07, 52.80 ± 0.18, 83.25 ± 0.32 and 61.11 ± 0.58 for Cd, Pb, Cu and Cr, respectively. Determine the metal concentrations in the seawater sample, expressed as $\mu g/L$.

SOLUTION:

The contents of heavy metals in seawater are usually very low, their analyses requiring currently some preconcentration steps. In addition, the analyte should be separated from the aqueous matrix due to the high complexity of the latter (large quantities of salts, biological products, etc.). For aqueous samples, liquid–liquid extraction is a widespread, suitable option. The experimental procedure described in this exercise involves many typical steps (Figure 5a).

While all these steps must be taken into account in the laboratory, for the calculations, they are reduced to a very simple scheme (Figure 5b).

The initial (V_i) and final (V_f) volumes of the sample aliquot can be considered in the same way as in a typical dilution/concentration calculation. Therefore, we can apply the well-known expression:

$$C_i \cdot V_i = C_f \cdot V_f$$

To start with (for instance) Cd, the experimental values yield:

$$C_i \cdot 100\,\text{mL} = 13.62\,\mu\text{g/L} \cdot 5\,\text{mL}$$

By algebraic arrangement:

$$C_i = \frac{13.62\,\mu\text{g/L} \cdot 5\,\cancel{\text{mL}}}{100\,\cancel{\text{mL}}}$$

After canceling the units, $C_i = 0.681 = 0.68\,\mu\text{g/L}$.

Before proceeding, you should verify that this result is logical. Comparing both concentration values, C_i, and C_f, it is seen that: $C_i = 0.68\,\mu\text{g/L} < C_f = 13.62\,\mu\text{g/L}$, which agrees with the preconcentration treatment performed (when concentrating, the final concentration of the analyte should be greater than in the original solution).

Now, the same calculations can be performed for the other metals, as well as for their confidence intervals. The following table shows the results:

Metal	Cd	Pb	Cu	Cr
Concentration (μg/L)	0.68 ± 0.004	2.64 ± 0.01	4.16 ± 0.02	3.06 ± 0.03

11. A soil sample is submitted to a microwave-assisted acid digestion based on the 3051EPA Method (Environmental Protection

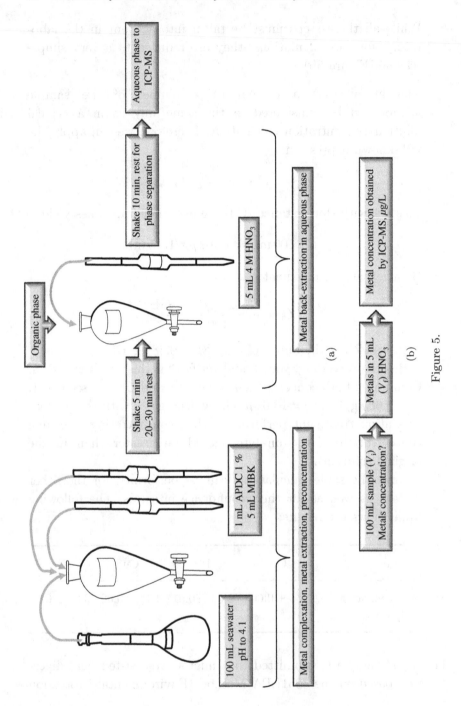

Figure 5.

Agency) for metal analysis. An aliquot of 0.5000 g of wet sample was digested with 10 mL of nitric acid in a closed reactor, microwave-heated until 170 °C during 10 min. The extract was filtered and made up to 20.00 mL with deionized water in a volumetric flask. This test solution was analyzed directly to determine Cd as a trace element; whereas iron was determined in a 20-fold diluted portion. The concentrations obtained were 8.24 μg/L and 2.55 mg/L for Cd and Fe, respectively. If a separate portion of soil was used to determine its moisture, which resulted to be 36 %, calculate the concentrations of Cd and Fe in the original sample, on both a wet and a dry basis.

SOLUTION:

The analytical procedure can be depicted (Figure 6a) and simplified to make calculations (Figure 6b).

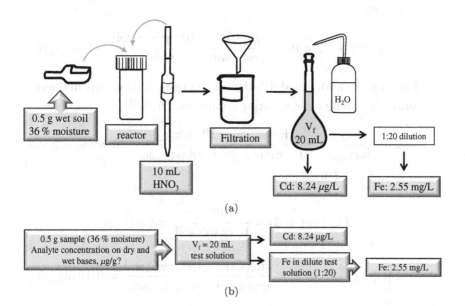

Figure 6.

Data	Given by enunciate:	Have to find out:
	0.5000 g wet soil	Cd, Fe concentration
	36 % moisture	on wet and dry
	V_f test solution 20 mL	bases.
	8.24 μg/L Cd	
	2.55 mg/L Fe in 1:20	
	solution	

To solve this problem, calculations should be carried out from the last working step to the starting point of the analytical procedure. For cadmium, calculations are presented first on a wet basis:

$$C_{Cd} \text{ (wet soil)} = 8.24 \frac{\mu g}{\cancel{L}} \cdot \frac{1 \cancel{L}}{10^3 \cancel{mL}} \cdot \frac{20 \cancel{mL}}{0.5 \text{ g wet soil}} = 0.33 \, \mu g/g$$

To convert to the dry basis, we need to take into account the percentage of moisture, which means that there are 36 g of water per each 100 g of wet soil, therefore, there will be $100 - 36 = 64$ g of dry soil.

$$C_{Cd} \text{ (dry soil)} = 0.33 \frac{\mu g \, Cd}{\text{g wet soil}} \cdot \frac{100 \text{ g wet soil}}{64 \text{ g dry soil}} = 0.52 \, \mu g/g$$

For iron, an additional dilution of the test solution was necessary since this element is a major component in soils.

$$2.55 \frac{mg}{\cancel{L}_{diluted}} \cdot \frac{1 \cancel{L}_{diluted}}{10^3 \cancel{mL}_{diluted}} \cdot \frac{20 \cancel{mL}_{diluted}}{1 \cancel{mL}_{test \, solut}} \cdot 20 \cancel{mL}_{test \, solution}$$
$$= 1.02 \, mg$$

$$C_{Fe} \text{ (wet soil)} = \frac{1.02 \, mg}{0.5 \text{ g wet soil}} = 2.04 \, mg/g$$

$$C_{Fe} \text{ (dry soil)} = 2.04 \frac{mg \, Cd}{\text{g wet soil}} \cdot \frac{100 \text{ g wet soil}}{64 \text{ g dry soil}} = 3.19 \, mg/g$$

As expected, the metal concentration in a dry basis is higher than on a wet basis.

12. A coal sample has a concentration of 24.3 μg/g of nickel. A 200 mg aliquot of this coal is digested in an acid medium, filtered and brought to a final volume of 50.00 mL (in a volumetric flask) with deionized water. Determine the concentration of Ni in the test solution if a 20 μL aliquot was analyzed.

SOLUTION:

This situation is somehow different from what we usually need in the laboratory, but sometimes it is indeed necessary to check what concentration would you obtain with a certain procedure for a given sample; for instance, when a certificated reference material with a known analyte concentration is studied (Figure 7a).

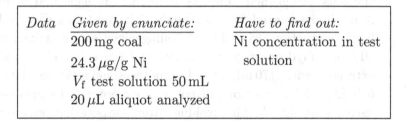

Data	Given by enunciate:	Have to find out:
	200 mg coal	Ni concentration in test
	24.3 μg/g Ni	solution
	V_f test solution 50 mL	
	20 μL aliquot analyzed	

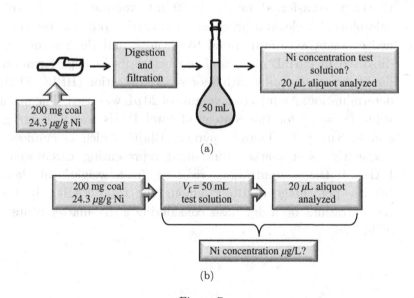

Figure 7.

It is worth noting that all nickel in coal has been dissolved and is in the 50 mL solution (test solution). The 20 μL aliquot analyzed of this solution, obviously, will exhibit the same Ni concentration (Figure 7b).

$$C_{Ni} = 24.3 \frac{\mu g}{g_{coal}} \cdot \frac{1 \, g}{10^3 \, mg} \cdot \frac{200 \, mg_{coal}}{50 \, mL_{test \, solution}} = 0.0972 \, \mu g/mL$$

Now, the result can be expressed in more appropriated units.

$$C_{Ni} = 0.0972 \frac{\mu g}{mL} \cdot \frac{10^3 \, mL}{1 \, L} = 97.2 \, \mu g/L$$

13. Polycyclic aromatic hydrocarbons (PAHs) are genotoxic carcinogens being their contents in foodstuffs legislated strictly. A method reported for their analysis in apples employs solid–liquid extraction using 20 g of homogenized sample mixed with 80 g of anhydrous sodium sulfate, and extracted in a Soxhlet extractor with 170 mL of a hexane–acetone mixture (1:1) for 6 h [2]. The extraction solvent was evaporated to dryness in vacuum at 40 °C. The residue after evaporation was quantitatively transferred into a 10.00 mL volumetric flask with chloroform. A cleanup procedure was carried out and the purified extracts were evaporated to dryness, and the residue was dissolved in 0.5 mL of acetonitrile before its high performance liquid chromatography with fluorescence detection (HPLC-FLD) determination. An injection volume of 20 μL was used. The mean value ($n = 3$) for the content of total PAHs was 9.30 μg/kg, and 0.03 ng/g for benzo(a)pyrene, B[a]P, which is employed frequently as a marker compound representing carcinogenic PAHs. If the concentrations refer to 'fresh weight' calculate: (a) the percentage of B[a]P among the total PAHs, (b) the concentrations on a dry basis considering a dry matter content of 14.7 %.

SOLUTION:

Data	Given by enunciate:	Have to find out:
	20 g apple	B[a]P percent
	PAHs 9.30 μg/kg	Concentrations on a dry basis
	B[a]P 0.03 ng/g	
	14.7 % dry matter	

Question a:

First, it is necessary to express both concentrations in the same units. In some cases, like this, they seem different but they represent the same analyte/sample ratio.

$$9.30 \frac{\mu g}{kg} \cdot \frac{1 \, kg}{10^3 \, g} \cdot \frac{10^3 \, ng}{1 \, \mu g} = 9.30 \, ng/g$$

Note that many units used frequently are interchangeable:

$$\frac{mg}{kg} = \frac{\mu g}{g}$$

$$\frac{\mu g}{L} = \frac{ng}{mL}$$

To calculate the percentage of B[a]P:

$$(\%) \, B[a]P = \frac{0.03 \, ng_{B[a]P}}{g_{apple}} \cdot \frac{g_{apple}}{9.30 \, ng_{PAHs}} \cdot 100 = 0.32 \, \%$$

Question b:

$$C_{B[a]P} \, (dry \, matter) = 0.03 \frac{ng \, B[a]P}{g \, fresh \, apple} \cdot \frac{100 \, g \, fresh \, apple}{14.7 \, g \, dry \, matter}$$

$$= 0.20 \, ng/g$$

$$C_{PAHs} \, (dry \, matter) = 9.30 \frac{ng_{PAHs}}{g \, fresh \, apple} \cdot \frac{100 \, g \, fresh \, apple}{14.7 \, g \, dry \, matter}$$

$$= 63.26 \, ng/g$$

EXERCISES PROPOSED TO THE STUDENT

14. Select the most appropriate units to express the concentration of a solution prepared by dissolving 125 mg of copper (II) sulfate in 500 mL of water.

    Solution: 0.25 g/L.

15. How would you prepare 50.00 mL of a 3 % (m/v) solution from 100 % pure sodium chloride?

    Solution: weigh 1.5 g of solute, dissolve it in water and make up to 50.00 mL.

16. 8.00 g of calcium chloride is dissolved in 25.00 mL of distilled water. (A) Set the concentration of this solution in mol/L. (B) What are the molar concentrations of both the chloride and calcium ions?

    Solution: (a) 2.12 M, (b) 4.24 M and 2.12 M for Cl^- and Ca^{2+}, respectively.

17. How many milliliters of an ammonia stock solution are needed to prepare 200 mL with a 5 % volume?

    Solution: 10 mL.

18. Determine the volume of each of the following solutions that contains 0.05 mol of potassium ion: (a) 0.200 M KNO_3, (b) 0.350 M $K_2Cr_2O_7$, (c) 1.5 M $KHC_8H_4O_4$

    Solution: (a) 0.25 L, (b) 71.4 mL, (c) 33.3 mL.

19. You have a 0.65 g/mL HNO_3 solution. Express its concentration in terms of both molarity and normality.

    Solution: 10.32 M, 10.32 N (*Hint*: this acid gives 1 mol of H^+).

20. How much sodium hydroxide is required to prepare 1.5 L of a 2.5 % solution, whose density is 1.126 g/cm³? What is its normality?

    Solution: 42.2 g, 0.7 N.

21. Calculate the molarity and normality of a 1 % $KMnO_4$ solution used as titrant (assume that its density is that of water at 25 °C).

    Solution: 0.063 M, 0.316 N (*Hint*: 5 moles of electrons per mole of MnO_4^- ion).

22. How would you prepare 50.0 mL of a 1 : 3 ethanol–water mixture

    Solution: 12.5 mL ethanol plus 37.5 mL H_2O.

23. How many milliliters of a commercial solution of nitric acid (69 % purity; 1.41 g/cm^3 density) are needed to prepare 100 mL of a 10 % solution?

    Solution: 10.28 mL (the density of the diluted solution is estimated as that for water at 25 °C).

24. How many mL of an HCl stock solution (35 % purity, specific gravity of 1.19) would be needed to prepare 1 L of a 0.2 N solution?

    Solution: 17.50 mL.

25. A maximum 1000 mg/L level of phosphate in milk (expressed as P_2O_5) has been regulated. What volume of milk (in milliliters) contains 50 mg of P?

    Solution: 114.5 mL.

26. Some countries had adopted salt fluorination schemes instead of water fluorination to prevent dental caries, mainly in young population [3]. A fixed amount of a fluoride compound (mostly NaF) is added to a fixed amount of refined common salt to achieve a maximum concentration of 250 mg/kg of fluoride. WHO recommends a maximum intake of 5 mg of common salt per day. If a maximum of 250 mg of fluoride per kg of salt is the established value, you are asked to calculate: (a) the mass percent of fluoride in salt, (b) what mass of fluoride should be ingested daily (at maximum)? (c) how much sodium fluoride should be added per each kilogram of salt?

    Solution: (a) 0.025 %, (b) 1.25 μg, (c) 553 mg NaF.

27. Phenolphthalein is widely employed as an acid–base indicator. For this, a 0.1 % (m/m) solution is used. Taking into account that this solute is insoluble in water, an alcoholic solvent must be used, e.g., ethanol (density 0.789 g/cm^3). You are asked to report the calculations to prepare 50.00 mL of the 0.1 % required solution. Also calculate the concentration of this solution in g/L.

Solution: 39.45 mg, 0.79 g/L (the density of the solution is considered as that for pure ethanol, at 25 °C).

28. To prepare a pH = 4 buffer solution, 5.10 g of potassium hydrogen phthalate ($KHC_8H_4O_4$) were dissolved in 250 mL of deionized water, 0.50 mL of 0.10 M hydrochloric acid added, and then the mixture was diluted to 500.00 mL. Calculate the molarity of both the salt and the acid in the buffer solution.

    Solution: 0.05 M $KHC_8H_4O_4$, 0.1 mM HCl.

29. To prepare a pH = 9.2 buffer solution, a solution containing 0.1 mol/L of ammonium chloride and 0.1 mol/L of ammonia is needed. How would you prepare 50 mL of such a solution? (for ammonia, consider 25 % purity and 0.91 g/cm^3).

    Solution: 0.37 mL NH_3, 0.27 g NH_4Cl.

30. A fuel oil sample is analyzed to determine its concentration of Ni. A 0.5000 g aliquot was weighed accurately and dissolved in 2.00 mL of toluene. A test portion of 200 μL was used to prepare 1.00 mL emulsion with ethanol, nitric acid and xylene. After the instrumental analysis, a 2.35 μg/L Ni concentration was obtained in the 20 μL analyzed aliquot. What is the concentration of Ni in the fuel oil sample?

    Solution: 47 ng/g (*Hint*: remember the meaning of 'aliquot').

31. Calcium hypochlorite is used widely for drinking water disinfection. Commercial products are available as bleaching powder or tablet forms. The disinfection 'power' of the various forms of presentation of calcium hypochlorite was expressed as its available chlorine, which ranged between 25 % (for bleaching powder) and 70 % (for HTH, high test hypochlorite). How would you prepare 10 L of a solution containing 50 g/L of available chlorine from bleaching powder and from HTH?

    Solution: 2 kg of bleaching powder, 0.7 kg for HTH.

EXERCISE REFERENCES

[1] Satyanarayanan, M.; Balaram, V.; Gnaneshwar, T.; Dasaram, B.; Ramesh, S.L.; Mathur, R.; Droila, R.K. (2007). Determination of trace

metals in seawater by ICP-MS after preconcentration and matrix separation by dithiocarbamate complexes, *Indian Journal of Marine Sciences*, 36(1): 71–75.

[2] Jánská, M.; Hajslová, J.; Tomaniová, M.; Kocourek, V.; Vávrová, M. (2006). Polycyclic aromatic hydrocarbons in fruits and vegetables grown in the Czech Republic, *Bulletin Environmental Contamination Toxicology*, 77: 492–499.

[3] Marthaler, T. M.; Petersen, P. E. (2005). Salt fluoridation — an alternative in automatic prevention of dental caries, *International Dental Journal*, 55: 351–358.

CHAPTER 2

BASIC DATA ANALYSIS

José Manuel Andrade-Garda and
María Paz Gómez-Carracedo

OBJECTIVES AND SCOPE

The main objective of this chapter is to train the student in some basic statistical concepts that are employed almost daily in laboratories. However, students must be aware that its intention is neither to deal with statistics *senso stricto*, nor to demonstrate the equations nor to consider every mathematical aspect in detail. Rather, it is about applying tools to address some current needs in laboratories. The techniques presented here should constitute a sound basis for objective and efficient decision-making in typical relevant laboratory issues, such as those related to precision, comparability of two average results, calibration or standardization, etc. Special emphasis was placed on the latter topic because it is of paramount importance in almost all instrumental techniques of analysis. Many tests discussed here will be applied in exercises posed in subsequent chapters.

1. INTRODUCTION

This chapter addresses a series of items which are not easy to label with just a brief title. Here, it is intended to learn how to perform some basic analyses of short series of analytical results and how to correctly study a calibration (standardization line). Nevertheless, the chapter does not dive deeply into statistical or chemometrical issues. Rather, it was attempted to give some conceptual, pragmatic ideas on several basic tasks that will have to be addressed in any laboratory. In case the title looks a bit pretentious to the reader, please, bear in mind that the authors tried to introduce undergraduates to some fundamental calculations that they will have to face almost daily in any professional laboratory job.

Chemometrics emerged strongly during the late 1970s to deal with objective methodologies to implement and optimize analytical methods and to gather information from (large) datasets. At present, it is a strong branch within analytical chemistry which, somehow, has revolutionized its epistemology. We strongly encourage readers to take a look on any of the excellent textbooks on chemometrics that are available nowadays, at least to overview the broad scope — and interest — it has for analytical chemists.

If we accept that the final mission of any analytical laboratory is to offer its customers trustworthy information, it is quite obvious that we must include within our skills some knowledge on basic data evaluation. Just for instance, we must be able to objectively decide whether a suspicious datum can be excluded from upcoming calculations or whether a calibration (the very fundamental corner stone of most analytical measurements) seems adequate.

In the following, several tools will be presented to address some routine questions that arise in current laboratory work when handling experimental data. Although only a few issues are discussed, they are indeed a part of the basic toolbox that any analyst must master. Due to space restrictions in this textbook and, not less important, lack of time in the undergraduate courses on instrumental analyses, some other relevant chemometric tools were not considered here (e.g., analysis of the variance, non-parametric tests, etc.). Deep statistical explanations will be avoided and readers and students are kindly forwarded to more specific textbooks (papers), some of which were used to develop this chapter.

Although the authors tried to restrict the number of different issues to be explained, and resumed and schematized main ideas, the theoretical part of this chapter is obviously larger than the others. This gets justified because, in our experience, undergraduates have a low statistical background.

Two reflections are in order here and they may well be a matter of discussion in a class or in a practical seminar. First, about the use of computers. They are so broadly available and used nowadays that students do not really care about questioning them. However, we all experienced software 'bugs' that in some circumstances might compromise our calculations. Indeed international guides describing quality in laboratories state clearly that it is the laboratories' responsibility to check (and demonstrate) that software works properly.

The correct functioning of computers and software is under the laboratory's direct responsibility. Correct data treatment constitutes also an unavoidable task of analytical chemists.

Second question is about 'good results'. Do 'perfect' results exist? Students tend to think that, without doubt, experimental results are good *per se*. To face them with reality, we usually devote a practical seminar after their laboratory classes to compare and discuss their results. Some groups made the 'same' experiments and — of course — they claim that they did what they should have done. Therefore, they do not find reasons to justify why their results did not agree.

Quality of the analytical results can only be defined by the customer. She/he has a need for information and '*if and only if* the information given by the laboratory is good enough to fit that need, the data can be said of being of appropriate quality. Here fit-for-purpose is the key word. The same analytical procedure yields results that may be excellent for a given purpose (e.g., infrared (IR) determination of the overall amount of hydrocarbons in waste water) but useless for others (the same determination in tap water).

Hence, there is no universal, undebatable level of laboratory quality. That is something that has to be gained at each assignment and for each client. Of course, there are indicators that can be used as objective figures of quality (broadly known as figures of merit; or analytical method performance characteristics, as recommended by Eurachem [1]): Limit of Detection (LOD) and Limit of Quantitation (LOQ), sensitivity, trueness and precision, etc. But also overall cost, speed, customer attention, health and security issues, etc.

We would like to pose another interesting discussion to the students. Following Oscar Wilde's best seller (*The Importance of Being Earnest*), we could say 'the importance of being precise in language'. This is about the importance of keeping clear concepts in mind and being able to justify our work and our calculations. Nowadays, most laboratories are entitled to deliver their analytical results with a statement of their uncertainty.

Uncertainty is defined as a parameter associated with the result of a measurement that characterizes the dispersion of the values that could reasonably be attributed to the measurand (in analytical chemistry we would say most times *the analyte*). This definition and much more information can be obtained freely at www.eurachem.org.

However, many analytical chemists experienced troubles when trying to explain this term — and mostly, its use — to current users. Not only that, pupils can enquire their (non-scientific) friends and relatives about the feeling the word inspires in them. Almost always they will answer that the word uncertainty has negative connotations. But scientifically speaking, it is exactly the opposite! An example will clarify this.

Most people like the term confidence interval and they argue that this creates confidence on the receiver/user of the laboratory information. Probably, you will never talk about 'lack of confidence interval', would you? Some people in some conferences discuss about this issue and propose the use of the word certainty instead of uncertainty. In our view, laboratories will gain from such a minimal change but the standard jargon is so deeply established, mostly in the physical and metrological fields, that it is unlikely that this happens. Therefore, students have to keep very clear ideas on basic concepts and how they must be used ... or not used. For instance, it has not much sense to calculate an average and — definitely, not at all — a standard deviation (SD) from just two or three results ... but this is so common even in top-rank scientific journals!

A final note has to be stated. To develop this chapter, it was assumed that undergraduates have some basic knowledge on what a distribution of data is and, therefore, that they have notions about the normal or Gaussian distribution. It was also accepted that they have some knowledge on the meaning and use of the confidence levels and how a hypothesis test can be used to accept or reject a null hypothesis. Although some brief notes will be given on these issues for the sake of clarity, their study is out of the scope of this monograph.

2. HOW TO DESCRIBE A SERIES OF DATA

Whenever a series of data is available (repeats of an analytical measurement, production figures, clinical data from several patients, etc.), we need to describe them without citing each and every datum. For two or three data, this might sound useless, but think about the data generated during a year in an industrial laboratory or in a production facility. It is impossible to enumerate all values in a discussion or for decision-making. Hence, we need to resume them using a reduced set of parameters. A quite intuitive approach would be to plot the data and try to describe the figure that arose there. This idea is behind the formal statistical definitions of the data distribution functions, among which the Gaussian one outstands (named after Carl Friedrich Gauss who systematized it around 1800). In many occasions, it is referred to as the normal distribution because it was found to be the current or 'normal' situation in many scientific fields, or as the bell shape (after Esprit Jouffret named it in a publication at 1903).

The core idea is that if the appearance of the graphical distribution of a series of data can be described using just some *descriptors* (*parameters* or *statistics*), then the overall data become described accurately. Four statistical descriptors can be used to unambiguously describe the distribution of the data. First, the experimental results have to be located in a scale (this can be done calculating different descriptors; typically, the arithmetic mean (or average) and/or the median); next, a parameter to describe their dispersion is needed (e.g., the range, the SD) and, finally, the appearance of the graphical shape has to be described using the skewness (symmetric appearance) and the kurtosis (existence of side tails).

2.1. Central trend of the data: the average

The arithmetic average, average value, average or just mean, is the numerical value obtained after dividing the sum of all the numerical values by the number of values. It represents the central trend of the distribution of the experimental data and it can be used to roughly represent them ('on average the values are around ...'). Equation (1) shows how it is calculated from a series of n experimental values, $x_1, x_2, \ldots, x_n$.

$$\bar{x} = \frac{x_1 + x_2 + \cdots + x_n}{n} = \frac{\sum x_i}{n} \tag{1}$$

Here, the Greek symbol sigma denotes the overall sum ($\sum x_i$), where the index 'i' extends from x_1 to x_n and the bar over the x indicates 'arithmetic mean'.

It is of paramount importance to remember that the mean is highly sensitive to outliers (data which once analyzed do not belong to the inherent distribution of the experimental series of values). Indeed, the breakdown point for the mean is one; i.e., just one wrong data will bias the calculations.

Therefore, if somebody suggests that the mean of 2 and 52 is 27, be critical and think about the scarce number of values and whether 2 and 52 can be considered close enough to undertake some calculations.

It is common practice to deliver the mean with a significant figure more than the individual values. This reflects the greater confidence we have on a (correctly calculated) mean than on the individual data. Clearly, the dimensional units of the mean are those of the original data.

The use of the mean is so widespread that it can yield to important mistakes when the average of several means is required. To calculate the 'grand mean', we intuitively tend to sum the averages and divide by the number of averages. Unfortunately, this is true only if the same number of values were used to calculate each and every individual average, otherwise, we should use Equation (2):

$$\bar{\bar{x}} = \frac{\sum (n_1 \cdot \bar{x}_1 + n_2 \cdot \bar{x}_2 + \cdots + n_n \cdot \bar{x}_n)}{n_1 + n_2 + \cdots + n_n} \qquad (2)$$

2.2. Central trend of the data: the median

The median ($\tilde{x}$) is the central value of a series of ordered results (from lowest to highest), so that 50 % of them are higher and 50 % are lower than it. In case of even results, the median is the arithmetic value of the two central ones. The median is a highly robust statistic because it will not be modified unless more than 50 % of the values are wrong. If we approach such an extreme situation, common sense would of course suggest that no calculation should be made but the experimental work reviewed carefully.

The median can be reported with an additional significant figure, as for the mean. Its dimensional units are those of the original data.

2.3. Measuring the dispersion: range or interval

A rough approach to evaluate the dispersion of the results is to report on the difference between the highest and the lowest values. However, the amount of information it yields is quite poor. The dimensional units of the range are those of the original data.

2.4. Measuring the dispersion: standard deviation

This is the usual statistic employed to inform on the dispersion of the data, it can be represented by SD or, simply, s. It is defined as the average deviation of the data from the central trend (Equation (3a)) and is used, typically, to estimate the analytical precision (i.e., how close the data are among themselves)

$$s = \sqrt{\frac{\sum (x_i - \bar{x})^2}{n - 1}} \qquad (3a)$$

Nevertheless, it has to be interpreted with some caution because in analytical chemistry, we need to think in an 'inverse way'. In effect, we consider that a high SD represents a poor performance, whereas a low SD represents good performance. As for the mean, a simple outlier will invalidate the calculations. The denominator $(n - 1)$ is called the 'degrees of freedom' (dof) and it is used to approximate the sample-derived SD to the population-related (or 'true') SD (which is denoted with the σ Greek letter). It is interesting to bear in mind that the sample-derived variance (the squared standard deviation) is an unbiased estimator of the true variance but, unfortunately, the sample-derived SD underestimates the true one. This is only serious when very few values are considered (between 2 and 5) and, so, a reliable SD (precise and unbiased) can be obtained when a very minimum of six values is used [2].

A note of caution when using pocket calculators or software. Many times, two keys (options) are available for the dof: $n - 1$ or n. They can be denoted anything like s_{n-1} and s_n or σ_{n-1} and σ_n (depending on the commercial brands). If less than 30 values (or 30) are available, then select the first option $(n-1)$, otherwise, select the second (n).

It is common practice to report it with two additional significant figures more than the original data. Of course, the dimensional units are those of the original data. Note that the SDs cannot be employed in arithmetic operations directly. If you need to sum them, divide them, etc., the corresponding variances, SD^2 (s^2), must be considered instead.

In case a pocket calculator is not available, Equation (3b) can be applied to perform the calculations by hand more straightforwardly than using Equation (3a).

$$s = \sqrt{\frac{n \sum (x_i^2) - \left(\sum x_i\right)^2}{n \cdot (n - 1)}} \qquad (3b)$$

Finally, remember to keep all digits in your calculator when performing calculations with the SD or the variance. If rounding is performed too early, the results will be quite different from the correct ones. This, of course, does not mean that you have to report a value with an excessive number of digits.

2.5. Measuring the dispersion: relative standard deviation

Derived from the previous statistic, an old denomination for this simple calculation was coefficient of variation (CV). It is used frequently to compare the scatter of the values at different levels of the mean (Equation (4)).

$$\text{RSD}(\%) = 100 \times \frac{s}{\bar{x}} \tag{4}$$

2.6. Measuring the symmetry: skewness

This statistic (denoted as g_1) describes the symmetry of the distribution of the data. Values around zero point out toward a symmetric (Gaussian-type) distribution while clearly negative or positive values indicate an accumulation of values to the left or right sides of the average (= tail); see Equation (5) and Figure 1 (a note of caution: in the figure, the three distributions are displaced to simplify their visualization but the student has to consider that their means are the same; i.e., without a horizontal displacement).

$$g_1 = \frac{\sum (x_i - \bar{x})^3}{n \cdot s^3} \tag{5}$$

2.7. Measuring the shape: kurtosis

Denoted as g_2, it indicates whether the experimental distribution has long tails ($g_2 < 0$), almost no tails ($g_2 > 0$) or 'normal' tails (considering a Gaussian distribution, $g_2 \approx 0$), see Equation (6) and Figure 1. In some textbooks, 3 is not subtracted and, therefore, the

Skewness

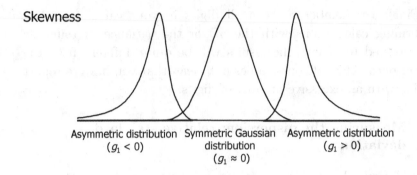

Asymmetric distribution　Symmetric Gaussian　Asymmetric distribution
$(g_1 < 0)$　　　　distribution　　　　$(g_1 > 0)$
　　　　　　　　　$(g_1 \approx 0)$

Kurtosis

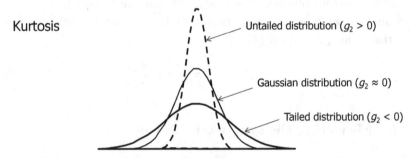

Untailed distribution $(g_2 > 0)$

Gaussian distribution $(g_2 \approx 0)$

Tailed distribution $(g_2 < 0)$

Figure 1.　Graphical concepts of skewed and tailed distributions.

Gaussian distribution should have $g_2 \approx 3$.

$$g_2 = \frac{\sum (x_i - \bar{x})^4}{n \cdot s^4} - 3 \tag{6}$$

3.　ARE THERE OUTLIERS?

This is a fundamental issue that must be considered carefully before proceeding with further studies. Indeed, it was mentioned in the previous section that the two most common descriptors of the data (the mean and the SD) are definitely affected by the presence of just a single outlier.

Suspicious data correspond to too low or too high values. They seem to pertain to a different distribution than the remaining experimental data do. Sometimes, they are detected simply after visual inspection of the data but in some other occasions, they do not appear as obviously. Our recommendation is to apply always a formal test to the extreme values of a series of data in order to unravel their presence.

Two statistical tests are common when looking for a single outlier; namely, the Dixon's and the Grubbs's tests, of which the latter is preferred. Here, both are presented and the student can check whether they lead to the same final conclusions in some of the proposed exercises (which, unfortunately, does not happen in every situation).

Note that the two tests are employed to check whether the lowest or the highest value of a series has an outlying behavior. Usually, this is done recursively; i.e., the test is applied first to an extreme and, then, to the other extreme. This is not totally correct although it is common practice. Some more comments will be given below. Throughout this textbook, it will be considered that only an outlier is present in each side of the series of data.

Both tests evaluate whether the suspicious result can be ascribed to an underlying data distribution. Most frequently, the analyst assumes that the underlying distribution is a Gaussian one. This is very reasonable when replicates are being performed although it may pose some doubts when the series of values correspond to different samples (e.g., from environmental studies). However, there exist statistical tests which do not assume any distribution, they are the non-parametric tests. Some of them became very popular because of their widespread availability in many software, as the box-and-whisker test. However, we will not consider them here.

All statistical tests benefit greatly from a relatively high number of values and, on the contrary, their performance degrades (sometimes dramatically) when a low number of values is available. Therefore, be careful when drawing decisions from small series of data (which, unfortunately, is the common situation in analytical laboratories).

When an analyst rejects a value from a logbook, such a figure must still be readable because in regulated and quality systems, data cannot simply disappear. This means that correction liquids or dramatic cross out (which impede the visualization of the underlying values) cannot be employed at all in laboratory logbooks. When informatic systems are used, electronic traceability of the electronic files is a must (in fact, corrected files are not deleted because they must remain in the system). In addition, when an outlier appears, it is good practice to investigate why it appeared and try to avoid the problem(s) that caused it.

Roughly speaking, the Dixon's and Grubbs's tests evaluate whether the suspicious datum lies far from the main bulk of results. The former considers how far the suspicious datum is from its closest neighbor and compares that distance to the range of values (although this must be corrected slightly in mathematical terms). The Grubbs's test compares the standardized value (i.e., distance between the suspicious value and the mean, divided by the SD), against a critical value.

The low number of values that is usually available in the laboratory makes selection of the confidence level at which the tests have to be used complex. This is due to the difficulty in estimating correctly the underlying distribution of the population from a very small sample (this word is used here not in the chemical sense, but in the statistical one). Although most times, a 95 % confidence level (5 % Type I error, or risk of false positives) is selected for many tests, a 99 % confidence interval (1 % Type I error) might be a safer suggestion for outlier recognition.

3.1. Dixon's test

The Dixon's test (denoted usually as Q, or τ) uses slightly different equations to take account of the number of data (see Equation (7)). It tests whether the null hypothesis (H_0: the distance between the suspicious datum and the remaining data is not significant) can be accepted or rejected. The working procedure is simple:

1st: Order your data, $x_1 < x_2 < \cdots < x_{n-1} < x_n$.
2nd: Decide which is the suspicious value (it is recommended to test the lowest and the highest).

3rd: Use the appropriate form of Equation (7) to calculate the experimental ratio, the Q statistic:

Number of data	x_n is suspicious	x_1 is suspicious
$3 \leq n \leq 7$	$(x_n - x_{n-1})/(x_n - x_1)$	$(x_2 - x_1)/(x_n - x_1)$
$8 \leq n \leq 10$	$(x_n - x_{n-1})/(x_n - x_2)$	$(x_2 - x_1)/(x_{n-1} - x_1)$
$11 \leq n \leq 13$	$(x_n - x_{n-2})/(x_n - x_2)$	$(x_3 - x_1)/(x_{n-1} - x_1)$
$14 \leq n \leq 25$	$(x_n - x_{n-2})/(x_n - x_3)$	$(x_3 - x_1)/(x_{n-2} - x_1)$

$$(7)$$

4th: Compare the experimental Q with the critical value (see Table 1), at the selected confidence level.

5th: If the experimental ratio is larger than the critical value, there are evidences to reject the null hypothesis (i.e., we accept the datum does not pertain to the underlying distribution of the remaining data).

3.2. Grubbs's test

This test is the preferred choice to detect single outliers and it uses a unique equation (as opposed to the Dixon's one). The working procedure is again quite simple:

1st: Order your data, $x_1 < x_2 < \cdots < x_{n-1} < x_n$.

2nd: Decide whether the highest or the lowest is suspicious, if it is not evident, check both.

3rd: Calculate the mean and the SD using all data.

4th: Calculate the experimental statistic (denoted as T or G), according to Equation (8):

$$G = \frac{|x_{\text{suspicious}} - \bar{x}|}{s} \qquad (8)$$

5th: Compare the experimental ratio to the critical values (Table 2); if the critical value is exceeded, there are evidences to reject

Table 1. Critical values for the Dixon's one-side test for the detection of a single outlier ($n =$ number of data values in the series, the percentages refer to the confidence level).

n	95 %	99 %
3	0.94	0.99
4	0.77	0.89
5	0.64	0.79
6	0.56	0.69
7	0.51	0.74
8	0.48	0.68
9	0.44	0.64
10	0.41	0.60
11	0.52	0.68
12	0.49	0.64
13	0.47	0.62
14	0.49	0.64
15	0.47	0.62
16	0.45	0.59
17	0.44	0.58
18	0.42	0.56
19	0.41	0.55
20	0.40	0.54

Source: F. E. Grubbs (1969). *Technometrics*, 11(1): 1–21.

the null hypothesis (H_0: all observations in the series come from the same normal population) and, so, the value must be deleted from further studies (it is assumed that the suspicious datum does not belong to the normal population described by the experimental average and SD).

An important advantage of the Grubbs's test over others is that it is flexible enough to consider other common situations, including the existence of several outliers simultaneously.

When we want to check simultaneously for outliers in both sides but we do not believe that both sides have wrong data (we expect only an outlier in one of the sides), calculate the Grubbs's test as in Equation (8) for each side but compare the maximum of these

Table 2. Critical values for the Grubbs's one-side test for the detection of a single outlier (n = number of data values in the series, the percentages refer to the confidence level).

n	95 %	97.5 %	99 %
3	1.15	1.15	1.15
4	1.46	1.48	1.49
5	1.67	1.72	1.75
6	1.82	1.89	1.94
7	1.94	2.02	2.10
8	2.03	2.13	2.22
9	2.11	2.22	2.32
10	2.18	2.29	2.41
11	2.23	2.35	2.48
12	2.28	2.41	2.55
13	2.33	2.46	2.61
14	2.37	2.51	2.66
15	2.41	2.55	2.70
16	2.44	2.59	2.75
17	2.47	2.62	2.79
18	2.50	2.65	2.82
19	2.53	2.68	2.85
20	2.56	2.71	2.88

two experimental ratios to the critical value. The critical value now corresponds to a 2-tail table. However, Table 2 can still be employed in the following manner: to work at a 95 % confidence level, and 2-tail, consider the critical values of the column headed as 97.5 % (i.e., 2.5 % Type I error); to work at a 99 % confidence level, select the critical values of the 95 % column.

A situation with two possible simultaneous outlying observations (the greatest and the smallest values) can also be addressed. This requires the calculation of a different statistic: the ratio of the range of the values to the calculated SD. This statistic requires dedicated critical values and some additional work because when the critical value is exceeded, we might think that both values are outliers. That would be only true when both the doubtful values are displaced from the mean by approximately the same extent. If this is not true, some further testing has to be done to decide whether to reject only the lowest, only the highest or both extreme values.

Grubbs considered also the case where two extreme values are checked for their outlying behavior simultaneously (i.e., the two largest or the two lowest values). A new statistic is defined so that it calculates the ratio of the sum of squares when the two doubtful values are omitted to the sum of squares when the two doubtful values are included. The sum of squares (or sum of squared deviations) is defined as *sum of squares* $= \sum(x_i - \bar{x})^2$.

If the calculated statistic is *lower* than the critical values (see Table 3), the two values are considered as outliers.

A note is in order here: Significance tests may be one-sided (one side) or two-sided (two sides). In short, and put in simple terms, to determine which one should be applied, think about the real question you want to solve (the null hypothesis). For instance, if you are comparing two average values and there is no additional information available, the test should be two-sided because mean A might be either larger or smaller than mean B. However, if you

Table 3.   Critical values for the Grubbs's one-side test for the detection of two outliers in one side of the data series (two largest or two lowest values).

n	95 %	99 %
4	0.0008	0.0000
5	0.0183	0.0035
6	0.0565	0.0186
7	0.1020	0.0440
8	0.1478	0.0750
9	0.1909	0.1082
10	0.2305	0.1415
15	0.3818	0.2859
20	0.4804	0.3909

Source: Grubbs, F. E. (1969). *Technometrics*, 11(1): 1–21.

need to check whether mean A is larger than mean B, this should be a one-sided test (because you do not consider mean $A <$ mean B). Also, the F-test to be studied later is a one-sided test because we organize the variances so that their ratio is always greater than one. One side and two sides are equivalent to 1-tail and 2-tail, which are two frequently used terms, as well.

3.3.   Calculate once you have eliminated outliers

Only after outliers have been deleted from your datasets (eventually, their causes studied and/or justified) should you proceed to the final calculations.

Most times *accuracy* will be the first question to be addressed. This term includes two concepts: *precision* and *trueness*. The former corresponds to a mutual agreement between the experimental results whereas the latter needs a 'true' value to which the average value of the series has to be compared with. Bias will denote the difference between the average and the 'true' value (its sign should not be discarded).

Precision can be estimated as repeatability (r) and reproducibility (R). For US National Institute of Standards and Technology (NIST) and International Vocabulary of Metrology (VIM), R is the closeness of the agreement between the results of measurements of the same measurand carried out under changed conditions of measurement; for IUPAC, it is the closeness of agreement between independent results obtained with the same method on identical test material but under different conditions (different operators, different apparatus, different laboratories and/or after different intervals of time). Similarly, r is the closeness of the agreement between the results of successive measurements of the same measurand carried out under the same conditions of measurement (NIST and VIM) or the closeness of agreement between independent results obtained with the same method on identical test material, under the same conditions (same operator, same apparatus, same laboratory and after short intervals of time) — IUPAC. R and r are calculated as SDs from series of measurements performed under the conditions specified in the definitions.

Some organizations (like American Society for Testing and Materials, ASTM) define R as the value below which the absolute difference between two single test results on identical material obtained under conditions implying different operators, different apparatus and reagents, and different environmental conditions, may be expected to lie with a specified probability. On the other hand, r is the value below which the absolute difference between two single test results obtained with the same operator, the same instrument and reagents, and identical environmental conditions, may be expected to lie with a specified probability. Although these definitions seem different from the NIST or IUPAC ones, they are not because when an SD is multiplied by a factor (for $n = 2$, this is approximately $2 \cdot \sqrt{2}$) a corresponding range is obtained.

> Precision values have to be compared to the requisites posed by the customer (legislation, etc.) to state whether they are good or not (for a given purpose). On the contrary, it is possible to state in a more general sense whether an analytical procedure yields 'true' results because a statistical test can be applied to check for the significance of the bias (which is the difference between the experimental and the true values). This is the Students' t-test. Incidentally, Student is the pseudonym used by William Sealey Gosset, a chemist, to publish his findings in 1908.

4. PRACTICAL APPLICATIONS OF THE STUDENT'S t-TEST

In the following, some common applications of the Student's t-test will be reviewed briefly. As a preliminary issue, recall that the Student's distribution describes how a reduced number of values distribute under conditions of quasi-normality. Therefore, when very few data exist (quite common in laboratories) the Student's, not the Gaussian distribution is the correct one. His seminal paper (of enormous importance in Statistics) was about the probable error of the mean (when calculated using few data points) and demonstrated that it is possible to bracket the range of values where the true mean of a series of data would be. This means that although for *this* particular series of values (consider your own ones) the best estimation of the true mean (i.e., the mean of the overall population) is given by xx (put here your own experimental average), you have to accept that it would be any other value between yy and zz (the two extreme values calculated as the confidence interval around the experimental mean), at a given probability level (usually, 95 %).

This simplified explanation is not absolutely correct from a historical viewpoint because the confidence intervals were defined by Neyman and Pearson around 1950, not by Student himself. Nevertheless, we felt that was a good way to understand easily the importance of the Student's t-test.

A Student's t-test is any statistical hypothesis test in which the test statistic follows a Student's t-distribution when the null hypothesis is supported. Most t-test statistics can be formulated as $t_{\text{experimental}} = \frac{\hat{\mu}-\mu}{S}$, μ being a population parameter, $\hat{\mu}$ being an estimator of μ and s being the standard error of the estimator (or, equivalently, an estimation of the SD of the estimator).

The confidence interval for a mean calculated with less than 30 values is given by Equation (9). Should more data be available, the Gaussian distribution has to be used.

$$\mu = \bar{x} \pm t \cdot \frac{s}{\sqrt{n}} \tag{9}$$

The $\frac{s}{\sqrt{n}}$ term is referred to as the standard error of the mean; μ stands for the true mean, and t is the Student's factor which, somehow, 'corrects' the Gaussian distribution for the number of values (outliers excluded!). The t-factor (t-value or, just, t) is obtained from statistical tables (Table 4) selecting the probability level and the dof (i.e., dof = $n-1$). The concept of dof is not trivial and for the purposes of this textbook, it would suffice to state that

Table 4. Selected values for the Student's t-value (2-tail and 1-tail tests).

$n-1$	2-tail test		1-tail test	
	95%	99%	95%	99%
1	12.706	63.657	6.314	31.821
2	4.303	9.925	2.920	6.965
3	3.182	5.841	2.353	4.541
4	2.776	4.604	2.132	3.747
5	2.571	4.032	2.015	3.365
6	2.447	3.707	1.943	3.143
7	2.365	3.499	1.895	2.998
8	2.306	3.355	1.860	2.896
9	2.262	3.250	1.833	2.821
10	2.228	3.169	1.812	2.764
11	2.201	3.106	1.796	2.718
12	2.179	3.055	1.782	2.681
13	2.160	3.012	1.771	2.650
14	2.145	2.977	1.761	2.624
15	2.131	2.947	1.753	2.602
16	2.120	2.921	1.746	2.583
17	2.110	2.898	1.740	2.567
18	2.101	2.878	1.734	2.552
19	2.093	2.861	1.729	2.539
20	2.086	2.845	1.725	2.528
30	2.042	2.750	1.697	2.457
40	2.021	2.704	1.684	2.423
50	2.009	2.678	1.676	2.403
60	2.000	2.660	1.671	2.390
70	1.994	2.648	1.667	2.381
80	1.990	2.639	1.664	2.374
90	1.987	2.632	1.662	2.368
100	1.984	2.626	1.660	2.364

is a mathematical restriction that needs to be considered when a statistic is estimated from another estimation. In general, it can be calculated as the number of values minus the number of estimated parameters, here $n-1$.

Several interesting conclusions can be derived from Table 4:

- 1st: When less than 4 values (dof = 3) are available, the *t*-values are so big that its use to draw conclusions should be questioned (has 2 ± 4 a logical explanation?).
- 2nd: The *t*-values decrease only marginally when dof > 7 (6 can be accepted, as well). This means that you will gain a lot when repeating up to seven or eight times an experiment (the factor decreases quite importantly and so the confidence intervals) but it is not worth to expend resources and time in many more experiments.
- 3rd: When the number of values is large (say, more than 20), the *t*-values converge to the *z*-distribution (the Gaussian one).
- 4th: As expected, 99 % probability level yields larger confidence intervals than the 95 % probability one.

4.1. Does my result coincide with a given one?

To address the common issue whether the average result of an experimental setup is statistically equal to a 'true' value derived from some theory, certified reference material (CRM), legal statement, target of a production plan, etc., the null hypothesis (H_0: the difference between the average value and the stated one is not significant) must be tested. The statistic to be used can be derived straightforwardly from Equation (9) and is given in Equation (10):

$$t_{\text{experimental}} = [(\bar{x} - \mu) \cdot \sqrt{n}]/s \qquad (10)$$

As long as the absolute value of the experimental statistic ($t_{\text{experimental}}$) is lower than the tabulated value ($n - 1$ dof and a given confidence level), the null hypothesis should not be rejected. In general, a 95 % confidence level is used and the test is valid for less than 30 replicates.

4.2. Comparing two series of data

Two series of (normal) data usually need to be compared in scattering (SD) and in average. Four different situations can occur, from different scattering and different averages to equal scattering and equal averages. Although this issue is resumed in many textbooks, it is far from trivial and the readers are kindly encouraged to take a glimpse on a more detailed review to get a feeling of the practical problems behind this ubiquitous test [3].

In order to proceed, the two variances must be compared first to decide which variety of the Student's t-test must be used. This is done by applying a Fisher's (or Fisher's and Snedecor's) F-test (H_0: the variances are not different):

$$F_{\text{experimental}} = s_1^2/s_2^2 \qquad \cdot \, (11)$$

The quotient is organized so that the largest numerical value is placed at the numerator (this means that the experimental F-value is always greater than one and, so, we will use 1-tailed distributions). If the experimental value is lower than the critical one ($n_1 - 1$ and $n_2 - 1$ dof for the numerator and denominator, respectively), at a given confidence level (99 % probability is recommended, although 95 % is selected many times), the null hypothesis cannot be rejected. Critical values are given in Tables 5 and 6.

Now, we can compare the averages of the two series. We can resource to a Student's t-test albeit somehow modified.

The t-test consists of evaluating whether the difference between the population means is statistically zero or, stated in other words, whether the confidence interval associated to the subtraction of the population means includes zero. The t-test contrasts the null hypothesis $H_0 \, : \, \mu_1 \, = \, \mu_2 \, (H_1 \, : \, \mu_1 \, \neq \, \mu_2)$. Both data series should proceed from normal populations with means μ_1 and μ_2 and variances σ_1^2 and σ_2^2, respectively. Of course, the populations are unknown and, so, the calculations must be based only in statistics derived from the experimental data: the $\bar{x}_1$ and $\bar{x}_2$ sample means (more specifically in their subtraction, $D = \bar{x}_1 - \bar{x}_2$, which estimates the true difference between the means) and in their sample variances (which are estimated usually by s_1^2 and s_2^2). We know that D ($= \bar{x}_1 - \bar{x}_2$) also follows a normal distribution whose true mean is $\mu_1 - \mu_2$ and its true variance is $\frac{\sigma_1^2}{n_1} + \frac{\sigma_2^2}{n_2}$, being n_1, n_2 the sample sizes for each sample data. This yields the general test posed by Welch in

1938 and depicted in Equation (12).

$$t_{\text{experimental}} = \frac{(\bar{x}_1 - \bar{x}_2) - (\mu_1 - \mu_2)}{\sqrt{\dfrac{\sigma_1^2}{n_1} + \dfrac{\sigma_2^2}{n_2}}} \tag{12}$$

In most textbooks, the theoretical means μ_1 and μ_2 do not appear in the previous equation because the null hypothesis assumes usually that their difference is zero (this is a common assumption but it may be other value) and, thus, their subtraction is not written explicitly. Here, we will maintain that criterion to simplify readability and to allow for a direct comparison with other texts. Recall that the denominator corresponds to the standard error of the difference.

The problem now is that we have to estimate simultaneously the averages and the variances of two *a priori* unknown populations, from a — usually — indeed very limited number of laboratory experimental data. Even worse, it is a well-known fact that the true variance of a population is underestimated when small sample sizes are used, see Section 2.4 (as it usually happens in analytical laboratories due to time and resource constraints). In order to proceed, we need to estimate as accurately as possible the standard error of the difference and, also very important, the effective dof.

In case the variances are comparable (the null hypothesis of the F-test cannot be rejected), we can pool them using the Welch–Satterthwaite's Equation (13):

$$s_{\text{pool}}^2 = \frac{(n_1 - 1) \cdot s_1^2 + (n_2 - 1) \cdot s_2^2}{(n_1 + n_2 - 2)} \tag{13}$$

The t-test statistic is

$$t_{\text{experimental}} = \frac{(\bar{x}_1 - \bar{x}_2)}{s_{\text{pool}}\sqrt{\dfrac{1}{n_1} + \dfrac{1}{n_2}}} \tag{14}$$

As long as the experimental test yields values lower than the tabulated ones ($n_1 + n_2 - 2$ dof, at a given confidence level), the null hypothesis holds on.

Table 5.   Selected values for the Fisher's F-test (1-tail test), 95 % significance level (in rows, dof of the denominator).

	Dof of the numerator →									
	1	2	3	4	5	6	7	8	9	10
1	161.4	199.5	215.7	224.6	230.2	234.0	236.8	238.9	240.5	241.9
2	18.51	19.00	19.16	19.25	19.30	19.33	19.35	19.37	19.38	19.40
3	10.13	9.55	9.28	9.12	9.01	8.94	8.89	8.85	8.81	8.79
4	7.71	6.94	6.59	6.39	6.26	6.16	6.09	6.04	6.00	5.96
5	6.61	5.79	5.41	5.19	5.05	4.95	4.88	4.82	4.77	4.74
6	5.99	5.14	4.76	4.53	4.39	4.28	4.21	4.15	4.10	4.06
7	5.59	4.74	4.35	4.12	3.97	3.87	3.79	3.73	3.68	3.64
8	5.32	4.46	4.07	3.84	3.69	3.58	3.50	3.44	3.39	3.35
9	5.12	4.26	3.86	3.63	3.48	3.37	3.29	3.23	3.18	3.14
10	4.96	4.10	3.71	3.48	3.33	3.22	3.14	3.07	3.02	2.98
11	4.84	3.98	3.59	3.36	3.20	3.09	3.01	2.95	2.90	2.85
12	4.75	3.89	3.49	3.26	3.11	3.00	2.91	2.85	2.80	2.75
13	4.67	3.81	3.41	3.18	3.03	2.92	2.83	2.77	2.71	2.67
14	4.60	3.74	3.34	3.11	2.96	2.85	2.76	2.70	2.65	2.60

15	4.54	3.68	3.29	3.06	2.90	2.79	2.71	2.64	2.59	2.54
16	4.49	3.63	3.24	3.01	2.85	2.74	2.66	2.59	2.54	2.49
17	4.45	3.59	3.20	2.96	2.81	2.70	2.61	2.55	2.49	2.45
18	4.41	3.55	3.16	2.93	2.77	2.66	2.58	2.51	2.46	2.41
19	4.38	3.52	3.13	2.90	2.74	2.63	2.54	2.48	2.42	2.38
20	4.35	3.49	3.10	2.87	2.71	2.60	2.51	2.45	2.39	2.35
25	4.24	3.39	2.99	2.76	2.60	2.49	2.40	2.34	2.28	2.24
30	4.17	3.32	2.92	2.69	2.53	2.42	2.33	2.27	2.21	2.16
40	4.08	3.23	2.84	2.61	2.45	2.34	2.25	2.18	2.12	2.08
50	4.03	3.18	2.79	2.56	2.40	2.29	2.20	2.13	2.07	2.03
60	4.00	3.15	2.76	2.53	2.37	2.25	2.17	2.10	2.04	1.99
70	3.98	3.13	2.74	2.50	2.35	2.23	2.14	2.07	2.02	1.97
80	3.96	3.11	2.72	2.49	2.33	2.21	2.13	2.06	2.00	1.95
90	3.95	3.10	2.71	2.47	2.32	2.20	2.11	2.04	1.99	1.94
100	3.94	3.09	2.70	2.46	2.31	2.19	2.10	2.03	1.97	1.93

(Continued)

Table 5.   (*Continued*)

	Dof of the numerator →									
	11	12	13	14	15	16	17	18	19	20
1	243.0	243.9	244.7	245.4	245.9	246.5	246.9	247.3	247.7	248.0
2	19.40	19.41	19.42	19.42	19.43	19.43	19.44	19.44	19.44	19.45
3	8.76	8.74	8.73	8.71	8.70	8.69	8.68	8.67	8.67	8.66
4	5.94	5.91	5.89	5.87	5.86	5.84	5.83	5.82	5.81	5.80
5	4.70	4.68	4.66	4.64	4.62	4.60	4.59	4.58	4.57	4.56
6	4.03	4.00	3.98	3.96	3.94	3.92	3.91	3.90	3.88	3.87
7	3.60	3.57	3.55	3.53	3.51	3.49	3.48	3.47	3.46	3.44
8	3.31	3.28	3.26	3.24	3.22	3.20	3.19	3.17	3.16	3.15
9	3.10	3.07	3.05	3.03	3.01	2.99	2.97	2.96	2.95	2.94
10	2.94	2.91	2.89	2.86	2.85	2.83	2.81	2.80	2.79	2.77
11	2.82	2.79	2.76	2.74	2.72	2.70	2.69	2.67	2.66	2.65
12	2.72	2.69	2.66	2.64	2.62	2.60	2.58	2.57	2.56	2.54
13	2.63	2.60	2.58	2.55	2.53	2.51	2.50	2.48	2.47	2.46
14	2.57	2.53	2.51	2.48	2.46	2.44	2.43	2.41	2.40	2.39

15	2.51	2.48	2.45	2.42	2.40	2.38	2.37	2.35	2.34	2.33
16	2.46	2.42	2.40	2.37	2.35	2.33	2.32	2.30	2.29	2.28
17	2.41	2.38	2.35	2.33	2.31	2.29	2.27	2.26	2.24	2.23
18	2.37	2.34	2.31	2.29	2.27	2.25	2.23	2.22	2.20	2.19
19	2.34	2.31	2.28	2.26	2.23	2.21	2.20	2.18	2.17	2.16
20	2.31	2.28	2.25	2.22	2.20	2.18	2.17	2.15	2.14	2.12
25	2.20	2.16	2.14	2.11	2.09	2.07	2.05	2.04	2.02	2.01
30	2.13	2.09	2.06	2.04	2.01	1.99	1.98	1.96	1.95	1.93
40	2.04	2.00	1.97	1.95	1.92	1.90	1.89	1.87	1.85	1.84
50	1.99	1.95	1.92	1.89	1.87	1.85	1.83	1.81	1.80	1.78
60	1.95	1.92	1.89	1.86	1.84	1.82	1.80	1.78	1.76	1.75
70	1.93	1.89	1.86	1.84	1.81	1.79	1.77	1.75	1.74	1.72
80	1.91	1.88	1.84	1.82	1.79	1.77	1.75	1.73	1.72	1.70
90	1.90	1.86	1.83	1.80	1.78	1.76	1.74	1.72	1.70	1.69
100	1.89	1.85	1.82	1.79	1.77	1.75	1.73	1.71	1.69	1.68

Table 6. Selected values for the Fisher's F-test (1-tail test), 99 % significance level (in rows, dof of the denominator).

	Dof of the numerator →									
	1	2	3	4	5	6	7	8	9	10
1	4052.2	4999.5	5403.4	5624.6	5763.6	5859.0	5928.4	5981.1	6022.5	6055.8
2	98.50	99.00	99.17	99.25	99.30	99.33	99.36	99.37	99.39	99.40
3	34.12	30.82	29.46	28.71	28.24	27.91	27.67	27.49	27.35	27.23
4	21.20	18.00	16.69	15.98	15.52	15.21	14.98	14.80	14.66	14.55
5	16.26	13.27	12.06	11.39	10.97	10.67	10.46	10.29	10.16	10.05
6	13.75	10.92	9.78	9.15	8.75	8.47	8.26	8.10	7.98	7.87
7	12.25	9.55	8.45	7.85	7.46	7.19	6.99	6.84	6.72	6.62
8	11.26	8.65	7.59	7.01	6.63	6.37	6.18	6.03	5.91	5.81
9	10.56	8.02	6.99	6.42	6.06	5.80	5.61	5.47	5.35	5.26
10	10.04	7.56	6.55	5.99	5.64	5.39	5.20	5.06	4.94	4.85
11	9.65	7.21	6.22	5.67	5.32	5.07	4.89	4.74	4.63	4.54
12	9.33	6.93	5.95	5.41	5.06	4.82	4.64	4.50	4.39	4.30
13	9.07	6.70	5.74	5.21	4.86	4.62	4.44	4.30	4.19	4.10
14	8.86	6.51	5.56	5.04	4.69	4.46	4.28	4.14	4.03	3.94

15	8.68	6.36	5.42	4.89	4.56	4.32	4.14	4.00	3.89	3.80
16	8.53	6.23	5.29	4.77	4.44	4.20	4.03	3.89	3.78	3.69
17	8.40	6.11	5.18	4.67	4.34	4.10	3.93	3.79	3.68	3.59
18	8.29	6.01	5.09	4.58	4.25	4.01	3.84	3.71	3.60	3.51
19	8.18	5.93	5.01	4.50	4.17	3.94	3.77	3.63	3.52	3.43
20	8.10	5.85	4.94	4.43	4.10	3.87	3.70	3.56	3.46	3.37
25	7.77	5.57	4.68	4.18	3.85	3.63	3.46	3.32	3.22	3.13
30	7.56	5.39	4.51	4.02	3.70	3.47	3.30	3.17	3.07	2.98
40	7.31	5.18	4.31	3.83	3.51	3.29	3.12	2.99	2.89	2.80
50	7.17	5.06	4.20	3.72	3.41	3.19	3.02	2.89	2.78	2.70
60	7.08	4.98	4.13	3.65	3.34	3.12	2.95	2.82	2.72	2.63
70	7.01	4.92	4.07	3.60	3.29	3.07	2.91	2.78	2.67	2.59
80	6.96	4.88	4.04	3.56	3.26	3.04	2.87	2.74	2.64	2.55
90	6.93	4.85	4.01	3.53	3.23	3.01	2.84	2.72	2.61	2.52
100	6.90	4.82	3.98	3.51	3.21	2.99	2.82	2.69	2.59	2.50

(Continued)

Table 6.   (*Continued*)

Dof of the numerator →

	11	12	13	14	15	16	17	18	19	20
1	6083.3	6106.3	6125.9	6142.7	6157.3	6170.1	6181.4	6191.5	6200.6	6208.7
2	99.41	99.42	99.42	99.43	99.43	99.44	99.44	99.44	99.45	99.45
3	27.13	27.05	26.98	26.92	26.87	26.83	26.79	26.75	26.72	26.69
4	14.45	14.37	14.31	14.25	14.20	14.15	14.11	14.08	14.05	14.02
5	9.96	9.89	9.82	9.77	9.72	9.68	9.64	9.61	9.58	9.55
6	7.79	7.72	7.66	7.60	7.56	7.52	7.48	7.45	7.42	7.40
7	6.54	6.47	6.41	6.36	6.31	6.28	6.24	6.21	6.18	6.16
8	5.73	5.67	5.61	5.56	5.52	5.48	5.44	5.41	5.38	5.36
9	5.18	5.11	5.05	5.01	4.96	4.92	4.89	4.86	4.83	4.81
10	4.77	4.71	4.65	4.60	4.56	4.52	4.49	4.46	4.43	4.41
11	4.46	4.40	4.34	4.29	4.25	4.21	4.18	4.15	4.12	4.10
12	4.22	4.16	4.10	4.05	4.01	3.97	3.94	3.91	3.88	3.86
13	4.02	3.96	3.91	3.86	3.82	3.78	3.75	3.72	3.69	3.66
14	3.86	3.80	3.75	3.70	3.66	3.62	3.59	3.56	3.53	3.51

15	3.73	3.67	3.61	3.56	3.52	3.49	3.45	3.42	3.40	3.37
16	3.62	3.55	3.50	3.45	3.41	3.37	3.34	3.31	3.28	3.26
17	3.52	3.46	3.40	3.35	3.31	3.27	3.24	3.21	3.19	3.16
18	3.43	3.37	3.32	3.27	3.23	3.19	3.16	3.13	3.10	3.08
19	3.36	3.30	3.24	3.19	3.15	3.12	3.08	3.05	3.03	3.00
20	3.29	3.23	3.18	3.13	3.09	3.05	3.02	2.99	2.96	2.94
25	3.06	2.99	2.94	2.89	2.85	2.81	2.78	2.75	2.72	2.70
30	2.91	2.84	2.79	2.74	2.70	2.66	2.63	2.60	2.57	2.55
40	2.73	2.66	2.61	2.56	2.52	2.48	2.45	2.42	2.39	2.37
50	2.63	2.56	2.51	2.46	2.42	2.38	2.35	2.32	2.29	2.27
60	2.56	2.50	2.44	2.39	2.35	2.31	2.28	2.25	2.22	2.20
70	2.51	2.45	2.40	2.35	2.31	2.27	2.23	2.20	2.18	2.15
80	2.48	2.42	2.36	2.31	2.27	2.23	2.20	2.17	2.14	2.12
90	2.45	2.39	2.33	2.29	2.24	2.21	2.17	2.14	2.11	2.09
100	2.43	2.37	2.31	2.27	2.22	2.19	2.15	2.12	2.09	2.07

In case the variances of the two series are different (the null hypothesis of the F-test can be rejected), there is no exact solution to the comparison of the two means (this is the so-called Fisher–Behrens problem).

Equation (15) presents how to calculate the experimental t-value in this case. The dof must be approximated from the variances and the number of experimental points (Equation (16)). An alternative to the use of Equation (16), which sometimes is a bit more convenient, is to estimate directly the critical t-value using Equation (17). There t_1 and t_2 are the Student's tabulated values for each series of data at a given significance level (usually 95 %), 2-tail and $n_i - 1, i = 1, 2$ dof. If the number of data points of each series is the same, then $t_{\text{critical}} = t_1 = t_2$.

$$t_{\text{experimental}} = \frac{(\bar{x}_1 - \bar{x}_2)}{\sqrt{\dfrac{s_1^2}{n_1} + \dfrac{s_2^2}{n_2}}} \tag{15}$$

$$\text{dof} = \frac{\left[(s_1^2/n_1) + (s_2^2/n_2)\right]^2}{\left[\dfrac{(s_1^2/n_1)^2}{n_1 - 1} + \dfrac{(s_2^2/n_2)^2}{n_2 - 1}\right]} \tag{16}$$

$$t_{\text{critical}} = \frac{t_1 \cdot s_1^2 + t_2 \cdot s_2^2}{s_1^2 + s_2^2} \tag{17}$$

5. CALIBRATION

Quantifying an analyte in the final aliquot under analysis is usually the primary objective of any laboratory. For this, analytical chemists employ a suite of instruments whose electronic signals must be related mathematically to the amount of substance or property we are interested in. Such methods are termed relative methods of analysis, in contrast to the absolute methods of analysis (which do not require a mathematical relationship because what we measure is what we are looking for). Recall here that although the examples presented in this book refer to 'concentration' as the parameter to be measured, any other physico-chemical property may be of interest and, so, measured; e.g., viscosity, strength of a material, gravity, octane number of a gasoline, percentage of sand, etc.

Producing a 'calibration model' or 'calibration function' is defined as the verification of the response of an instrument to a material of known properties and, if necessary, the correction by a factor to take the instrument to the corresponding mark. Frequently, this term is used interchangeably with the concept of standardization, although the latter is a distinct idea. Standardization means to characterize the response of an instrument according to the known properties of the material, and this is usually done by the 'calibration curve' (which should be called 'standardization curve'). Indeed, most present-day instruments consist of different systems and their complexity makes it difficult to really calibrate them *senso stricto*. In fact, what is usually done in laboratories is standardizing the response by means of a series of samples of known concentrations (or any other property to be quantified).

Accordingly, standardization ('calibration') in analytical chemistry is the operation that determines the functional relationship between measured values (signal intensities, y) and the chemical properties that characterize analytes and their amounts or, usually, concentrations (x). This operation includes the selection of the model, the estimation of the parameters and, in addition, their errors and their validation. The terms calibration and standardization will be used synonymously throughout this book because this is common practice in most laboratories.

It is worth noting the empirical nature of standardization because it is a working process that is performed under some 'specified conditions'. This remarkably means that a system would have a straight-line behavior for me (right now) but a similar system could show a different behavior in 'your' laboratory.

The working procedure is conceptually rather simple:

- 1st: Fix an adequate setup for your instrument. Check for its (statistical) stability.
- 2nd: Prepare a suite of mixtures where the property to be measured varies within your control (e.g., a suite of solutions with

increasing amounts of an analyte). These are termed currently calibrators, standards or, a bit more correctly, standardization (or calibration) solutions (if you measure liquid phase). A critical issue to develop these solutions is to match their matrix to the matrix you expect in the aliquots derived from the sample treatment (for this, sometimes, you would add acids, salts, etc.). If possible, treat standards as you will treat samples, including sample preparation steps.

- 3rd: Measure the standards, in the same manner as you will measure the sample aliquots.
- 4th: Set an empirical mathematical relationship between the signals and the values of the property you are interested in. This is the standardization (calibration) function.
- 5th: Measure the sample aliquots and evaluate the amount of property by interpolation. Do never extrapolate! Scientists should never 'invent' what would happen outside the experimental conditions they can study.

However, in practice, many problems can arise due to practical difficulties in adhering to these principles. Just for instance:

(a) Instrumental fluctuations, like drifts, pressure variations (in chromatography), source temperature functioning, etc. They can be compensated for using the *internal standard calibration method*. This consists of adding another species (the internal standard), different from the analyte, to all the solutions that are to be measured. The ratio between the signals attributed to the analyte and the internal standard is calculated for each solution and plotted against the concentration of analyte. The internal standard must fulfill several relevant characteristics to be of use, the most important ones being:

1. similar physical and chemical behavior as the analyte in the measurement device,
2. range of concentrations similar to that of the analyte,
3. should not be present either in the test samples or in the calibrators.

(b) It is commonly difficult to match the matrix of the working aliquots (obtained from the treatment of the samples) to the matrix of the standardization solutions employed to standardize the instrument (prepared from pure standard solutions). Hence, it might be difficult to accept that the behavior of the standards and the sample aliquots is the same. If this is so, ... well, all your work might reasonably be questioned. There is no room in this chapter to deal with the so many difficulties that may occur when performing standardization (calibration). As a frequently applied alternative, we will focus on the typical standard addition method (SAM) calibration that was reviewed recently [4].

5.1. Basic calibration calculi

The term 'calibration curve' or 'calibration line' is used ubiquitously despite its obvious ambiguity and lack of rigor. However, when an analytical chemist says 'linear calibration,' it usually should be understood as a 'straight-line' calibration function of the form signal = offset + constant · property, or just: $y = a + b \cdot x$. It is worth noting that this function is a model scientists deploy to understand the physical or chemical system but it is not real (!). This means that a model tries to represent reality, but it is not reality by itself.

From the experimental pairs of values (y_i, x_i), the a and b constants ('intercept' and 'slope', respectively) are calculated by means of the least squares criterion. Such a criterion represents a common approach to adjust a function to a set of experimental points so that the function goes through all the points, as close as possible (the sum of the residuals is zero). The least squares fit is implemented nowadays in almost any software and any basic pocket calculator. Therefore, we will not present the equations here.

On the contrary, it is important to stress the three basic, mathematical requisites to apply this fundamental criterion correctly:

1. The experimental errors accumulate in the signal domain; i.e., the x values do not have error (or, at most, errors in x values are

negligible compared to the errors in the y values). This statement can be problematic in trace analysis.

2. The errors in the signals do have a Gaussian distribution. In general, this can be accepted providing our instrument is stable and is under statistical control.

3. The errors in the signals are independent on the magnitude of the x values. This is called homoscedasticity. The opposite situation is called heteroscedasticity and, unfortunately, is quite common in analytical chemistry and it appears many times when the standardizations are studied seriously.

If you experience some difficulties in fulfilling these three basic statements, be aware of your calibration. It might not be correct and it might yield wrong or unjustifiable quantitations. You should always spend several minutes in studying your model and, eventually, trying to solve its problems. A quite simple approach to detect these problems graphically is presented in the next section.

It is intuitive for any person in a laboratory to accept that the measured signals do have some unavoidable errors (lets think here only about random errors). Thus, it should not be surprising that, according to the error propagation rules, they propagate throughout the calculi to the intercept and the slope. This means that the least squares criterion yields a and b values which, in fact, correspond to the most probable values that we could obtain under our experimental conditions in case we would repeat the experiments many times. They constitute the average values of their associated random distributions. Nevertheless, we have to accept that their true values (consider the term true in an abstract, 'philosophical' sense; i.e., population of every possible value) may not totally agree with the average ones. We need to calculate intervals of values for the intercept and the slope where we reasonably can accept that the true values will be.

The Student's t-test is of much help here and, so, a standardization (calibration) function should be expressed correctly as $y = (a \pm t \cdot s_a) + (b \pm t \cdot s_b) \cdot x$. Here, s_a and s_b are the standard errors of the intercept and slope, respectively, and t is the Student's tabulated t-value for $n - 2$ dof and a given confidence level (usually 95 %).

Although s_a and s_b are given automatically in most software, pocket calculators do not offer them. This is not a major problem as they can be calculated straightforwardly

$$s_a^2 \cong s_{y/x}^2 \left(\frac{1}{N} + \frac{\bar{x}^2}{\sum_{i=1}^{N} (x_i - \bar{x})^2} \right) \tag{18}$$

$$s_b^2 \cong \frac{s_{y/x}^2}{\sum_{i=1}^{N} (x_i - \bar{x})^2} \tag{19}$$

$$s_{y/x}^2 = \frac{\sum_{i=1}^{N} e_1^2}{N-2} = \frac{\sum_{i=1}^{N} (\hat{y}_i - y_i)^2}{N-2} \tag{20a}$$

$s_{y/x}$ is called the standard error of the fit and the lower it is, the better the function will fit the experimental points. The e values are called the residuals and they represent the amount of experimental information which is not explained by the model. They should have a random distribution (this is linked to the homoscedasticity condition given above). The 'hat' ($^\wedge$) over a letter means 'predicted value using a model'.

It is worth noting that it is immediate to check whether the intercept or the slope is statistically zero, either by performing a t-test or by observing if zero is within the corresponding confidence intervals. These are independent tests. However, when you have to test statistically whether the intercept equals a value (here, zero) and the slope equals a value (here, one) *simultaneously* (e.g., when comparing two analytical methods), you need to perform an F-test because both parameters are not independent, see Equation (20b) (with 2 and $n-2$ dof for numerator and denominator, respectively, the x values refer to the abscissas).

$$F_{\text{experimental}} = \frac{a^2 + 2 \cdot \bar{x} \cdot (0-a) \cdot (1-b) + \left(\sum x_i^2 / n \right) \cdot (1-b)^2}{2 \cdot s_{y/x}^2 / n} \tag{20b}$$

A final reasoning leads us to an important conclusion: if the intercept and the slope inherently have a (Gaussian) distribution, any interpolation will do (again, due to error propagation). Thus, any interpolated value should be given with its associate confidence interval (Equation (21a))

$$\hat{x}_0 \pm t \cdot \frac{s_{y/x}}{b} \cdot \sqrt{1 + \frac{1}{N} + \frac{(y_0 - \bar{y})^2}{b^2 \sum_{i=1}^{N}(x_i - \bar{x})^2}} \qquad (21a)$$

where N is the number of standards, and y_0 is the signal obtained for the measured aliquot wherefrom x_0 is interpolated. In case replicated measurements are made for the standards and the sample aliquot (denoted by q), the equation should be modified slightly (Equation (21b)).

$$\hat{x}_0 \pm t \cdot \frac{s_{y/x}}{b} \cdot \sqrt{\frac{1}{q} + \frac{1}{N} + \frac{(y_0 - \bar{y})^2}{b^2 \sum_{i=1}^{N}(x_i - \bar{x})^2}} \qquad (21b)$$

These equations are not calculated manually in general and they can be implemented easily in any popular spreadsheet. Note that the t-values correspond to 2-tail statistics because a symmetric interval around a value is considered (which, in fact, it is an approach itself!).

5.2. Validate your model

The calculations above are not too complex and, therefore, there is an unfortunate generalized trend to assume that they are 'perfect'. Without doubt, the numerical values themselves will be fine unless some programming error occurs (and recall that according to international quality guides, it is under the laboratory's responsibility to demonstrate that this does not occur). But this does not mean at all that the model is correct. It is your duty to demonstrate that the calibration is acceptable beyond any reasonable doubt. This is not a trivial task and several aspects can be addressed. Here, we will only resume some conceptual ideas in order to provide a basic background for undergraduates but, please, bear in mind that they are not exhaustive.

The very first and critical step to validate a calibration model is to detect and eliminate outliers. There are some potential statistical tests for this but in order to get sound and reliable results, they require quite a lot of experimental data. This is not usually the case in analytical laboratories. However, anyone can plot their experimental values (e.g., absorbance versus concentration) and assess whether a point departs from the overall trend. Most of the times this was caused by some experimental error and can be corrected, otherwise just delete the point. It is of paramount importance to recall that even a single outlier might totally bias the calibration straight line and, so, invalidate your work. Be specially cautious with possible outliers at the extremes of the calibration range because they induce a strong rotational effect on the slope and intercept (this jargon was coined by Ellison and Thompson [5]).

Calibration plots are very useful but sometimes they do not allow either a perfect view or a sound decision-making. Somehow we need to 'zoom in' to visualize trends best. This can be done easily by plotting the residuals of the fitted function. Residuals are, simply, the difference between your experimental signals and the signals proposed by the model ($y_{\text{pred}} - y_{\text{experimental}}$, or just, $\hat{y}_i - y_i$). Two key aspects should be studied:

1. There are no trends. Remember that residuals should be random and of a similar magnitude (see Figure 2). Figure 2a occurs very frequently and it means that our model is wrong because we tried to fit a straight line to something that, indeed, should be a parabola. Figure 2b is not so frequent but it can appear when there is a systematic problem (e.g., with the pipettes, be careful when performing dilutions), the model is again invalid because it

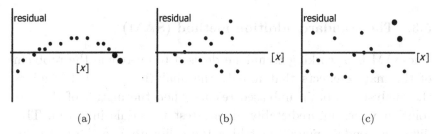

Figure 2. Examples of calibration residuals, (a) distribution when a parabola was fit (incorrectly) by a straight line, (b) distribution when a proportional error is present on the calibration, (c) typical homoscedastic distribution.

does not fit the basic assumptions of the least squares fit criterion. Figure 2c is what should be obtained.

2. There is no large residual associated to a data point. If that is the case, that point is likely an outlier. Problem here is to define large objectively. A pragmatic and very simple criterion consists of plotting the standardized residuals. Thompson and Lowthian [6] recently proposed to use a simplified standardized residual, which is just the result of ratioing each residual by the SD of all the residuals. Hence, any residual out of the ±3 boundaries should be associated to an outlier. These calculations are trivial nowadays using any popular spreadsheet.

Finally, it is of utmost importance to emphasize that the correlation coefficient (denoted as r) is not a valid criterion by itself to decide on the validity of a calibration model. This is an extended, common mistake in many published works, likely because of the availability of software and a lack of a true understanding of this parameter. Even worse, some authors discuss about the correlation coefficient but they present the coefficient of determination (denoted as R^2) which has a different meaning.

Note that a large correlation coefficient does not imply a straight-line behavior, it could be a very nice parabola. This has been demonstrated many times and it is still a very common misunderstanding.

5.3. The standard addition method (SAM)

The SAM is a working technique devised to overcome the problem of the matrix effects that modify the analytical signal. It enables the analyst to obtain unbiased results when the matrix of the test solution varies unpredictably among test materials in a run. This difficulty renders matrix matching the calibrators to the test solutions impossible to apply. The sensitivity of the method is affected unpredictably from test solution to test solution [5]. In essence, the method comprises three steps:

- 1st: Measure the analytical response produced by the test solution.
- 2nd: Spike the test solution with one or more known amounts of analyte (usually in the form of an aqueous solution without concomitants) to get corresponding solutions, and measure the new responses.
- 3rd: From the responses calculate a straight-line fit of the experimental data and from that evaluate the concentration that produced the response obtained from the untreated test solution.

IUPAC recommends adding amounts of analyte x_i, equimolar as expected for the test solution, so that $x_1 \sim x_0$; $x_2 \sim 2 \cdot x_1$; ... $x_p \sim p \cdot x_1$ (p is the number of levels of the addition; frequently $p = 3$ or 4). A preliminary measurement of x_0 would be needed [7].

However, one must be certain on the straight-line behavior of the calibration otherwise it should not be applied. One more typical problem of SAM is that frequently it is not possible to get reference samples without the analyte to evaluate the recovery. Finally, SAM has traditionally been used as an extrapolation method whose predictions may be obviously affected by such an inadequate practice

When extrapolating, the quantity of analyte in the test solution, $\hat{x}_0$, is obtained by prolonging the line to the abscissa: $y = 0$, $x_0 = -(a/b)$. Its standard error is calculated using Equation (22a) [8], in many textbooks, the minus sign in the (a/b) term is avoided, which leads to a relevant error

$$s_{\hat{x}_0} = \frac{s_{y/x}}{b} \cdot \left(1 + \frac{1}{N} + \frac{\left(-\left(\frac{a}{b}\right) - \bar{x}\right)^2}{\sum_{i=1}^{N}(x_i - \bar{x})^2}\right)^{1/2} \tag{22a}$$

where N is the total number of calibration solutions used in the standardization step (i.e., including replicates of the spiked and unspiked calibrators). The term '1' within the parenthesis stems from error propagation.

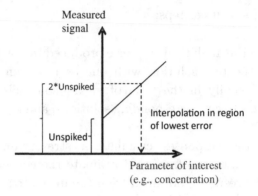

Figure 3. Graphical concept of interpolation in the SAM.

We strongly recommend interpolation, which is as simple as depicted in Figure 3. The calculation of the standard error of the interpolated value is obtained from Equation (22b), which is usual Equation (21a) considering $2 \cdot y_0$ as the signal to be interpolated. This very simple approach leads to smaller confidence intervals than extrapolation and is more correct [9]. Note that y_0 is the signal measured for the test aliquot, not the intercept of the calibration line [4, 9]. Recall that the t-values should be two-sided.

$$\hat{x}_0 \pm t \cdot \frac{s_{y/x}}{b} \cdot \sqrt{1 + \frac{1}{N} + \frac{(2 \cdot y_0 - \bar{y})^2}{b^2 \sum_{i=1}^{N}(x_i - \bar{x})^2}} \tag{22b}$$

5.4. The addition calibration method

A well-known drawback of SAM is that it cannot be used for more than one test solution. Hence, a new standard addition calibration has to be developed for each test solution. This implies usually very large workloads in laboratories.

A strategy to overcome this consists of using the so-called *addition calibration method*. The key idea here is to develop a unique SAM calibration and use it for several test solutions whose matrices are essentially similar. If such a similarity can be accepted (e.g., different samples from a same production process), the SAM calibrations should only differ in their intercepts, not in the slopes. It is possible, then, to predict the concentration of each sample by adapting the intercept that is considered in the SAM regression equation [10,11]. Given an intercept (ordinate 1, O1) calculated from a first SAM calibration developed with a characteristic test solution (sample 1), subtract the signal of the unspiked test solution ($S1$) to get the background and, then, add the signal of the next unspiked test solution (e.g., sample 2, $S2$); in brief, the calculations are resumed as:

Estimated concentration $= (|O1 - S1 + S2|)/b$, where the absolute value is considered ($||$) and b is the slope of the initial SAM calibration. This method is particularly suited for replicate analysis of samples. Numerical exercises are given in Chapter 5.

5.5. Comparing calibration lines

A relevant question to be addressed is how can we decide whether SAM should be applied instead of current (aqueous-based) calibration. In principle, it is not possible to state *a priori* an answer and some work has to be done at the laboratory to get an *ad hoc* final decision. It is commonly accepted that a good way to evaluate whether the concomitants of a sample matrix interfere with the analyte signal consists of comparing the slopes of the two regression straight lines obtained with classical (aqueous-based) and standard addition calibrations. If they do not differ, it is accepted that the matrix does not modify the sensitivity of the system under those particular experimental conditions and, so, the SAM is not mandatory. Otherwise, the SAM will be required.

Comparing the slopes of two regression (straight) lines is a particular case of comparing two averages (studied in Section 4.2 in this chapter), although with some changes due to the use of lines instead of replicated series of values.

First, the variances must be compared (Equation (23)) using an F-test

$$F_{\exp} = \left(s^2_{(y/x),\text{large}} / s^2_{(y/x),\text{small}} \right) \qquad (23)$$

which is compared to the critical value ($F_{\text{tabulated}}$, $n_1 - 2, n_2 - 2$, % confidence level). While the experimental value is lower than the tabulated one, H_0 cannot be rejected and, so, the variances are comparable. In this situation, Equations (24) and (25) have to be applied:

$$t_{\text{experimental}} = \frac{(b_1 - b_2)}{\sqrt{s^2_{(y/x),\text{pool}} \left(\dfrac{1}{\sum (x_{i,1} - \bar{x}_1)^2} + \dfrac{1}{\sum (x_{i,2} - \bar{x}_2)^2} \right)}} \tag{24}$$

$$s^2_{(y/x),\text{pool}} = \frac{(n_1 - 2)s^2_{(y/x)1} + (n_2 - 2)s^2_{(y/x)2}}{n_1 + n_2 - 4} \tag{25}$$

Otherwise, we face again the Fisher–Behrens problem and Equation (26) must be used:

$$t^*_{\text{experimental}} = \frac{(b_1 - b_2)}{\sqrt{s^2_{b_1} + s^2_{b_2}}} = \frac{(b_1 - b_2)}{\sqrt{\dfrac{s^2_{(y/x)1}}{\sum (x_{i1} - \bar{x}_1)^2} + \dfrac{s^2_{(y/x)2}}{\sum (x_{j2} - \bar{x}_2)^2}}} \tag{26}$$

The tabulated t-value is derived either from Equation (17) (considering the s_b^2 of each regression line) or from the typical tables, for a given % probability level and dof calculated by Equation (27):

$$\frac{1}{\text{dof}} = \left[\left(\frac{c^2}{(n_1 - 2)} \right) + \left(\frac{(1 - c)^2}{(n_2 - 2)} \right) \right]$$

$$\text{with} \quad c = \frac{s^2_{(y/x)1} / \sum (x_{i1} - \bar{x}_1)^2}{\left(\dfrac{s^2_{(y/x)1}}{\sum (x_{i1} - \bar{x}_1)^2} + \dfrac{s^2_{(y/x)2}}{\sum (x_{j2} - \bar{x}_2)^2} \right)}; \quad \text{and } n_1 \leq n_2 \tag{27}$$

Although the equations above seem hard to apply, they are not because calculations like $s^2_{(y/x)1} / \sum (x_{i1} - \bar{x}_1)^2$ are obtained immediately from any spreadsheet or statistical package under the heading 'squared standard error of the slope' ($s^2_{b_1}$).

5.6. Limits of detection and quantitation

To calibrate an instrument may be cumbersome and time-consuming and, so, it is advisable to extract as much information as possible from it. In particular, two relevant performance parameters (this is the new term for the traditional denomination *analytical figures of merit*) can be derived from an optimized calibration: the LOD and the LOQ. Although their fundamental concepts are logical and broadly accepted, few calculations have been so controversial in analytical chemistry.

In this section, we will present both the classical and the updated definitions (they will be applied also in other chapters of this book), along with a brief discussion on the conceptual ideas behind them. A good example of the controversy associated to the calculations involved with these parameters is that many recent papers in relevant journals and laboratory reports still maintain the 1970s-classical, old definitions... although IUPAC, ISO and the European Union set new common definitions by 1995 which superseded the previous ones (the EU set its guidance a bit later, in 2002). Without doubt, changes need time to settle down, they raise questions and discussions and must gain acceptance by demonstrating they are good enough (or, at least, that they solve some previous problems).

The slow pace in adopting the new definitions might be explained by usual practices associated to some analytical techniques (e.g., in atomic spectrometry, it is current practice to report a value plus and minus an SD, instead of a true confidence interval), the different ways to evaluate methods within the analytical chemistry field and, worst, there is still no universal agreement on the definition and use of some terms [1]. It is also important to bear in mind that, after all, we must work in a given environment, with its own customs, prejudices or preferences and it might happen that even with the conceptual background being more or less accepted, no single universal experimental protocol can be agreed upon [12]. To complicate things further, a very popular textbook amongst analytical chemists emphasizes the arbitrariness in defining the LOD [13] and that this term is open to alternative definitions for particular purposes. Another difficulty is that although everyone accepts that validation is a must, they do not agree on the criterion (criteria) to which it has to be done. This is particularly so for the LOD and LOQ, which are seldomly stated by the laboratories' customers and, so, many laboratories develop their own strategies. Finally, it has to be acknowledged that the relevant IUPAC's report from 1995 is not of straightforward understanding/use and, likely, this is a barrier to its broad acceptance. The ISO and EU approaches, however, seem more pragmatic.

Despite the previous comments, the basic ideas for these performance characteristics (in the 1970s and nowadays) are almost universally accepted by analytical chemists because they do respond to current limitations, common scientific knowledge and normal way of working. Laboratories cannot detect any amount of analyte at any intended level. Probably, we can measure concentrations easily in percentage or parts per million (e.g., $mg \cdot L^{-1}$) ranges. However, parts per billion (e.g., $\mu g \cdot L^{-1}$) or less would not be feasible because our instruments, sample treatment, reagents, stability of the electric supply, etc. impose limits that impede us to detect such tiny amounts.

Further, even if you were able to detect a given substance at a given level, are you sure you can quantify it beyond reasonable doubt? Likely, you will have no problems at large concentrations but what happens at very low concentrations?

5.6.1. Traditional superseded definitions

In classical terms (set by IUPAC in 1975 and reaffirmed by ACS in 1980), the LOD is the lowest concentration of analyte which yields a signal whose magnitude is statistically different from that of the blank, at a stated probability level, usually 99 %. In simple terms, a value from a signal will not be confounded with a blank.

Here, a relevant misunderstanding arose as the LOD was considered (incorrectly) a boundary to state whether a sample 'has' or 'does not have' analyte and this wrong idea is sustained frequently even today.

An important ambiguity of the definition above is that the LOD can be calculated in different ways depending on how the 'blank' is defined experimentally, and on the selection of the multiplication factor which represents the probability level. Such a blank may be, for instance, the intercept of the calibration line, a solution made from reagents, measured several times, or a true procedural blank, which are just three different options. The traditional equation is formally the same:

$$LOD_{\text{signal domain}} = (\bar{y}_{\text{blank}} + 3 \cdot s_{\text{blank}}) \tag{28a}$$

$$LOD_{\text{concentration domain}} = (\bar{y}_{\text{blank}} + 3 \cdot s_{\text{blank}})/b \tag{28b}$$

where $\bar{y}_{\text{blank}}$ is the average of the blank signals or the intercept of the calibration line, s_{blank} is their SD (standard error of the intercept); the constant 3 corresponds approximately to the 99 % confidence limit of the theoretical (population) Gaussian distribution, and b is the slope of the calibration line. If the average value of the blank is statistically zero, the equation can be simplified and that was a common approach in many cases.

Observe that the value thus calculated in the concentration domain (Equation (28b)) can only be a true reflection of the LOD when the slope is well defined and the intercept is statistically zero [14]. Hence, it was common practice to background correct the signals (again, problem is which blank should be used) and, so, the simple classical equation appeared:

$$LOD_{\text{concentration domain}}$$

$$= (LOD_{\text{signal domain}} - \bar{y}_{\text{blank}} + 3 \cdot s_{\text{blank}})/b = 3 \cdot s_{\text{blank}}/b \tag{28c}$$

A common approach considered the ordinate of the calibration line and its 99 % one-side confidence interval, calculated with the Student's t-value and the adequate dof (Equation (28d)):

$$LOD_{\text{concentration domain}} = (a + t_{(99\%,1\text{-tail,dof})} \cdot s_a)/b \tag{28d}$$

The traditional LOQ was defined as the lowest concentration of analyte which yields a signal which cannot be confounded statistically with that from a blank and, besides, can be quantitated without risk (usually, this has been interpreted as being able to get a value with less than 10 % relative standard deviation (RSD); but this was not mandatory). In other words, it is the lowest concentration of analyte that can be identified and quantified. It indicates quite intuitively that the LOQ has to be higher than the LOD because we want not only to identify a compound but also to be reasonably sure that we can quantitate it acceptably (e.g., with less than 10 % RSD). As a rule of thumb, the LOQ is approximately three times the LOD and it represents a more realistic limit to establish the lowest boundary for the correct use of an analytical method. In that way, the probability for a signal to be considered as a blank (i.e., to say that its concentration is null when indeed it is not zero) is really negligible. Its simplest form can be calculated as:

$$LOQ_{\text{signal domain}} = (\bar{y}_{\text{blank}} + 10 \cdot s_{\text{blank}}) \qquad (29a)$$

$$LOQ_{\text{concentration domain}} = 10 \cdot s_{\text{blank}}/b \qquad (29b)$$

Again, the background was subtracted from the signals to get Equation (29b).

As a token of the controversy that these classical definitions carry, we can reflect on two practical issues:

(a) if no more details are given in a report, one has to assume that the calculations associated to these definitions correspond to the so-called *instrumental* LOD and LOQ as they derive from the calibration measurement stages. To offer more reliable values, possible dilutions your samples experienced and the overall sample processing must be considered. In such a case, you should consider the corresponding dilution factors and calculate the (estimated) *method-related* LOD and LOQ.

Even better, if you can prepare a representative blank for the overall analytical process (which your sample aliquots suffer during their preparation, handling, treatment, measurement, etc.), you should use such a blank to perform these calculations, as it would be a true and good representation of your 'noise' and, so, a good basis to calculate what you can really measure with your overall process.

(b) The equations above were designed for a large number of replicates (not less than 20, so that the (statistical) sample approaches its population), which are not usually done due to, economical, time, manpower or resources limitations. This means that factor '3' in the LOD definition should be substituted by the corresponding 1-tail Student's t-value and that the SD has to be corrected for its bias (the SD derived from a low number of values underestimates the true one).

As a conclusion, the traditional equations — in particular, the LOD ones — suffer from many problems, definition ambiguities and they are correct only in very special circumstances (which are not met in common practice!).

5.6.2. Rationale for the modern definitions, a bit of history

The conceptual idea associated to LOD (regardless of its confusion with the decision limit) has been present almost always in analytical chemistry (a 100-year review was presented recently [15]). Historically, IUPAC tried to systematize it first time in 1975 after Kaiser's studies [16], who defined the LOD in a somehow slightly subjective way and gave rise to the 'traditional' equations presented in the previous section. The LOQ was initially defined in 1995 [17] by the American Chemical Society (ACS) as a means to address the general view that the LOD was not good enough for quantitative analysis [18]. Although, the LOD appeared quite natural and intuitive, and simple to understand and to use, its definition was adopted very slowly and not without difficulties, as Winefordner and Long reported in a classical paper [14]. Hence, it seems that it has always been difficult to convince the analytical community to adopt/change standards. Nowadays, we face the same problem and most scientific works still report LODs according to the classical calculations.

Almost 40 years after the 1975 IUPAC's definition, analytical chemistry witnessed enormous advances and faces new working fields and challenging arenas and, besides, includes a new discipline, chemometrics (which is focused on applying statistics to analytical problems). Analytical chemists now face critical problems everywhere and the complexity of the consequences of their scientific decisions has increased. As a matter of example, we would like the student to reflect on two simplified problems he/she might face in their professional lives:

Let us assume that a student gets a scholarship from a food company (XYZ Co.) to perform his/her Bachelor's project at its facilities. XYZ has good expansion opportunities as it developed a wonderful recipe to produce bottled sausages. They are really tasteful and their acceptance by final customers has exploded almost exponentially. The student is entitled to work with the laboratory staff as they are experiencing a lot of job. His/her task is to perform analyses to ascertain absence of pathogenic bacteria on the sampled sausages according to a given analytical protocol. The method is based on measuring fluorescence signals. Ideally, the signal should be 'zero' (no harmful bacteria) or, more correctly, be around an average value of zero (they established a confidence interval around the average zero signal value).

One day, the student discovers that a sample yielded a signal out of the confidence interval and immediately rings the alarm which makes the factory line to stop working. A task force is set to investigate the problem, new samples are taken and while the analyses are repeated, the task force researches cleanliness, raw materials, storage containers, etc. Finally, nothing unusual is discovered and the new analyses report a correct situation. After 24 h, the task force concludes that the alarm was in fact a false alarm, the factory line is restarted and the product delivered again.

In another occasion, the Senior Manager receives an urgent call from the Governmental Consumers' and Inspection Office. Several people are in hospitals with acute gastroenteritis. Preliminary epidemiological research pointed out that all patients ate the same brand of sausages. The production batches are reported to the CEO and the corresponding analytical certificates are retrieved urgently. Experimental values are within usual values, nothing anomalous appears. The corresponding samples stored in the laboratory freezer are taken, the analyses repeated and results were as previously (the confidence intervals agreed largely). Therefore, it has to be concluded that there are no pathogenic bacteria in the samples. Aliquots of the samples are sent to governmental laboratories where more sensitive analytical methods will be applied. There, a gastroenteric bacteria is discovered which might affect people.

There are many statistical issues that can be considered in this example, but we want to focus now on what is called the *error Type I* (error of the first kind, or alpha — α-error, or false positives) and the *error Type II* (error of the second kind, or beta — β-error, or false negatives). In production environments and in quality control, they are sometimes known as risk for the producer and risk for the consumer, respectively.

In statistical terms, a Type I error is the incorrect rejection of a true null hypothesis (a *false positive*). In the first example, the true fluorescence signal belonged to the population of zero values, although the value obtained by the student did not seem to belong to it. A Type II error is the failure to reject a false null hypothesis (a *false negative*). In our second example, the experimental value seemed within the normal distribution of the zero values when in fact it was out. It is also commonly said that a Type I error consists of detecting an effect that is not present, while a Type II error consists of failing in detecting an effect that is present.

Some students best visualize another way of thinking: the error of the first type consists of accepting the alternative hypothesis (e.g., analyte is present) when that is wrong, whereas the error of the second type consists of accepting the null hypothesis (e.g., the analyte is absent) when that is wrong.

It is of most importance to notice that these two errors are inherent to the use of any statistical test and that, unfortunately, they are strongly related. We cannot reduce one of them unless we increase the other! We commented slightly on this issue in this chapter when reviewing the tests to detect outliers. Recall here that a statistical test checks for the probability of a statement (the null hypothesis, H_0) to be true. Table 7 presents a mnemotechnic rule to remember the two types of errors easily. As an additional note, the *power of the test* is defined as the probability of correctly accepting the alternative hypothesis, given α; it is calculated as $1 - \beta$.

Historically, the risk of false positives (Type I error) has been considered in most statistical applications in analytical chemistry

Table 7. Mnemotechnic rule to differentiate the Type I and Type II errors associated to a statistical test.

		TRUTH (unknown)	
		H_0 true (accept)	H_0 false (reject) (accept H_1)
MY DECISION (using calculations with my experimental data)	H_0 true (accept)	OK	Type II (β) False negative Error to the client
	H_0 false (reject) (accept H_1)	Type I (α) False positive Error to the maker	OK

although the Type II error has been neglected. This has been an unfortunate issue because although the Type I error is fixed (usually) at 5 % probability, the corresponding Type II error is not considered at all and, thus, it can amount up to around 50 % (almost a chance out of two to deliver false negatives!).

Therefore, it is necessary to take into account these statistical difficulties for decision-making and, so, new definitions for LOD and LOQ are required which take account of the Type I and Type II errors [17, 19, 20]. It is also important to avoid the wrong use of the LOD as a criterion to which a particular signal (concentration) is compared against in order to decide whether detection/no detection was achieved.

Before digging into the calculations, it is worth emphasizing that the LOD and LOQ are intended to serve as performance characteristics of the analytical process and, therefore, it makes no sense to attempt their calculation unless we have written the working protocol in advance and its performance is in a state of statistical control (i.e., the analysts have experience in using it and its precision has been tested and it was found satisfactory and stable enough). Both performance characteristics are used to objectively characterize the analytical procedure and to decide whether it can be used for a specific task (e.g., the LOD might be useful to decide if a procedure can measure a given pollutant in a set of samples whose expected level is around ...).

In general, each time an analysis is performed, it is also needed to decide whether the signal (concentration) obtained in such analysis can be discriminated against the background noise (i.e., decide if the signal associated to the analyte was 'detected' or was 'not detected'). This — natural — concept is supported by a new term, the *critical value* (or *decision limit*), which is associated to each series of measurements. Following, the LOD and LOQ are method-dependent, whereas the critical value may be different for various measurement series of a particular method.

The wealth of different definitions and criteria (which affected clinical analyses, commercial transactions, etc.) made IUPAC and ISO meet and harmonize these concepts. This collaboration started in 1993. The level of agreement was very high although the denominations of the terms still differ slightly (interested readers are forwarded to a Currie's study where relevant differences between the two organizations are discussed and treated in mathematical terms, which is out of scope here [21]).

In the following, numerical calculations related to the critical level and the LOD adhere to the ISO and European Union standards because they are based on the straight-line statistics studied in previous sections of this chapter. They are also focused on setting values in the concentration domain. LOQ ideas are extracted from IUPAC documents because neither ISO nor EU guidelines consider it.

Potential readers and students are warned, however, on the complexity of this theme and they are forwarded to specific literature (some relevant references are covered in this chapter and they form an acceptable starting point).

5.6.3. Calculations and current agreed definitions

For the sake of undergraduate training, in the next discussions, it is accepted that the experimental values fit a series of assumptions. Otherwise, the derived calculations are not valid and must be reviewed (ISO and IUPAC references discuss how to do that). The assumptions are:

1. the calibration is defined correctly by a straight-line function.
2. the variance (SD) is constant throughout the calibration (or, at least, in the range between the zero concentration (the blank) and the LOQ). This is commonly taken for granted but it must be assessed experimentally.
3. the underlying distribution of values (i.e., the population) associated to any response (signal) follows a Gaussian distribution.

The *critical value*, or *decision limit* or *critical value of the net state variable*, x_c, (detection decision, critical level, detection limit, or similar), or CC_α (in the European Union Directives), is the minimum significant value of an estimated net signal or concentration, applied as a discriminator against background noise.

From a practical viewpoint, the net state variable (as used by ISO) is the variable we are interested in (e.g., the concentration of the analyte). The adjective *net state* reinforces an interesting point any scientist should bear in mind and that ISO and IUPAC (implicitly) pointed out and took into consideration to set definitions and calculations: in reality, an absolute zero is not possible (your zero might not be so if I can measure your sample with a cutting-edge instrument). Hence, any experimental value can only be correctly characterized in terms of difference from the blank (the so-called basic state); i.e., in terms of the net state. In practice, the basic state is set (maybe artificially) to (your own) zero and, therefore, any results are reported in terms of supposed concentrations. This is why setting a good blank is so important in analytical chemistry.

The *critical value*, x_c or CC_α, is used only to decide whether a signal does not belong to the distribution of a blank. In its definition, only the Type I error is fixed (in general, $\alpha = 0.05$) whereas the Type II error is not considered at all (like in the old LOD definition); i.e., *ca.* 50 % probability of false negatives. Formally, it can be defined as '*the lowest concentration level of the analyte that can be detected in a sample with a probability α of a false positive decision; otherwise stated, the signal above which it can correctly be concluded, with a given α probability, that a sample does not belong to the blank* (*or stated value*)' [20]. Hence, x_c or CC_α measures the ability of the method to not give false positive results by focusing on samples that contain the target analyte with a concentration below or at a given level.

Taking into account that we will usually handle a calibration straight line and a low number of replicates to evaluate the concentration, x_c or CC_α can be calculated as in Equation (30):

$$x_c = t_{(0.95,1\text{-tail},\nu)} \cdot \frac{s_{y/x}}{b} \cdot \sqrt{\frac{1}{q} + \frac{1}{N} + \frac{\bar{x}^2}{\sum_{i=1}^{N}(x_i - \bar{x})^2}} \qquad (30)$$

where the terms have the same meaning as in previous sections (mainly, those related to calibration). Observe that N extends to all standards and their replicates (N = number of standards · number of (true) replicates per standard) and q is the number of (true) replicates of the studied 'test aliquot' (to simplify q = number of (true) replicates per standard). The Student's factor is determined at 95 % confidence, 1-tail table and $N - 2$ dof. The t-test here is one-sided because, roughly speaking, we do not consider either negative net analyte signals or negative concentrations.

Note that classical LOD definitions almost corresponded to this definition and this is why confusion on its correct use arose frequently. It is of maximum importance to realize that the critical level is a way to decide that we have measured 'something' (a signal) which, with a 5 % probability of Type I error, is different from noise (or background) — i.e., we reject the null hypothesis H_0: the measured signal does not differ from the background distribution. However, it might well happen that even when this signal is produced by a 'real' concentration, we conclude around 50 % of the times that our signal belongs to the blank (i.e., that the 'real' concentration is not present).

This means that the critical level is not good enough for sound and comprehensive decision-making, although it is a good starting point to foresee that such a low signal is due to some non-null analyte concentration. When more certainty is required, as it is usually the case, we have to resort to the 'LOD' (this term has actually been superseded to avoid confusion, but we can keep it here).

The *capability of detection* (CC_β), according to European Union, *minimum detectable net concentration* (x_d), following ISO, or *minimum detectable value* (or *detection limit*), as suggested by IUPAC, is *the true net concentration of the analyte in the material to be analyzed which lead, with probability* $1 - \beta$, *to the correct conclusion that the concentration in the analyzed material is different from that in the blank material* [20] (i.e., we correctly accept H_1). In other words, it is the minimum concentration of analyte that can be discriminated from the blank controlling the risks of false positives and false negatives (it is common to set both at $\alpha = \beta = 0.05$, or 95 % confidence, although other possibilities are also acceptable as long as they are clearly indicated in the associated report). It can be calculated approximately as in Equation (31):

$$x_d = \delta_{(\alpha=0.95, \beta=0.95, \nu)} \cdot \frac{s_{y/x}}{b} \cdot \sqrt{\frac{1}{q} + \frac{1}{N} + \frac{\bar{x}^2}{\sum_{i=1}^{N} (x_i - \bar{x})^2}} \quad (31)$$

The $\delta_{(\alpha=0.95, \beta=0.95, \nu)}$ parameter describes a non-central t-distribution. It depends on the two types of errors and ν dof (dof $= \nu = N - 2$). This term replaces the well-known Student's t-test because we have to take into account two critical values. In an ideal situation, they are the $t_{(1-\alpha)}$ critical value of the blank and the $t_{(1-\beta)}$ critical value of the distribution of values of the estimated quantity (in this case, a concentration exactly at the LOD, which is assumed to have the same variance as the blank, as we stated at the beginning of this section).

When specific tables are not available (as it is common), δ can be estimated roughly as $2t$ (providing the α and β errors are set equal, and t corresponds to the Student's t statistic, 1-tail $(1 - \alpha)$ probability, and ν dof). For small ν (around 4), this approach has about 5 % error and, so, it is preferable to estimate it as $\delta \approx 2t \cdot \frac{4\nu}{4\nu+1}$, which yields less than 1 % error [21].

Note that CC_β is calculated upon CC_α as it seems logical. Finally, recall that these calculations correspond to the final measurement stages and, therefore, method CC_α ($=x_{c,method}$) and method CC_β ($=x_{d,method}$) should refer to the original sample problem.

This means that we should consider the dilutions (concentrations, treatments, etc.) suffered by the test solution of the sample.

It was discussed above that the calculation of the LOD and LOQ is complex. As a token of this, just consider the following issue: it has been reported that the ISO definitions that we have considered so far might be biased and might yield too many false negatives because of problems with the non-central parameter of the non-central t-distribution [15, 22]. In our view, this is not of major relevance for our purpose here — undergraduate training. Likely, and unfortunately, there would be many more relevant problems affecting the calculations and we prefer to concentrate on the agreed international definitions.

As a measure of the inherent *quantitation capability* of an analytical procedure, the *minimum quantifiable (true) value* or *quantitation limit* (x_q) is defined as the analyte (true) value that will produce estimates having a specified (subjective) relative SD. This boundary has traditionally been used to denote the ability of a measurement process to adequately quantify an analyte [17].

If a common 10 % RSD is considered (and remember that this was merely a pragmatic decision), this leads to the classical equation seen above (Equation (29b), $LOQ_{signal\ domain} = 10 \cdot s_{blank}$). Some other authors tried to avoid this subjective RSD value and proposed to consider three times the LOD (considering the old definition, this yields a $9 \cdot s_{blank}$ simplified criterion) [18]. This proposal has some logic but it was not successful. Interestingly, the $10 \cdot s_{blank}$ criterion equals (given the basic assumptions are correct) $3.04 \cdot LOD$.

The calculation of x_q is not straightforward because one has to take into account the error associated to b (the slope) for error propagation; this is not simple and a comprehensive mathematical development was presented elsewhere [21]. For our purposes here, the simplified calculations will be considered. Once more, bear in mind that we are assuming homoscedasticity in the range between the blank and the LOQ and that we are using a calibration line (here, a major problem arises because the ordinate and slope are correlated and their errors are related).

To get the LOQ, we resort to an approach given by Currie [21], which calls for an iterative process because the term we are looking for (x_q) must enter into both sides of the equation (see Equation (32)) due to the error (variance) inherent to the prediction of a value using the calibration line. A reasonable starting value for x_q might be $3 \cdot x_d$ or the rough classical LOQ, this is introduced in the right side of the equality, calculations are done, the result is introduced again in the right part of the equality, calculations repeated and so forth until the solution converges. This is a very fast process using any popular spreadsheet (no additional laboratory experiments are needed to calculate it). Note that Equation (21a) cannot be used here because we do not know the true signal at x_q.

$$\text{LOQ} \equiv x_q \approx 10 \cdot \frac{s_{y/x}}{b} \cdot \sqrt{\frac{1}{q} + \frac{1}{N} + \frac{(x_q - \bar{x})^2}{\sum_{i=1}^{N} (x_i - \bar{x})^2}} \qquad (32)$$

5.6.4. Additional practical issues for calculations

Reporting data

A problem for many laboratory managers is how to translate the performance limits into their reports. IUPAC established that no data should be censored and that all values have to be reported, even when they are lower than the decision limit. In that case, report the signal and the estimated quantity, along with their uncertainties [23]. ISO agreed on that and it also recommends to add the comment 'not detected' in case the value is lower than the decision limit. It also discourages the inclusion of comments like 'smaller than the minimum detectable value'.

However, laboratory managers cannot disregard common practices in their own field. As a matter of example:

(a) For analyses of residues of pesticides in food, the EU demands that individual analytes below the reporting limit (the LOQ) are reported as < the numerical value of the LOQ (mg/kg) [24].

(b) The Clinical and Laboratory Standards Institute and the Association of Public Health Laboratories recommend reporting [12, 25]:

a. If result < LOB (equivalent to the decision limit), provide the LOB value.

b. LOB ≤ result < LOQ, 'detected (experimental value) < LOQ', provide LOQ value.

c. If result > LOQ, report the quantitative result.

Likely, the key idea here is not to hide the experimental value obtained in the laboratory and simplify its understanding by indicating how meaningful such value is. Problems will appear whenever the experimental value is lower than the LOQ because the closer the value is to the decision limit, the closer it is to the random variability of the measurements. For some users, it will be difficult to understand that a value may not be too reliable below a certain level.

Inclusion of blanks in calibration

Before advancing further in this point, it has to be absolutely clear that we should never force the least squares fit to pass through the origin (this is an option available in many software). If that is done, the fit would be biased and, further, some properties of the classical least squares straight line might be lost.

Whether the signal of the blank should be included in the calibration is a matter of discussion because it can greatly alter the calibration fit. In effect, the extreme position of this point might seriously affect the straight line if its behavior does not match that of the other standard solutions. Therefore, be sure that your blanks are truly representative of your calibration standards (likely, of the test samples). If doubts appear, do not include that point into the least squares fit and check its position afterwards. That will help you decide whether it will be safe to include it in the fit. This issue is exemplified in exercises 14 and 15 in this chapter. An extreme example illustrates it:

In atomic absorption spectrometry, calibration solutions are frequently prepared in an acidic medium. If the amount of acid is not considered in the blank, the signal would be in general very low and, so, when that value is included in the least squares fit, the straight line calibration will be affected greatly (likely, this point will pull down for the line toward it, which is not correct). ISO 11843-2 and EU recommend including the blank in the calibrations, although ISO stresses that care must be placed on assuring its adequacy (e.g., passing it throughout the overall analytical process, as for the standards).

As a corollary, when the decision limit, the capability of detection (minimum detectable value) or the LOQ method performance characteristics are to be calculated, it is mandatory that all analytical steps (sampling, sampling preparation, dilutions, purifications, etc.) are considered in deriving those values [23].

Unfortunately, it is not always easy (possible?) to carry them out in laboratories dealing with complex samples, such as those related to environmental studies, food, clinical, studies etc.

Usage of blanks to evaluate performance characteristics

The guides mentioned above allow the analyst another possibility to evaluate the decision limit, capability of detection (minimum detectable value) and, even, the LOQ. This is based on measuring a large number of blanks under different experimental trials. Again the key point here is to represent as accurately as possible the overall analytical procedure with the blanks. Many times, this is far from simple due to the complexity of the procedure, lack of time or resources to carry out studies under different experimental conditions, lack of samples without the analyte (typically in clinical studies), etc.

If good/representative blanks are not available or we cannot perform a large number of measurements (different days, different calibrations, different lots, etc.) or we feel that our blanks do not accurately represent the overall analytical procedure ('overall' is again the critical key word), then probably it is safer not to try deriving performance parameters from them and resort to the equations presented in the previous sections. They take into account the errors associated to the calibrations, which otherwise might remain disconsidered.

Remember that a series of blanks is not repeating the measurement of a blank solution several times. On the contrary, the experimental conditions must be varied to account for 'normal variability in the measurement process'. In clinical analyses, not less than 60 replicates per reagent lot are needed [12]. In trace analyses, it is even more difficult to estimate accurately the signal for the blank. Note that at least 50 replicated observations are needed to estimate a value with a 10 % RSD [23]. The EU relaxes the demands to only 20 blanks [20] but without detailed justification. Further, the overall procedure should be performed for each calibration solution, which is not usually done. This approach has not been considered applicable in general [21].

Another difficulty is how to estimate a proper blank. To illustrate this theme, we would like to extract some comments from a nice paper from Van der Voet [26]. As the performance limits are a function of the signal attributed to zero concentration (the zero of the net state variable, in ISO's words) and the signals obtained at very low concentrations, the calculated limits may be different depending on whether we consider a part of the analytical process or the method as a whole. At least, three possibilities can occur:

(a) Repeatability conditions: calculate the limits using calibration data from several measurement series and report the median of individual values. There is no guarantee that the median is representative of other measurement processes.

(b) Within-laboratory reproducibility conditions: they take into account, precisely, the last comment in the previous paragraph. Reproducibility here means that the several measurement series used to calculate the limits bracket the usual experimental conditions and they were considered (e.g., different calibrations, different reagents, personnel, etc.); i.e., between-series variability, under within-laboratory conditions. Formal experimental designs (not considered in this book, please consult more comprehensive textbooks) can be of help here. Although initially the limits here will be larger than when only a calibration is considered

(= repeatability conditions), the more values we accumulate, the smaller the statistical uncertainties will be.

(c) Between-laboratory reproducibility conditions: the limits are derived from interlaboratory exercises. These limits would be of relevance for legal and trade regulations.

When repeatability or within-laboratory conditions are considered, different levels can be considered in the operating procedure intended to calculate the performance limits:

(a) Only the instrument measurements (not very reliable).
(b) Sample preparation plus measurements (include here, standards), quite affordable in most cases.
(c) Sampling plus sample preparation plus measurement, not always affordable in complex problems.

This means that the most convenient, affordable option is to estimate the blank from the intercept of the calibration line (if the calibration itself is representative of the samples; i.e., the matrices of the standards and the samples really match). However, recall that real measurement processes include sampling, sampling treatment, sample transportation, recoveries from sample processing, etc. [21]. So, we end up discussing again about the suitability of the standards.

Anyway, it is very important that the laboratory reports clearly on which level the calculations were performed because the performance parameters will depend on them. Be honest and do not let users imagine how the calculations were performed.

Estimate reliable performance parameters out of your calculations

As a consequence of many of the previous discussions, you may have already noticed that the particular values of the decision limit, capability of detection and LOQ may vary when different experimental trials are considered. This is true because of random error and inherent laboratory variability. Therefore, the experimental x_c (CC_α), x_d (CC_β) or LOQ limits, which are intended to estimate the method-related ones, should not depend so strongly on the trials you made.

A reasonable estimate of the method-related limits is by considering the median of a series of individual x_c, x_d or LOQ values [19], although other approaches exist.

Pragmatic considerations to address the numerical exercises

The overall complexity associated to the so-called LOD and LOQ, which was partially simplified in the previous sections — although we tried not to lose rigor — leads us to pose several issues to solve the numerical exercises in a very pragmatic way. It is expected that the guidances we present here can serve as a reasonable approach to foresee how real problems may be addressed. Numerical exercises are intrinsically simpler than many real problems and although we tried to follow international guides, some issues could not be found there. For instance, how to consider in a same measurement series the procedural blank and the reagents blank. This is a quite common situation and we tried to apply common sense but it was not possible for us to ascertain whether statistical flaws are hidden in those approaches. Feedback from our colleagues will be, thus, indeed welcome.

As a general rule in laboratory work, procedural blanks are almost always included within each series of sample aliquots treated to get the final test solutions. They pass all the analytical procedures and are a good means to take account of contamination and impurities, manipulation, etc. Calibration blanks are more related to the use of reagents and concomitants (when the latter are added to simulate the sample matrix) to prepare the calibration standards. They seldom pass through the overall analytical process (although ISO 11843-2 recommends so). Currently, average signals associated to procedural blanks are larger than those of calibration blanks (in simple analytical methodologies they are of the same order) and it can be assumed that, roughly, the former 'include' the latter. This seems logical but it implies a relevant problem in many practical situations when using the calibration to interpolate the signals.

In effect, we have to take into account that somehow the background signal associated to a (treated) sample test solution is different from that of the calibration standards. How to minimize this problem is far from trivial. A general-purpose solution, although without full-success guarantee, might be to perform a standard addition calibration per sample, providing you have resources and enough time and your analyte does not exhibit complex behavior.

A pragmatic approach more suited for current daily work consists of subtracting always the signal of the procedural blank from the gross signal of the sample test aliquots. In addition, the average signal of the calibration blanks has to be subtracted from those of the calibration standards (not from the signals of the test samples) if the intercept of the calibration line is statistically different from zero. The rationale of this approach is to avoid oversubtracting backgrounds to the signals of the calibration standards and of the samples and work always with net analyte signals. This will be exemplified in Exercises 14 and 15 here and in Chapter 5.

REFERENCES

[1] Magnusson, B.; Örnemark, U. (eds.) (2014). *Eurachem Guide: The Fitness for Purpose on Analytical Methods — A Laboratory Guide to Method Validation and Related Topics*, 2nd ed. Available at: www.eurachem.org.

[2] Vandeginste, B. G. M.; Massart, D. L.; Buydens, L. M. C.; De Jong, S.; Lewi, P. J.; Smeyers-Verbeke, J. (1998). *Handbook of Chemometrics and Qualimetrics (Part A and Part B)*. Elsevier, Amsterdam.

[3] Andrade, J. M.; Estévez-Pérez, G. (2014). Statistical comparison of the slopes of two regression lines: a tutorial, *Analytica Chimica Acta*, 838: 1–12.

[4] Andrade-Garda, J. M.; Carlosena-Zubieta, A.; Soto-Ferreiro, R.; Terán-Baamonde, J.; Thompson, M. (2013). Classical linear regression by the least squares method, in *Basic Chemometric Techniques in Atomic Spectroscopy*, 2nd ed. RSC (UK).

[5] Ellison, S. L. R.; Thompson, M. (2008) Standard additions: myth and reality, *Analyst*, 133: 992–997.

[6] Thompson, M.; Lowthian, P. J. (2011). *Notes on Statistics and Data Quality for Analytical Chemists*. ICP, London.

[7] Danzer, K.; Currie, L. A. (1998). Guidelines for calibration in analytical chemistry. Part 1: Fundamentals and single component calibration, *Pure and Applied Chemistry* 70: 993–1014.

[8] Ortiz, M. C.; Sánchez, S.; Sarabia, L. (2009). Quality of analytical measurements: univariate regression, in S. Brown, D.; Tauler, R.; Walczack, B. (eds.), *Comprehensive Chemometrics: Chemical and Biochemical Data Analysis* (Volume 1). Elsevier, Amsterdam.

[9] Andrade, J. M.; Terán-Baamonde, J.; Soto-Ferreiro, R.; Carlosena, A. (2013). Interpolation in the standard addition method, *Analytica Chimica Acta*, 780: 13–19.

[10] Cal-Prieto, M. J.; Carlosena, A.; Andrade, J. M.; Muniategui, S.; López-Mahía, P.; Fernández, E.; Prada, D. (1999). Development of an analytical scheme for the direct determination of antimony in geological materials by automated ultrasonic slurry sampling-ETAAS, *Journal of Analytical Atomic Spectrometry*, 14: 703–710.

[11] Perkin-Elmer, Atomic Absorption Laboratory Benchtop. User's Guide, 1991. Perkin Elmer, Überlingen (Germany).

[12] Pierson-Perry, J. F. (2012). Understanding detection capability for quantitative *in vitro* assays, Webinar Training course. Available at http://eo2.commpartners.com/users/APHL/downloads/588-610-11-3 SlidesPerPage.pdf. (Last accessed April 2016).

[13] Miller, J. C.; Miller, J. N. (2010). *Statistics and Chemometrics for Analytical Chemistry*, 6th ed. Pearson, UK.

[14] Winefordner, J. D.; Long, G. L. (1983). Limit of detection: a closer look at the IUPAC definition, *Analytical Chemistry*, 55(7): 712A–724A.

[15] Belter, M.; Sajnog, A.; Baralkiewicz, D. (2014). Over a century of detection and quantitation capabilities in analytical chemistry — Historical overview and trends, *Talanta*, 129: 606–616.

[16] Kaiser, K. (1947). Die Berechnung der nachweisempfindlichkeit, *Spectrochimica Acta*, 3: 40–67.

[17] Currie, L. A. (1995). Nomenclature in evaluation of analytical methods including detection and quantification capabilities, *Pure and Applied Chemistry*, 67(10): 1699–1723.

[18] Mocak, J.; Bond, A. M.; Mitchell, S.; Scollary, G. (1997). A statistical overview of standard (IUPAC and ACS) and new procedures for determining the limits of detection and quantification: application to voltammetric and stripping techniques, *Pure and Applied Chemistry*, 69(2): 297–328.

[19] ISO 11843-2-2007, Capability of detection — Part 2: methodology in the linear calibration case, ISO (Geneve), 2000 (Corrigendum 2007).

[20] Commission Decision of 12 August 2002, implementing Council direc-
     tive 96/23/EC concerning the performance of analytical methods and
     the interpretation of results (2002). *Official Journal of the European
     Communities*, 8–36.
[21] Currie, L. A. (1997). Detection: international update, and some
     emerging dilemmas involving calibration, the blank, and multiple
     detection decisions, *Chemometrics and Intelligent Laboratory Systems*,
     37: 151–181.
[22] Montville, D.; Voigtman, E. (2003). Statistical properties of limit of
     detection test statistics, *Talanta*, 59: 461–476.
[23] Currie, L. A. (1999). Detection and quantification limits: origins and
     historical overview, *Analytica Chimica Acta*, 391: 127–134.
[24] SANCO/12571/2013, Guidance document on analytical quality con-
     trol and validation procedures for pesticide residues analysis in food
     and feed, SANCO (rev. 0).
[25] Beckman Coulter, Information Bulletin, Understanding detection
     capability: LoB, LoD and LoQ in the clinical laboratory, 2013.
[26] Van der Voet, H. (2012). *Detection Limits, Encyclopedia of Environ-
     metrics*, 2nd ed. John Willey & Sons.

WORKED EXERCISES

Note: There are several criteria to fix the number of decimal places
(i.e., significant figures) to be reported in the final results. We adhere
here to a simple common one, which stems from the American
Society for Testing and Materials (ASTM), of broad application in
the industry. The average mean and the SD consider one and two
additional significant figures, respectively, with regard to the original
values.

For practical reasons, more numerical exercises dealing with
calculations associated to regression lines and performance characteristics are presented in Chapter 5, which is devoted to atomic
absorption spectrometry.

1. Compare the experimental average and median of the two series
   of three analytical replicates each made on two corresponding
   treated sample aliquots (values represent mg/L): (a) 2.8; 3.0; 5.1
   (b) 1.9; 3.1; 50.0.

SOLUTION:

To perform three measurement replicates of the aliquot being measured is a quite common practice (in some instrumental devices, this is set as a typical default). Most scientists would think that this is a way to assure their reported values but, unfortunately, you can see in this simple example that it is not so.

Series (a): $\bar{x} = 3.63$; $\tilde{x} = 3.00$

Series (b): $\bar{x} = 18.33$; $\tilde{x} = 3.10$.

The medians are practically the same (they are a robust parameter), whereas the averages are totally different. So what about the 50.0 value? Well, the analyst was wrong in copying the result (this happens very often).

Critical mind is a keystone when making calculations.

2. The following results correspond to Fe in fresh water (mg/L). Decide whether there are outliers using the Dixon's and Grubbs's tests. Data set: 10.45, 10.47, 10.47, 10.48, 10.49, 10.50, 10.50, 10.52, 10.53, 10.58.

SOLUTION:

In this example, both tests are requested for the student to practice their basic concepts. In next examples, the Grubbs's test will be requested (in case, no explicit comment is given, just consider that is the default to be applied). The student is encouraged to apply the Dixon's one from time to time to practice it and check for coincidence.

When the data are ordered, it appears that the possible outlier may be 10.58 (the student can check for the lowest value, 10.45). Dixon's test:

1. Experimental ratio, equation 7 ($n = 10$), $Q = (X_n - X_{n-1})/(X_n - X_2) = (10.58 - 10.53)/(10.58 - 10.47) = 0.454$.
2. Table 1 shows that the tabulated values at 5 % and 1 % error probability ($n = 10$) are 0.41 and 0.60, respectively.
3. As a conclusion, there are evidences against accepting the null hypothesis at 95 % confidence level (and, so, the value should

be rejected from the series) but it cannot be rejected at 99 % confidence (the value should be kept for calculations).

This is a typical situation where the analyst (or somebody else, the laboratory manager or the customer) has to decide carefully on the confidence level at which the values should be reported for a particular use. With very few values, a 99 % confidence level might be more reasonable.

Grubbs's test:

1. Experimental ratio: $G = (10.58 - 10.499)/0.037 = 2.17$.
2. Tabulated values (Table 2, $n = 10$) at 95 % and 99 % confidence levels = 2.18 and 2.41.
3. To conclude, there are no evidences to reject the null hypothesis at any of the two confidence levels and, so, the value should be retained.

This is a good example where both tests do not agree. In general, the Grubbs's one is to be preferred.

3. Sodium in physiological serum can be quantified using atomic absorption spectrometry. Measurements from several production batches yielded 102, 97, 99, 98, 101 and 106 mg/L. Calculate the 95 % and 99 % confidence intervals for the average production.

SOLUTION:

The summary of the batches is: $\bar{x} = 100.5$ mg/L; $s = 3.27$ mg/L. The critical values from Table 4 (5 dof), at 95 % and 99 % levels, are 2.57 and 4.03.

Therefore:

$$\mu = 100.5 \pm \frac{2.57 \cdot 3.27}{\sqrt{6}}$$

$$= (100.5 \pm 3.4)\,\text{mg/L} \ (95\,\%\ \text{confidence level})$$

$$\mu = 100.5 \pm \frac{4.03 \cdot 3.27}{\sqrt{6}}$$

$$= (100.5 \pm 5.4)\,\text{mg/L} \ (99\,\%\ \text{confidence level})$$

Observe that the 99 % confidence interval is clearly larger than the 95 % one; i.e., we accept that the true value may be within a larger range around the most probable value (the average).

Note: When working with series of data, it is highly recommended to always check for outliers first. This was not done in this example in order to focus on the confidence interval concept and the importance of selecting the confidence level. The student can verify that the largest value (106) is not an outlier, according to Grubbs's test.

4. A physical measurement of the diameter of sausages in a Galician factory yielded 38.9 mm, 37.4 mm and 37.1 mm before cooking. Assuming that the target value was established as 38.9 mm, were they adhering to the production plan?

SOLUTION:

This is a typical question that can be formulated in a wealth of forms. However, the answer can be found quite straightforwardly using Equation (10).

$$t_{\text{experimental}} = [(37.80 - 38.9) \cdot \sqrt{3}]/0.964 = 1.98$$

From Table 4, the Student's critical value (95 % confidence, 2-tail, 2 dof) is 4.30. As the experimental value is clearly lower than the critical value, the null hypothesis should not be rejected and, so, the average and the target values are not statistically different.

This exercise is important to emphasize the erroneous common belief of the goodness of performing three replicates. Observe that only 2 dof yield an enormous tabulated value (for 99 % confidence, this would be 9.92!) and, so, it would be difficult to visualize differences between the mean and the target.

5. Two analytical methods, flame atomic absorption spectrometry (FAAS) and graphite furnace atomic absorption spectrometry (GFAAS), are being evaluated in order to apply one of them

for a particular type of samples. Different aliquots of a typical homogenizate were analyzed in different runs. The following table presents the results and their summary. Decide whether the two methods lead to different averages (95 % confidence).

FAAS	4.99 4.99 5.00 5.01 5.00 5.00; $\bar{x}_1 = 4.998$ $s_1^2 = 5.67 \cdot 10^{-5}$
GFAAS	5.01 4.95 5.03 4.96 5.02 4.99; $\bar{x}_2 = 4.993$ $s_2^2 = 1.07 \cdot 10^{-3}$

SOLUTION:

Assuming that outliers are not present in any series, the first step to check both averages is to test whether their variances are comparable using the Fisher–Snedecor's test.

$F_{experimental} = 1.07 \cdot 10^{-3}/5.67 \cdot 10^{-5} = 18.8$

$F_{tabulated(99\%, 5, 5, 1\text{-tail})} = 10.97$ (5.05 at 95 %, Tables 5 and 6).

Here, the experimental F is larger than any of the two tabulated values and, so, the null hypothesis should be rejected (the variances are not equivalent). Following, the Student's t-test (Welch's mode) is to be applied (Equations (15) and (16)).

$$t_{experimental} = \frac{5.000 - 4.993}{\left[\left(\dfrac{5.67 \cdot 10^{-5}}{6}\right) + \left(\dfrac{1.07 \cdot 10^{-3}}{6}\right)\right]^{1/2}} = 0.73$$

$$dof = \left[\frac{(5.67 \cdot 10^{-5}/6 + 1.07 \cdot 10^{-3}/6)^2}{\left[\dfrac{(5.67 \cdot 10^{-5}/6)^2}{6+1}\right] + \left[\dfrac{(1.07 \cdot 10^{-3}/6)^2}{6+1}\right]}\right] - 2 \sim 6$$

From Table 4, the critical value (2-tail, 95 %, 6) = 2.45.

As the experimental value is lower than the tabulated one, there are no evidences to reject the null hypothesis, i.e., the experimental averages are equal (although their precisions are not).

6. An applicant for a job in a laboratory is asked to perform a typical analysis to determine sulfate ion in tap water. She receives

a sample (which is in fact an interference-free 5.00 mg/L sulfate standard) and is requested to make the analysis following a standard operating procedure of the laboratory. Her results were 4.94; 4.88; 5.06; 4.93; 5.10; 5.00; 5.10 and 5.18 mg/L of sulfate. The manager has to decide: (a) Has the candidate a bias in her results? (b) Calculate the confidence interval of the average (95 % confidence) and, so, the maximum spread of the results given by the candidate (c) What number of measurements would be required for this candidate to get a ±0.03 mg/L confidence interval?

SOLUTION:

Similar strategies to that shown in the enunciate can be deployed to monitor the performance of the staff in a laboratory (in addition to the use of certified reference materials (CRMs) or participation in interlaboratory exercises), or to address similar issues related to quality control.

Question (a):

First, evaluate whether outliers are present in the data series. In this case, we checked the largest and the lowest values:

$G_{experimental}$ (lowest) = 1.39; $G_{experimental}$ (highest) = 1.52
($\bar{x}$ = 5.024 mg/L; s = 0.1031 mg/L)
$G_{tabulated(99\%, n=8)}$ = 2.22 ($G_{tabulated(95\%, n=8)}$ = 2.03)

Therefore, the null hypothesis cannot be rejected in any case, the values are not outliers.

Now, applying Equation (10):

$t_{experimental}$ = [(5.00 − 5.024) · $\sqrt{8}$]/0.1031 = 0.66, which is lower than $t_{tabulated\ (2\text{-tail},\ 95\%, n-1=7)}$ = 2.37

Therefore, the means are not statistically different and, so, the candidate has no bias.

Question (b):

The 95 % confidence interval (CI) around the average is $\bar{x} \pm$ CI = 5.024 ± 2.37 · 0.1031/$\sqrt{8}$ = (5.024 ± 0.086) mg/L. In other words, the analyst should not obtain values outside the 5.110–4.938 mg/L range, otherwise she might be biased.

Question (c):

The new confidence interval was already defined. However, we do not know n, so let us assume that she will need a lot of measurements to reduce her interval by more than a half. Let us, initially, consider a large n and, thus, the theoretical t-value (95 % confidence), 2:

$$\text{CI} = t \cdot s/\sqrt{n}; \quad 0.03 = 2 \cdot 0.1031/\sqrt{n};$$

so that $n \simeq 47$ measurements.

We could refine this approach a bit more by using the t-value associated to 47 measurements and repeating the calculations. However, n will not decrease substantially.

Recall that the candidate received the operating protocol first time when she was asked to make the analysis and, so, in general, we expect her to improve her performance with time and experience.

7. Instrumental methods of analysis rely critically on the correct preparation of the calibration standards. The least squares fit method requires the error associated to their concentrations be negligible. This justifies that many standards nowadays are prepared by weighting. Let us assume that a particular laboratory uses certified weights whose accuracy is 1.18 mg throughout the whole working range. To get acceptable errors associated to the calibration standards, the relative error in the weights should not exceed 0.8 %. Questions are: (a) calculate the minimum weight that can be measured in the balance to prepare the standards; (b) calculate the error associated to 0.2315 g and 10.00 mg.

SOLUTION:

Although the exercise may appear complex, it is not. Indeed, these calculations should be taken into account currently. You only need to reflect on the definition of absolute and relative errors. The term accuracy (which is used by most manufacturers) refers here to the concept of trueness (agreement amongst measured and true values; bias or (absolute) error).

Recall to use coherent units.

Question (a):

Relative error (in %) = absolute error · 100 /weighted value.
Using the values stated above: $0.8\% = 0.00118\,g \cdot 100/\text{weight} \rightarrow$ weight = 0.1475 g.

This is the minimum weight to get the stated relative error, higher weights yield lower relative errors, as desired. See next calculations.

Question (b):

Relative error (%) = 0.00118 · 100 / 0.2315 = 0.51 % and
Relative error (%) = 0.00118 · 100 / 0.0100 = 11.8 %.

8. A researcher needs to determine exactly the volume she uses for a non-aqueous titration to measure total acidity in kerosene fuels (aviation fuels). She, therefore, needs to check the volume of the pipette to be used to withdraw isooctane (ASTM motor grade, whose gravity was certified). She measured the weights of several volumes of isooctane and made the calculations, whose results were: 9.982; 9.998; 9.995; 9.991; 9.998; 10.003; 9.994; 9.980; 9.996 and 9.997 mL. Calculate the bias (if it exists), considering that the theoretical volume of the pipette is 10.000 mL.

SOLUTION:

First, determine the presence of outliers using the Grubbs's test (check the extreme values). The student can also apply the Dixon's one to practice it.

The statistics of the series are: $\bar{x} = 9.9934$, $s = 7.24 \cdot 10^{-3}$

$$G_{(\text{experimental, highest value})} = 1.33$$

and

$$G_{(\text{experimental, lowest value})} = 1.85$$

$$G_{(\text{tabulated, 99\%,}n=10)} = 2.41 \quad \text{and} \quad G_{(\text{tabulated, 95\%,}n=10)} = 2.18$$

Therefore, the null hypothesis cannot be rejected for any of the two confidence levels, the extreme values should not be considered as outliers.

To test whether the average and the theoretical values exhibit a significant difference, the Student's test is applied:

$$t_{\text{experimental}} = [(10.000 - 9.9934) \cdot \sqrt{10}]/7.24 \cdot 10^{-3} = 2.88$$

$t_{\text{tabulated}(95\,\%,\,2\text{-tail},n-1=9)} = 2.26$, therefore the null hypothesis can be rejected and, so, there is a significant difference (bias) between the values, which on average is $9.9934 - 10.000 = -0.0066$ mL; i.e., the pipette delivers less than 10.000 mL.

However, $t_{\text{tabulated}(99\,\%,2\text{-tail},n-1=9)} = 3.25$, in which case, the null hypothesis should not be rejected and, thus, the pipette has no bias.

This example puts forward two relevant issues: 1st: the need for a careful selection of the confidence level and, 2nd, the need for performing as much experiments as possible in order to get sound conclusions (in this example, 10 measurements were done but, likely, they still constitute a small set).

9. A new methodology is being implemented to measure palladium in groundwater. A control standard solution was measured on different days and yielded results of 0.51; 0.51; 0.49; 0.51; 0.52; 0.51; 0.48; 0.50; 0.51 and 0.53 μg/L. The theoretical value (according to the report from the laboratory manager, calculated after weighting and diluting) was 0.52 μg/L. To accept the method, the manager has to demonstrate the absence of bias (at least, with respect to this internal standard) and a precision equal or better than 0.005 μg/L. Can he accept the method (considering only these requisites)?

SOLUTION:

A preliminary evaluation of eventual outliers showed that none of the extreme values can be rejected ($G_{\text{experimental}}$ values were 1.90 and 1.62, for the lowest and highest values, respectively, which are smaller than the tabulated critical values at both 95 % and 99 % probabilities).

To study whether bias is present, we can calculate whether the theoretical and experimental average do not differ significantly. As the theoretical value has no error:

$t_{\text{experimental}} = [(0.52 - 0.507) \cdot \sqrt{10}]/0.0142 = 2.89$, which is significant at a 95 % confidence level (not significant at a 99 % level, 9 dof)

In general, industries work at a 95 % confidence level and, following, we must accept that there is a small bias of $-0.013\ \mu g/L$.

Further, the Fisher–Snedecor's test is used to compare two variances:

$$F_{\text{experimental}} = (0.0142)^2/(0.005)^2 = 8.04$$

which is clearly greater than

$$F_{\text{tabulated}(99\,\%,\,9,\,30)} = 3.07 (F_{\text{tabulated}(95\,\%,\,9,\,30)} = 2.21)$$

Note that the target SD is not an experimental value and, so, we selected an arbitrary high value for its dof.

As a consequence, the method does not fulfill the precision requirement (nor the trueness one) and some review/correction seems required.

10. In an environmental study to characterize industrial discharges to a river, Arsenic was selected as the monitoring metal. An issue that has to be addressed by the researchers is whether the amount of As(III) is the same as that of As (total) (in such a case, the most toxic form, As(V), would not be present in the discharges). Samples were withdrawn from the discharge point on several days and quantitated for As (see the following table). Can it be affirmed that As(V) is not significantly present in the discharge?

$C_{\text{As(III)}}$ ($\mu g/L$)	1.84	1.92	1.85	2.07	1.94	1.92	1.91
$C_{\text{As(total)}}$ ($\mu g/L$)	1.94	2.05	1.97	1.90	1.89	1.92	2.10

SOLUTION:

As(V) will not be present in the samples if the average values for As(III) and As(total) do not have a significant difference. Let us check this.

First, recall that outliers should not be present. In this example, the Dixon's and the Grubbs's tests do not agree completely. Let us consider the latter.

$\bar{x}(\text{As(III)}) = 1.921\,\mu\text{g/L}$, $s(\text{As(III)}) = 0.0756\,\mu\text{g/L}$
$\bar{x}(\text{As(tot)}) = 1.967\,\mu\text{g/L}$, $s(\text{As(total)}) = 0.0795\,\mu\text{g/L}$

— Series of As(III) values:

$G_{\text{experimental}}$ statistics were 1.07 and 1.97, for the lowest and highest values, respectively.

— Series of As(total) values:

$G_{\text{experimental}}$ statistics were 0.97 and 1.67, for the lowest and highest values, respectively.

Since

$$G_{(\text{tabulated},\,99\,\%,\,n=7)} = 2.10 \quad \text{and} \quad G_{(\text{tabulated},\,95\,\%,\,n=7)} = 1.94$$

the null hypothesis will not be rejected (a value might be rejected, but at 95 % confidence level, not too obvious).

Next, compare the variances:

$$F_{\text{experimental}} = (0.0795)^2/(0.07559)^2$$
$$= 1.11 < F_{\text{tabulated}(99\,\%,\,6,\,6)} = 8.47$$

The null hypothesis should not be rejected, the variances are similar and, so, we can pool them (Equation (13)):

$$s_{\text{pool}}^2 = [6 \cdot (0.07559)^2 + 6 \cdot (0.0795)^2]/(7 + 7 - 2)$$
$$= 6.02 \cdot 10^{-3}; \quad s_{\text{pool}} = 0.07757$$
$$\text{dof} = 7 + 7 - 2 = 12$$

So, we can apply Equation (14):

$$t_{\text{experimental}} = (1.921 - 1.967)/[0.07757 \cdot \sqrt{(1/6 + 1/6)}]$$
$$= 1.03 < t_{\text{tabulated}(95\,\%,\,12)} = 2.18$$

The difference between the two average concentrations is not significant, which means that the most toxic species, As(V), is not likely to be present.

11. A good way to control the performance of an FT-mid-IR device is to use a standard polystyrene film. Some researchers proposed to monitor a ratio between two bands. Two research groups performed spectra on different days and calculated the proposed ratio (see the following table). Is it possible to affirm that the results are statistically comparable? Does one group have better precision? Why? Suppose that from these data, a standard value should be obtained for the proposed ratio, calculate it (*Note*: operate in all steps at 95 % confidence level).

| Research Group 1 | 5.9 | 5.7 | 5.6 | 6.9 | 6.3 | 5.8 | 6.0 | 5.6 | 5.7 | 5.4 |
| Research Group 2 | 5.5 | 5.9 | 5.8 | 5.7 | 5.4 | 5.6 | 5.6 | 5.0 | 5.7 | 5.9 |

SOLUTION:

Grubbs's test indicates that there are two outliers, one at each series: 6.9 (Group 1) and 5.0 (Group 2) for which $G_{experimental} = 2.33$ and 2.28 (respectively) $> G_{tabulated(95\%, n=10)} = 2.18$. These values must be deleted before proceeding with the other calculations. Hence:

Group 1: $\bar{x} = 5.78$; $s_1 = 0.2635$
Group 2: $\bar{x} = 5.68$; $s_2 = 0.1716$.

Are the variances comparable?
$F_{experimental} = (0.2635)^2/(0.1716)^2 = 2.36 < F_{tabulated(95\%,8,8)} = 3.44$; i.e., the null hypothesis cannot be rejected and, so, the variances are comparable. This means that neither group is more precise.

Note that many students fail in answering this question because they compare the SDs just visually, which is not correct and leads to wrong conclusions frequently.

The t-test requires a pooled variance:

$$s^2_{\text{pool}} = [8 \cdot (0.2635)^2 + 8 \cdot (0.1716)^2]/(9 + 9 - 2) = 0.0494$$

$$s_{\text{pool}} = 0.2223$$

Therefore:

$$t_{\text{experimental}} = (5.78 - 5.68)/[0.2223 \cdot \sqrt{(1/9 + 1/9)}]$$

$$= 0.95 < t_{\text{tabulated}(95\,\%,16)} = 2.12$$

As a conclusion, we cannot affirm that the two research groups yield different average values. In this case, we can merge all the data and calculate the grand average for the ratio (this could be a tentative 'standard' value), with an associated 95 % confidence interval calculated from the overall set of 18 values) as $t_{\text{tabulated}(95\,\%,17)} \cdot s/\sqrt{18}$.

Thus, the tentative standard value would be $= 5.73 \pm 0.11$ (no units because it is a ratio).

12. A group of students made 18 measurements of the dissociation constant of 2-chloro-phenoxyacetic acid in aqueous solution (conductimetry at 25 °C). Out of them, nine were obtained considering a mathematical correction for the ion mobility, whereas the others were given without that correction (see the following table). Decide (with a 5 % significance) whether the mathematical correction affects the results within the experimental conditions of the study.

$pK_{(\text{uncorrected})}$	2.995	2.995	2.996	2.997	2.999	3.002	3.004	3.005	3.006
$pK_{(\text{corrected})}$	2.999	2.999	3.000	3.000	3.000	3.000	3.001	3.003	3.007

SOLUTION:

As usual, the first step should consist of ascertaining the presence of outliers in each series of data. Consider the Grubbs's test (remember that the student should practice the Dixon's one by him(her)self). To summarize:

1st series (uncorrected values): $G_{\text{lowest}} = 1.10$; $G_{\text{highest}} = 1.38$.
2nd series (corrected values): $G_{\text{lowest}} = 0.78$; $G_{\text{highest}} = 2.35$.

Considering that

$$G_{\text{tabulated}, 99\%, 9} = 2.32 \quad \text{and that } G_{\text{tabulated}, 95\%, 9} = 2.11$$

the highest value of the second series should be considered as an outlier (the null hypothesis is rejected for it at both confidence levels).

Note: Due to the very low number of experimental datapoints available usually, in this textbook, we prefer not to repeat the Grubbs's test to the remaining values once one of them was deleted. If, however, that were done, the second series in this example would show another outlier. Nevertheless, that would also mean that we should apply the Grubbs's test for two simultaneous outliers on one side of the series (not to perform two tests!). This was mentioned in the theoretical part of this chapter.

Before continuing, delete it and recalculate the mean and SD of the series.

To compare the means, decide the equation to use by studying first whether the variances are comparable:

$$F_{\text{experimental}} = (4.43 \cdot 10^{-3})^2 / (1.28 \cdot 10^{-3})^2$$
$$= 12.25 > F_{\text{tabulated}(99\%, 8, 7)} = 6.84$$

and, also, $> F_{\text{tabulated}(95\%, 8, 7)} = 3.73$

Hence, we must reject the null hypothesis and accept that there is a significant difference among the variances and, so, apply the Welch's equations:

$$t_{\text{experimental}} = \frac{2.9999 - 3.0003}{\left[\left(\dfrac{(4.43 \cdot 10^{-3})^2}{9} \right) + \left(\dfrac{(1.28 \cdot 10^{-3})^2}{8} \right) \right]^{1/2}} = 0.25$$

$$\text{dof} = \left[\frac{((4.43 \cdot 10^{-3})^2/9 + (1.28 \cdot 10^{-3})^2/8)^2}{\left[\dfrac{((4.43 \cdot 10^{-3})^2/9)^2}{9+1} \right] + \left[\dfrac{((1.28 \cdot 10^{-3})^2/8)^2}{8+1} \right]} \right] - 2 \sim 10$$

The critical levels are

$$t_{(\text{tabulated, 99\%,2-tail,10})} = 3.17 \quad \text{and}$$
$$t_{(\text{tabulated, 95\%,2-tail,10})} = 2.23$$

and, so, the null hypothesis should not be rejected and we conclude that the two averages are 'the same', wherefrom the grand mean (3.0001) can be considered as the 'true' pKa. Using the overall 17 values, the confidence interval (95 % level) can be calculated as:

$$pK_a = 3.0001 \pm 0.0017$$

13. In a study intended to evaluate whether a new equipment can be of use for the routine measurement of the specific gravity of diesel and fuel oils, a collection of typical samples was obtained from a petrochemical refinery. Their specific gravity was measured using both the classical (a glass densitometer) procedure and the new proposal (PRO). The statistical treatment of the data can be done in different ways but it was decided to consider a typical regression fit. What conclusions can you derive from the study of the straight-line regression model? (*Note*: Operate in all cases at 95 % probability).

Classical	PRO	Classical	PRO
0.9214	0.9215	0.9769	0.9765
0.9750	0.9742	0.9737	0.9732
0.9214	0.9216	0.9838	0.9834
1.0003	0.9989	0.9778	0.9767
0.9215	0.9215	0.9317	0.9307
0.9955	0.9954	0.9744	0.9738
0.9596	0.9595	0.9746	0.9746
0.9214	0.9215	0.9726	0.9719
0.9745	0.9740	0.9780	0.9780
0.9705	0.9705	0.9015	0.9014
0.9796	0.9795	0.9698	0.9693

SOLUTION:

When two methods are compared, an intercept equal to zero and a slope equal to one are expected if they provide the same (statistical) results. Therefore, this is what has to be checked.

A simple plot of the regression line presents an — apparently — good model (see Figure 4a). However, an objective and sound decision-making needs, at least, the study of the residuals of the calculated model (see Figure 4b). There it can be clearly seen that the straight-line model is affected by three

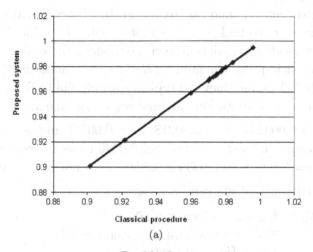

(a)

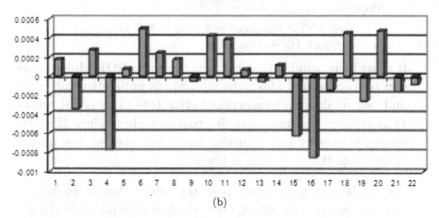

(b)

Figure 4.

experimental values whose residuals almost double the others and, worst, are in one side of the least squares straight line, likely 'pulling' it toward them. These three points (#4, #15 and #16) should be eliminated from the series of values and the model recalculated (the student can check that the standardized residuals for these points are very close to, or overpass, the typical warning limit, 2; although the decision for point #15 is not simple).

The new model, PRO $= 0.00531 + 0.9942 \cdot$ Classical ($r = 0.99996$) does not show obvious anomalous points, the residuals are homogeneous and no trends can be observed. Although the residual for the 2nd point is slightly higher than for the others, it was decided to maintain it in the model (studies done without it did not improve the model); besides, its standardized residual is only 1.9 (lower than the typical warning limit, 2).

The basic statistics of the final regression line are:

$S_{y/x} = 0.000264$; $S_a = 0.00218$; $S_b = 0.00227$ and, so:
Intercept $= a \pm t \cdot S_a = 0.00558 \pm 0.0046$ (does not contain zero),
Slope $= b \pm t \cdot S_b = 0.9939 \pm 0.0048$ (does not contain unity).

Alternatively, t-tests can be done:

Intercept: $t_{\text{experimental}} = |0 - a|/S_a$
$\qquad = 2.55 > t_{\text{tabulated, 95\%,2-tail,17}} = 2.11$
$\qquad$ reject $H_0 \rightarrow$ not zero

Slope: $t_{\text{experimental}} = |1 - b|/S_b$
$\qquad = 2.68 > t_{\text{tabulated, 95\%,2-tail,17}} = 2.11$
$\qquad$ reject $H_0 \rightarrow$ not unity

It must be concluded that the new system (PRO) exhibits some sort of 'background bias' (the intercept is not statistically zero) and, worst, shows a proportional error (the slope is not unity). In particular, the new system predicts higher values than the reference ones at low densities and lower ones at the highest range (note that the slope is lower than one).

Note: It can be argued that this study has a problem in its design. In effect, the student should note that there are very few experimental data at the lowest range of gravities. This might bias the final model. However, these are real data from production units and the number of batches which yielded those values was very limited. From a purely theoretical view, we could think of eliminating those points from the study and focus only on the upper part. However, this would be a suboptimal solution because if this were done in routine use, some samples would have to be measured using the classical method, whereas others would require another system, which is not a very convenient option.

Another comment is in order here. The exercise was focused on interpreting the meaning of the slope and the intercept. However, an F-test to check simultaneously whether intercept $= 0$ and slope $= 1$ should be applied instead (Equation (20b)). Here, $F_{\text{experimental}} = 10.53$ which is larger than the tabulated values $(F_{(99\%,2,17)} = 6.11)$, and so the intercept and slope are not simultaneously equal to zero and one, respectively.

14. An ion selective electrode is used to quantify ammonium in a water treatment plant. Calibration was done by diluting appropriate aliquots of a freshly prepared ammonium stock solution to 100.00 mL. The calibration data are shown below:

$C(NH_4^+)$ (mg/L)	0.15	0.30	0.40	0.60	0.80	1.00
Transformed signal (V)	0.026	0.052	0.069	0.106	0.135	0.170

The average signal for the calibration blank was $9 \cdot 10^{-4}$ V, with a corresponding $7.4 \cdot 10^{-4}$ V SD $(n = 10)$. The procedure to treat the samples consists of several steps: 5.0 mL are withdrawn from the 100 mL containers where the samples are taken. They are treated to adjust the ionic strength and the pH and the final volume is made up to 25.00 mL. A 3.00 mL aliquot is withdrawn and diluted to 10.00 mL, wherefrom 5.00 mL are taken to perform the final measurement. A procedural blank is prepared following the same stages, its average signal was 0.008 V, with a $2.0 \cdot 10^{-3}$ V SD. Questions to be addressed here are:

(a) Sensitivity of the analytical process.
(b) The ammonium concentration (mg/L) in two samples whose signals were 0.192 V and 0.060 V, each; along with their confidence intervals.
(c) Should the concentrations of the samples be reported as 'quantitated'?

SOLUTION:

The calibration looks quite acceptable, as it is seen next, but when the residuals and the standardized residuals are studied, it appears clear that the fourth point behaves anomalously. It is true that the standardized value does not overpass 2 but its difference with the other points is so clear that it seems more reasonable to discard it than to keep it. We decided to delete it from the studies. The calibration blank was not included in this preliminary calculations to avoid any further distortion (Figure 5).

Question a:

The new calibration line resulted: Signal $= 0.0012(\pm 0.0019) + 0.1684(\pm 0.0031) \cdot C(NH_4^+)$; and therefore the sensitivity of the calibration is $0.1684(\pm 0.0031)$ V/(mg/L).

When the calibration blank is introduced in the plot, the numerical value fits perfectly into the line and, so, we can introduce it safely as an additional calibration point. In this particular case, the calculations above are not modified (within rounding).

Question b:

It is very important to notice that the 0.182 V signal is out of the calibration range and, so, it is not correct to quantify it. It must be stated something like 'the sample cannot be quantitated because it is out of the calibration range'. In practical terms, we

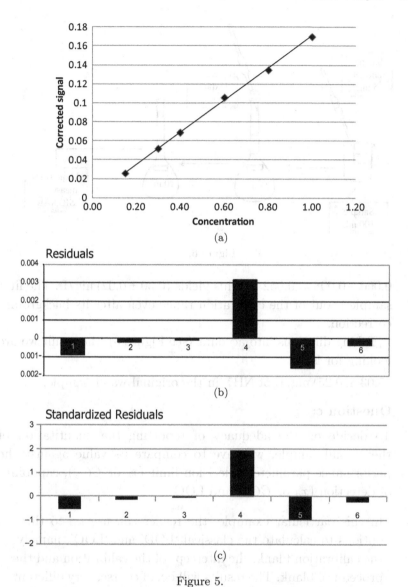

Figure 5.

would usually dilute it in the laboratory but nothing should be anticipated about its concentration.

The second sample needs to be background corrected using the procedural blank and, thus, the signal to be interpolated is

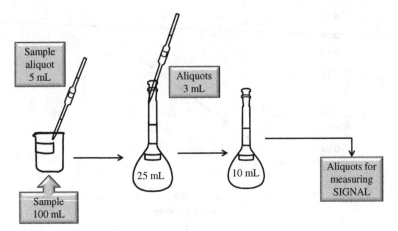

Figure 6.

$0.060 - 0.008 = 0.052$ V. This yields (0.30 ± 0.02) mg/L. The first sample is out of the calibration range even after its background correction.

Taking dilutions into account (see Figure 6), the result we are looking for is:

(5.03 ± 0.25) mg/L of NH_4^+ in the original water sample.

Question c:

To decide on the adequacy of reporting the quantitation of the second sample, we have to compare its value against the performance parameters: decision limit (x_c or CC_α), capability of detection (x_d or CC_β) and LOQ.

In this particular example, the reader can use up to three options to calculate the classical 'LOD' and 'LOQ', namely: the calibration blank, the intercept of the calibration and the procedural blank. The results will be, of course, very different and this exemplifies some of the problems encountered when using such definitions (the student is encouraged to perform such a comparison).

Using the intercept of the calibration line, classical definitions lead to (Equations (28d) and (29a) and associated explanations there):

$$\text{LOD}_{\text{concentration domain, classical definition}}$$

$$= (0.0012 + 3.75 \cdot 0.00061)/0.1684 = 0.021 \, \text{mg/L}$$

$$\rightarrow \text{taking dilutions into account} \rightarrow = 0.35 \, \text{mg/L of NH}_4^+$$

(considering the 99 % Student's 1-tail t-value, dof $= 4$, instead of simply using the theoretical value of 3).

$$\text{LOQ}_{\text{concentration domain, classical definition}}$$

$$= (0.0012 + 10 \cdot 0.00061)/0.1684 = 0.043 \, \text{mg/L}$$

$$\rightarrow 0.72 \, \text{mg/L of NH}_4^+ \text{ in original water}$$

This means that the sample can be quantitated without concerns.

Let us now consider the *updated definitions for the performance parameters* (Equations (30)–(32)):

$x_{\text{c,calibration}} = 0.009 \, \text{mg/L of NH}_4^+$ in the calibration line $\rightarrow$ considering dilutions:

$$x_{\text{c}} = 0.16 \, \text{mg/L of NH}_4^+$$

$x_{\text{d,calibration}} = 0.017$ in the calibration line $\rightarrow$ considering dilutions:

$$x_{\text{d}} = 0.29 \, \text{mg/L of NH}_4^+$$

$\text{LOQ}_{\text{calibration}} = 0.043$ in the calibration line (if $3 \cdot x_{\text{d}}$ is used as a first trial in the iterative calculations, only three iterations are required to get convergence) $\rightarrow$ considering dilutions:

$$\text{LOQ} = 0.72 \, \text{mg/L of NH}_4^+$$

This means that the sample can be quantitated without concerns.

In this example, there were not large differences between the 'old' and the 'new' definitions of the performance parameters. But this was just by chance and next example reveals how different they can be in other circumstances.

15. The next calibration table corresponds to anodic stripping polarographic measurements made to determine Cd in extracts of oceanic sediments. A convenient signal is the integration of the area under the peak obtained for aqueous standards. Determine: (a) the sensitivity of the calibration straight line; (b) the concentration of Cd (expressed as mmol/kg) in a sediment sample whose final integrated signal was 27.5 (units of integrated area). Such a signal was obtained after the following treatment: 5.0000 g of sediment are extracted (lixiviated) in a 50.0 mL acidic solution (pH = 2) and kept in darkness for several hours. Then, 3.50 mL are withdrawn and diluted to 100.00 mL. From the latter solution, 4 mL aliquots are sampled and finally measured.

(c) Evaluate the 'LOD' and 'LOQ' performance parameters considering the classical and the updated definitions.

[Cd] mmol/L	0.50	1.50	2.50	3.50	4.50	5.50
Polarographic area (integrated units)	2.71	8.11	13.98	19.37	28.45	30.92

Additional data: Signal of the calibration blank: 0.5 units ($s_{cal} = 1 \times 10^{-2}$ units, $n = 7$); Signal of the procedural blank: 5 units; ($s_{proc} = 3$ units, $n = 7$).

SOLUTION:

Question a:

A preliminary calibration plot to assess whether relevant outliers are present reveals that one of the points (the fifth) has indeed a very suspicious behavior (this task is let to the student's personal work). This is confirmed by visualizing a plot of the residuals or of the standardized residuals (see Figure 7a), which clearly reveals an outlying behavior (standardized residual close to −2 and far apart from other residuals).

This point is deleted and the calibration repeated, which yields: Area = −0.230 (±0.417) + 5.647 (±0.130) · [Cd], with a very good residuals pattern (homoscedasticity) (see Figure 7b),

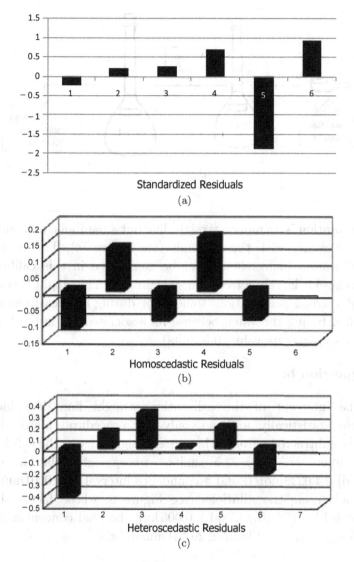

Figure 7.

and the requested sensitivity is:

5.647($\pm$0.130) units of integrated area/(mmol/L)

When the calibration blank is included in the fit as an additional calibration point, serious problems appear because the

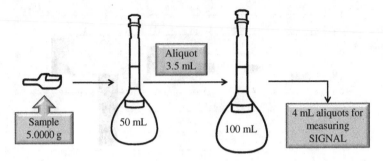

Figure 8.

calibration is no more a straight line but a parabola, as visualized very clearly with the residuals (see Figure 7c). Therefore, the calibration blank should not be considered in this calibration because it largely affects the straight line. This might be caused by a wrong preparation, some error during its measurement or when typing the value because its associated SD is too low (as it also seems its value, 0.5 units).

Question b:

The intercept of the calibration straight line does include zero statistically and, so, only the procedural blank should be subtracted from the gross signals of the unknown sample: $27.5 - 5 = 22.5$ units, whose interpolation yields $[Cd] = 4.02$ mmol/L and a confidence interval of ± 0.10 mmol/L. Considering the dilutions (see Figure 8; whose factor should be: $0.1\,L \cdot (50\,mL/3.5\,mL)/0.005\,kg$), the final concentration is: $C_{Cd\ sediment} = (1148.57 \pm 29.14)$ mmol/kg.

Question c:

We discarded the calibration blank already and, so, no calculations will be done with it.

- Using the intercept of the calibration line, classical definitions lead to:

$\text{LOD}_{\text{concentration domain, classical definition}}$

$$= (-0.230 + 4.54 \cdot 0.131)/5.647 = 0.065 \, \text{mmol/L}$$

$$\rightarrow \text{taking dilutions into account} \rightarrow = 18.45 \, \text{mmol/kg}$$

(considering the 99 % Student's 1-tail t-value, dof $= 3$, instead of using simply the theoretical value of 3).

$\text{LOQ}_{\text{concentration domain, classical definition}}$

$$= (-0.230 + 10 \cdot 0.131)/5.647 = 0.191 \, \text{mmol/L}$$

$$\rightarrow 54.64 \, \text{mmol/kg}$$

- Taking dilutions into account using the procedural blank, we would obtain for Cd: LOD' $= 942.09 \, \text{mmol/kg}$ and LOQ' $= 1770.71 \, \text{mmol/kg}$
- Taking the *updated definitions* into consideration:

$x_{\text{c,calibration}} = 0.085 \, \text{mmol/L}$ in the calibration line $\rightarrow$ considering dilutions:

$\underline{x_{\text{c}} = 24.39 \, \text{mmol/kg}}$

$x_{\text{d,calibration}} = 0.158 \, \text{mmol/L}$ in the calibration line $\rightarrow$ considering dilutions:

$\underline{x_{\text{d}} = 45.02 \, \text{mmol/kg}}$

$\text{LOQ}_{\text{calibration}} = 0.350$ in the calibration line (if $3 \cdot x_d$ is used as a first trial in the iterative calculations, only two iterations are required to get convergence) $\rightarrow$ considering dilutions:

$\underline{\text{LOQ} = 100.05 \, \text{mmol/kg}}$

Contrary to the previous numerical exercise, note the huge differences between the performance parameters. In this example, the procedural blank and its SD were high, probably due to the complexity of the samples. This yielded very high figures when the classical LOD and LOQ were computed using it.

EXERCISES PROPOSED TO THE STUDENT

Note: It is definitely critical to ascertain the existence of outliers before solving the exercises. If this study is not presented in the resolution, the exercise should be considered 'faulty'.

16. The laboratory of a factory determines nitrites in sausages daily (reported as ‰, m/m). The following table presents the measurements obtained from four batches. The target objective was established at 9.970‰ (to accomplish with legislation, quality preservation, etc.). Calculate the basic statistics (average, median, SD, relative SD), the absolute and relative errors.

Batch (nitrites, ‰ m/m)			
1	2	3	4
9.873	11.007	9.973	9.731
9.997	9.989	10.003	9.653
9.876	11.010	9.899	9.841
9.898	9.998	9.987	9.798

SOLUTION:

	Batch (nitrites, ‰ m/m)			
	1	2	3	4
$\bar{x}$ (‰)	9.9110	10.5010	9.9655	9.7558
SD (‰)	0.05841	0.58602	0.04599	0.08210
RSD (%)	0.59	5.58	0.46	0.84
$\bar{\bar{x}}$ (‰)		10.0333		
Overall median (‰)		9.9360		
E (‰)		+0.063		
E_r (%)		+0.63		

17. To test a balance used for practical classes, pupils are given a set of certified weights. They are told to measure them as precisely as possible. The results are given in the following table. Considering that decisions have to be taken at a 95 % significance level, can the pupils state that the balance is working correctly? (*Note*: accept Gaussian distributions for the series of data).

Certified value (g)	Average experimental value ($n = 10$) (g)	SD (g)
0.10	0.08	0.04
1.04	1.00	0.20
10.12	10.21	0.52
100.15	99.83	0.64

Solution: The exercise can be solved in different ways although the study of the regression straight line seems not too adequate because of the very low number of data points (only 2 dof). A quite simple possibility is to check whether each average weight includes the certified value (Student's t-test, 95 % significance). In this case, all certified values are included within the experimental confidence intervals and, so, the null hypothesis should not be rejected at any weight level, i.e., the experimental averages do not have significant differences with the certified values.

Note that the result of this and many other exercises in this chapter is an explanation — a decision — not a number (e.g., the experimental t-value, or any other).

18. A laboratory manager prepared a chloride standard solution (aqueous matrix) to assess the performance of the analysts in charge of the test. The experimental values reported by the analysts of several shifts were (non-consecutive days): 49.4, 49.8, 50.8, 49.3, 51.3, 50.0, 50.8 and 51.8 mg/L. Calculate the confidence intervals of the average value (95 % and 99 % significance).

Solution: (50.40 ± 0.76) mg/L (95 % significance) and (50.40 ± 1.13) mg/L (99 % significance).

There were no outliers (using both Dixon's and Grubbs's tests) at any significance level, 95 % or 99 %.

19. Suppose that the Technical Manager of your company states that your target yield (for a given process) should be 80 %, with an accepted 2 % variability (given as a confidence interval). The last batches obtained from your process were: 82, 84, 81, 85, 80, 79, 80, 76, 79, 82 and 89 (%). Explain clearly the conclusions you would present to your manager.

Solution: Grubbs's test shows that the null hypothesis should not be rejected either at 95 % or at 99 % significance (we conclude that there are no outliers). Application of the Student's t-test revealed that the null hypothesis cannot be rejected ($t_{experimental} = 1.42 < t_{tabulated(95\%, n-1=10)} = 2.23$) and, so, the conclusion is that your production batches are around the target value. However, you could have a slight problem with the dispersion of your values because the confidence interval around the mean was 2.35, a bit higher than the desired maximum variability. Somebody has to check the operational parameters in order to reduce such a variability.

20. Two research groups decided to compare the results they obtained using their proprietary analytical methods. They prepared 300 g of a homogenizate and distributed aliquots to each person in charge of the analyses. Each group performs 10 independent analyses (different analysts, days, etc.). Compare their precisions and average values. Is one of the laboratories yielding better precision? In case you need to report a true value for the material under study, what would such a value be? *Note*: operate at 95 % significance.

Group A (mg/L)	5.25	5.55	5.45	5.85	5.65	6.15	6.75	5.45	5.55	5.75
Group B (mg/L)	6.60	6.40	5.70	6.30	6.30	6.10	6.40	6.50	6.60	6.40

Solution: First, check for outliers in each series. The suspicious values are the highest and the lowest for series A and B, respectively. At 95 % significance, the null hypotheses can be rejected for them but at 99 % significance, they cannot. It is, therefore, not simple to take a decision. In most laboratories, the Grubbs's test is applied routinely at 95 % significance. In this case, this seems acceptable because we have 10 experimental values (with much less values we should operate at 99 % significance). So discard the two values and recalculate the basic statistics. According to the F-test, the variances are comparable and, so, any group is more precise (at any significance level, 95 % and 99 %). The Student's t-test yields: $t_{experimental} = 5.76 > t_{tabulated,95\%,2\text{-tail},16} = 2.13$.

Therefore, the groups are reporting different averages and, thus, we cannot estimate a true value for the material.

21. Kanamycin is a powerful antibiotic which is used to alleviate myopia and that can be determined using high performance liquid chromatography (HPLC). A governmental laboratory is in charge of monitoring commercial pills which contain this active pharmaceutical ingredient (API). For this, the laboratory purchases boxes of the commercial product in several drugstores. Considering that the pharma company declared 5 mg of API per pill, you have to determine: (a) Does each batch of pills fulfill that declaration? (b) Does one of the batches present a more homogeneous distribution of the API? (c) Do the two batches have the same API average composition? *Notes*: Perform calculations at 95 % significance; data represent mg of API/pill.

| Batch A | 5.45 | 5.15 | 7.71 | 5.55 | 4.75 | 5.32 | 5.53 | 5.09 | 5.70 | 4.42 |
| Batch B | 4.98 | 4.84 | 4.77 | 4.91 | 4.84 | 4.98 | 4.91 | 5.21 | 4.67 | 5.21 |

Solution: The highest value of Batch A behaves as an outlier (at both 95 % and 99 % confidence levels). The experimental Fisher–Snedecor's test value (=5.72) is larger than the critical values (at both 95 % and 99 % confidence levels) and, therefore, the API distribution in Batch B is more homogeneous than in Batch A (this is bad news for the pharma company). For the two batches, the confidence intervals (95 %) around the experimental averages include the target value (5 mg/pill); the Student's t-test to compare the two experimental averages yields a value (=1.92, 11 dof) which cannot be used to reject the null hypothesis, at 95 % probability, and we therefore accept that there are no significant differences between the averages of the two batches.

22. A premium-brand sherry importer measures several organoleptic and chemical parameters before accepting the products. Iron is quantified routinely because it is used to assess the quality of the original grapes, the geological origin of the grapes (by means of isotopic dilution techniques) and possible undue contamination processes. The next results show the analyses of several aliquots of sherry withdrawn from randomly selected bottles (total Fe concentration, mg/L) from two 'soleras' of the same maker. Can it be said that the two soleras have different total contents of iron (95 % significance)?

| Solera A, C_{Fe} (mg/L) | 49 | 44 | 70 | 50 | 58 |
| Solera B, C_{Fe} (mg/L) | 44 | 57 | 34 | 48 | 50 |

Solution: The Grubbs's test does not show any outlier; the variances are not different [$F_{experimental}$(=1.44) < $F_{tabulated(95\%,4,4)}$ (=6.39)]; the Student's t-test does not allow rejection of the null hypothesis and, so, there are not evidences against the equality of both averages [$t_{experimental}$(=1.29) < $t_{tabulated(95\%,2\text{-tail},8)}$ (=2.31)].

Note: It is true that the number of experimental points is low (each solera is obtained from a 500-L cask, approximately 400 bottles), but the cost of each bottle of premium-brand sherry is so high that the number of analyses should be kept at a minimum.

23. Galician mussels hold a worldwide reputation because of their large size and delicious taste. They can be exported either in refrigerated containers or, even, in containers with seawater

ponds to preserve their freshness and quality. To be exported, every batch is monitored by a European-accredited laboratory ruled out by the Galician government. One of the critical analytical parameters consists of determining the absence of Diarrhetic Shellfish Poisoning (DSP, a toxin that appears in seawater when blooms of dinoflagellate occur). There are two main analytical procedures to determine DSP, one is based on a bioassay performed on a suite of selected mice, whereas the other uses HPLC (nowadays, this is the method recommended by EU authorities). With the data below, and using linear regression, conclude whether the two methods yield statistically comparable results (values in μg/L; consider 95 % significance).

Bioassay	90	182	175	142	110	101	122	100
HPLC	75	155	143	122	90	85	98	86

Solution: The presence of outliers was not visualized in a plot of residuals. Recall that the results of the bioassay should be considered as the abscissa (because this has been the reference method for many years). Then, the regression between the two methods yields: HPLC $= -0.50(\pm 10.58) + 0.84(\pm 0.08) \cdot$ bioassay. It is obvious that the intercept is statistically zero (there is no background difference between the methods) although the slope is not unity (which suggests that there is a proportional bias).

One of the problems associated to the use of bioassays is the spread of the results, which might be the root problem here for the lack of comparability.

Note: application of the F-test (Equation (20b)) here might also be of interest and it should result in $F_{\text{experimental}} = 202.4$, which is clearly larger than the tabulated ones (2 and 6 dof, for numerator and denominator, respectively), the intercept and slope are not zero and one, simultaneously.

24. A study was made to evaluate the linearity of an IR deuterated triglycine sulphate detector (DTGS detector). Solutions of propanone were prepared and the baseline-corrected height of the

non-interfered peak at $1700\,\text{cm}^{-1}$ was measured. Considering a regression straight line for the data, evaluate the behavior of the detector.

[Propanone] (M)	0.50	1.50	2.50	3.50	4.50	5.50
Baseline-corrected height ($\times 10$)	3.76	9.16	15.03	20.42	25.33	31.97

Solution: There is an outlier (4.50, 25.33); rejecting it, the regression straight line is: Height $= 0.819 + 5.650 \cdot$ [Propanone]; with, $S_{y/x} = 0.1577$, $S_a = 0.1312$, and $S_b = 0.041$. The Student's t-test revealed that the intercept does not include zero (i.e., there is an offset; maybe because of some baseline shift — which has to be studied visualizing the spectra; $t_{\text{experimental}}$ ($=6.25$) $> t_{\text{tabulated}(95\,\%,2\text{-tail},3)}$). In addition, the slope does not include unity, $t_{\text{experimental}}(=137.8) > t_{\text{tabulated}(95\,\%,2\text{-tail},3)}$ ($=3.18$). A possibility to solve this problem is to restrict the working range (the last standard seems to offer a too low transmittance).

Application of the F-test to check both coefficients simultaneously yields $F_{\text{experimental}} = 24{,}399$ which is larger than the tabulated 95 % confidence value (2 and 3 dof), and leads to the same conclusion; for a 99 % confidence level that is not so.

25. An alternative method to quantify the flash point of light fuels (gasoline and kerosene) was proposed. A correlation using a suite of routine samples was done (see the following table). According to this information, are the two methods comparable?

Reference method (EC)	50	47	48	51	44	53	48	49	44
Proposed method (EC)	51	48	49	52	45	54	49	50	46

Solution: Note that an alternative way to study the problem is to perform a paired t-test (which was not detailed in this chapter); however, it is more common to perform a regression study (the reference values are the abscissas). The last point has to be deleted because of its outlying behavior. The resulting line is: proposed method $= 1.0 + 1.0 \cdot$ reference method; correlation coefficient $= 1.0000$. The slope is statistically unity but the intercept shows that there is a constant offset of $+1.0\,^\circ$C (the proposed method increases the 'true' value by $1\,^\circ$C.

26. To evaluate the presence of interferences when measuring benzo-a-pirene (BAP) using gas chromatography, a preliminary study was made using a typical sample. The following table collects the data associated to a traditional aqueous calibration and to an SAM. You have to (a) determine the presence of interferences; (b) estimate the concentration of BAP in a sample (ng/L) considering that it suffered the treatment detailed next: 250.00 mL of waste water are filtered and extracted through a commercial solid–liquid extraction cartridge. The cartridge is dried in vacuum (30 min). To elute polycyclic aromatic hydrocarbons (PAHs), including BAP, a mixture of 3 solvents is used. The eluate is dried and redissolved in 0.50 mL of hexane, just before the chromatographic injection.

Note: To simplify the calculation, consider that the calibrations do not have outliers.

C_{BAP} (mg/L)	Transformed signal	C_{BAP} added (mg/L)	Transformed signal
0	0.003	0	0.109
10	0.061	10	0.152
20	0.123	20	0.191
30	0.17	30	0.242
40	0.215	40	0.295
50	0.272	50	0.321

Solution: Strictly speaking, each calibration has an outlier but the exercise is devoted to focus on comparing the slopes of the calibration lines. First, the straight-line statistics have to be calculated and the standard errors of the fits compared with an F-test. It is found that $F_{experimental} = 1.42$ is lower than the tabulated values (at 95 % and 99 % probabilities). Following, they can be pooled ($s^2_{pool} = 4.25 \cdot 10^{-5}$). The dof can be estimated as 8. Finally, the experimental t-value applied to compare the slopes was 4.07, which is larger than the 95 % tabulated one.

As a conclusion, there is/are some interference(s) that modify the slope of the calibration lines and, so, the SAM must be used. The interpolation of the unspiked test solution leads to (24.92 ± 4.81) mg/L, which considering the dilutions yields the concentration we were looking for: (49.8 ± 1.0) ng/L.

CHAPTER 3

ULTRAVIOLET AND VISIBLE SPECTROSCOPY

Rosa María Soto-Ferreiro

OBJECTIVES AND SCOPE

This chapter covers the main aspects related to the quantitative applications of ultraviolet and visible (UV–VIS) spectroscopy: direct application of the Beer's law to calculate the concentration of an analyte; using multicomponent analysis for the determination of two or more components in a sample, even if their individual absorption spectra overlap partially; the derivative spectroscopy to avoid the interferences; and the different methods to determine the stoichiometry of a complex.

1. INTRODUCTION

1.1. Basic concepts

The region of the electromagnetic spectrum that spans from 160 to 780 nm, with the visible region starting at 400 nm gives rise to the so-called *spectroscopy in the ultraviolet and visible* (UV–VIS) *regions*. Such a spectroscopy is based on the measurement of the radiation absorbed by chemical species which experiment transitions between different electronic energy states. The absorbance spectrum is obtained by plotting the energy retained by the sample against the photon's energy or, more commonly, its wavelength.

These two spectral regions are currently studied altogether because their underlying physical processes are the same (electronic transitions). In addition, the instrumentation requirements are similar and, thus, the same instrument can be utilized for both regions, albeit with some minor modifications. Moreover, the procedural operations required for sample pretreatment, calibration, signal processing, etc. are mostly the same.

Quantitative applications are based on the Bouguer–Lambert–Beer law (in brief, Beer's law), which states that for a monochromatic electromagnetic radiation, the absorbance (A) is proportional to the optical path (usually, the sample thickness (b) and the concentration of each absorbing species (C) (Equation (1a)):

$$A = \sum a_i \cdot b \cdot C_i \tag{1a}$$

where a_i is the absorptivity of each species, i (a physical property of the species) independent of the concentration, whose unit is concentration$^{-1} \cdot$cm^{-1}, and C_i is the concentration of each absorbing species. If the concentration is expressed as molarity, then a is replaced by the molar absorptivity, ε, which has the unit $L/(\text{mol}\cdot\text{cm})$, and for a unique absorbing substance, Equation (1b) is applied:

$$A = \varepsilon \cdot b \cdot C \tag{1b}$$

Deviations from linearity have been frequently observed owing to the limitations of Beer's law, which can be divided into three categories:

(i) *Fundamental*: It is valid only for low concentrations of analyte and, in principle, only one absorbing species (so, be careful if you have more absorbing species);

(ii) *Chemical*: The analyte should not react with the solvent or another species present in the solution;

(iii) *Instrumental*: The radiation obtained in the instruments is not purely monochromatic (due to, e.g., the effect of the stray radiation, filters, etc.).

Therefore, the analytical chemist has to check carefully the experimental conditions in order for them to fulfill the requirements of the Beer's law, specifically when calibration and quantitation are performed.

Roughly, two types of instruments are available for UV–VIS spectroscopy. They are termed single beam and double beam, according to the physical path(s) the radiation follows within the instrument. When using a single beam instrument, it is necessary to measure the absorbances of the blank and the sample consecutively to get a blank (background) correction, while in double beam spectrophotometers, the beam from the lamp is divided into two beams that pass through the sample cell and the reference cell respectively, compensating for source fluctuations and blank (background) signal. The basic components of UV–VIS instruments are shown in Figure 1.

A full description of the instrumental aspects of UV–VIS spectroscopy is out of the scope of this book and readers are kindly forwarded to any of the textbooks describing the instrumental methods of analysis [1–4].

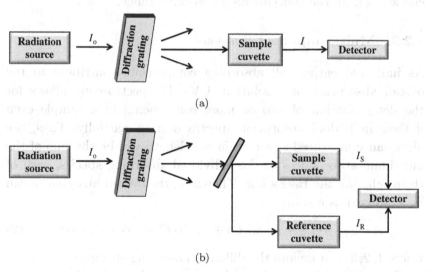

Figure 1. Schematic diagrams of (a) single beam and (b) double beam spectrophotometers.

1.2. Quantitative applications of the Bouguer–Lambert–Beer law

Some major quantitative applications of UV–VIS spectroscopy are resumed in this section. The basic principles are presented first for each application and then, in the next sections, typical case studies are solved, and several exercises proposed in order to help students practicing these topics.

1.2.1. Calculating the concentration of the analyte

The most common quantitative application of UV–VIS spectroscopy consists of calculating the concentration of an absorbing species using the linear relationship between absorbance and concentration, namely the Beer's law. For this purpose, as it was explained in Chapter 2, standard solutions of the analyte have to be prepared in the range of concentrations we are interested in, their absorbances measured at a wavelength where maximum absorbance was observed (usually, after preliminary studies) and, finally, a calibration function is deployed. The performance of the model (Beer's law) should be checked within the concentration's working range.

1.2.2. Multicomponent analysis

As indicated earlier, all absorbing components contribute to the overall absorption of a solution. UV–VIS spectroscopy allows for the determination of two or more components in a sample even if their individual absorption spectra overlap partially. Thus, the absorbance measured at a certain wavelength will be the sum of the individual absorbances of the individual absorbing species, each of them fulfilling the Beer's law. Therefore, the overall absorbance can be expressed as a sum:

$$A = \varepsilon_1 \cdot b \cdot C_1 + \varepsilon_2 \cdot b \cdot C_2 + \varepsilon_3 \cdot b \cdot C_3 + \cdots + \varepsilon_n \cdot b \cdot C_n \qquad (2)$$

where $1, 2, 3 \cdots n$ denote the different absorbing species.

To solve this equation, theory says that absorbance measurements must be performed at as many wavelengths as target compounds have to be quantified. However, from a practical point of view, the number of species to be quantified is usually restricted to two or three. This can be explained because of the large experimental errors incurred when more concentrations have to be calculated. Thus, when binary mixtures ought to be resolved, absorbance measurements at two selected wavelengths will be performed. Each wavelength should characterize a different substance (here, it is the main experimental difficulty to resolve systems with more than three components), and a system of two equations has to be solved, as it is shown below. The molar absorptivities of both species at the selected wavelengths must be as different as possible.

$$A_{\lambda_1} = \varepsilon_{1\lambda_1} \cdot b \cdot C_1 + \varepsilon_{2\lambda_1} \cdot b \cdot C_2$$
$$A_{\lambda_2} = \varepsilon_{1\lambda_2} \cdot b \cdot C_1 + \varepsilon_{2\lambda_2} \cdot b \cdot C_2 \tag{3}$$

1.2.3. Derivative spectroscopy

Derivative spectroscopy consists of calculating the first or superior order derivative of the spectrum. Derivative spectra often reveal details that are not visible in current spectra. The major problem in derivative spectroscopy is that the signal-to-noise ratio (S/N) degrades progressively with the order of the derivative.

One of the more interesting applications of derivative spectroscopy is the accurate determination of the concentration of a component when interferents are present. Calculations in this situation are based usually on signal measurements at the wavelength where the derivative spectrum of the interfering compound becomes zero (discarding those where the analyte becomes zero as well).

1.2.4. Determination of the stoichiometry of a complex

The stoichiometric composition of a complex can be ascertained using UV–VIS spectroscopic measurements since the absorbance can be determined without disturbing the equilibria under study. Three methods are commonly employed: the method of the continuous variations, the molar-ratio method and the slope-ratio method. The results obtained from the first two methods also provide information about the stability of the complex as its dissociation/formation constants can be determined easily; as a particular application, acidity constants can be evaluated too.

The methods are depicted below considering that the stoichiometry of a metal–ligand complex has to be determined (say, M_nL_m).

(i) Method of continuous variations

Aliquots of stock solutions of the cation and the ligand (for convenience, they may be of the same concentration) are mixed in different proportions, so that the total volume and the total amount of moles is constant in all mixtures. Just for instance, if we decided to handle stock solutions of the same concentration, we will mix 1 mL of metal with 4 mL of ligand; 2 mL with 3 mL, 3 mL with 2 mL, and so forth. The absorbance of each solution is measured at the wavelength where a maximum was observed after a preliminary study of the complex. Plotting the molar fraction of one of the reagents against the absorbance (providing that the reagents alone do not absorb at the selected wavelength) yields a characteristic triangular-shaped curve, shown in Figure 2. The straight portions of the triangle are extrapolated upwards until they cross each other. The

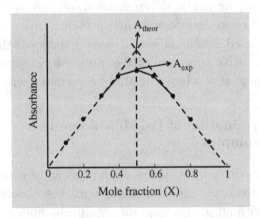

Figure 2. A typical plot obtained for the method of continuous variations.

intersection point corresponds to the molar fraction of the theoretical stoichiometry of the complex. For instance, a 0.5 molar fraction denotes an ML complex, whereas a 0.33 molar fraction denotes an ML_2 complex.

(ii) The molar-ratio method

Working solutions are prepared by mixing portions of separate stock solutions. The concentration of one of the reagents (usually, the cation) is kept constant throughout, while the concentration of the other varies continuously. The absorbance of the solutions is measured at a suitable wavelength (as determined from preliminary studies; in principle, only the complex should absorb at the selected wavelength) and plotted against the ratio between the moles of the ligand and the metal, yielding a curve as shown in Figure 3. The extrapolation of the straight-line portions, so that a crossing point can be visualized easily, shows the stoichiometric ratio of the complex.

As mentioned above, the data generated from both the method of continuous variations and the molar-ratio method can be utilized to determine the dissociation/formation constants. Consider as a starting point, the dissociation equilibrium of the complex:

$$M_nL_m \leftrightarrow nM^{m+} + mL^{n-}$$

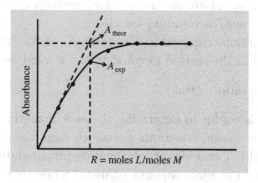

Figure 3. A typical plot obtained for the molar-ratio method.

Then, the dissociation constant is expressed as:

$$K_\mathrm{d} = \frac{[M^{m+}]^n \cdot [L^{n-}]^m}{[M_n L_m]}$$

Of course, the formation of equilibrium and the constant are expressed analogously using basic calculations:

$$nM^{m+} + mL^{n-} \leftrightarrow M_n L_m$$

$$K_\mathrm{f} = \frac{[M_n L_m]}{[M^{m+}]^n \cdot [L^{n-}]^m}$$

The determination of both K_d and K_f requires the knowledge of the concentrations of the different species in the equilibrium. These can be calculated (in fact, estimated) from the initial concentrations of the reagents and some basic considerations at the stoichiometric point. Thus, we can foresee the 'theoretical' concentration of the complex (theoretical, here, means that we estimate its concentration assuming that no dissociation occurs) from the absorbance value obtained by extrapolating the straight portions of the plots (in Figures 2 and 3, that was termed A_theor). Since dissociation does occur in a solution, we can now calculate the experimental concentration of the complex in the solution, using the experimental absorbance signal, $A_\mathrm{experimental}$, also measured at the stoichiometric point. Note that A_theor denotes the theoretical signal associated to the quantity of complex formed when exact stoichiometric amounts of metal and ligand ($[M_n L_m]_\mathrm{theor}$) are in the solution, without any consideration of dissociation and/or common ion effect.

The calculations required to determine K_d or K_f will be described in more detail in the section devoted to the worked exercises.

(iii) The slope-ratio method

This method is useful to determine the stoichiometry of complexes with high dissociation constants (in which case, the other methods would not work). Two series of solutions are prepared. In one of them, the flasks have the same amount of metal while that of the ligand increases steadily. In the other series, the opposite should occur; i.e.,

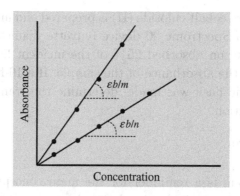

Figure 4. A typical plot obtained for the slope-ratio method.

the same concentration of ligand is mixed with increasing quantities of metal. The plot of the absorbances versus the concentrations of the reagent (for each series) yields two straight lines. The ratio of the slopes is the stoichiometric metal–ligand ratio, for a generic complex M_nL_m (Figure 4).

REFERENCES

[1] Douglas, A.; Skoog, F.; Holler, J.; Crouch, S. R. (2007). *Principles of Instrumental Analysis*, 6th ed. Thomson Brooks/Cole, Belmont, CA.

[2] Harvey, D. (2009). *Analytical Chemistry 2.0. An Electronic Textbook for Introductory Courses in Analytical Chemistry*. McGraw-Hill, New York.

[3] Robinson, J. W.; Skelly Frame, E. M.; Frame, G. M. II. (2014). *Undergraduate Instrumental Analysis*, 7th ed. CRC Press, Chichester, UK.

[4] Thomas, M. (1996). *Ultraviolet and Visible Spectroscopy*. John Wiley & Sons, New York.

WORKED EXERCISES

Preliminary notes:

(1) We strongly recommend to depict the analytical process in a figure in order to avoid typical missing calculations associated with sample treatment.

(2) Throughout this chapter, the optical path, b, will be considered as 1 cm, unless otherwise stated.

1. A solution of cobalt chloride (II) is prepared and measured immediately in a Spectronic 20 device (cuvette diameter 1 cm). This colored solution absorbed 25 % of the incident light at 642 nm. Determine the absorbance of the sample. If a 10-fold dilution in a volumetric flask was made, determine the transmittance (%) of this solution.

SOLUTION:

This problem is a reminder for the main concepts of radiation absorption: transmittance and absorbance.

Taking into account the fundamental definition of absorbance, the first question can be addressed easily:

$$A = -\log T$$

If 25 % of P_0 is absorbed, 75 % of the incident light is transmitted, hence:

$$A = -\log(0.75) = 0.125$$

The second question is resolved taking into account that the molar absorptivity is a physical property of a compound at a certain wavelength, independent of its concentration. So, two equations can be posed:

$$A_1 = \varepsilon \cdot b \cdot C_1 \qquad A_2 = \varepsilon \cdot b \cdot C_2$$

where $C_2 = C_1/10$, and A_2 corresponds to the unknown parameter to be calculated:

$$\frac{A_1}{A_2} = \frac{C_1}{C_1/10}$$

$$A_2 = \frac{0.125}{10} = 0.0125$$

Finally, using the absorbance definition:

$$A = -\log T = 0.0125; \quad T = 0.9716$$

And, so: $\% T = 97.16 \%$

2. Manganese is an oligoelement that plants need for good bloom-
   ing. To determine the content of this metal in a fertilizer, a
   complexation with 1–10-phenanthroline can be performed in
   order to get a colored complex which is suitable for absorbance
   measurements in the visible region. A 5.00 g aliquot of a fertilizer
   was digested, the complexing agent was added and the solution
   diluted to 250.00 mL. A 3 mL aliquot was withdrawn to measure
   its absorbance. Calculate the concentration (μg/g) of manganese
   in the fertilizer with the data supplied in the table below.

Solution	Absorbance
Blank (including reagents)	0.025
Mn^{2+} standard solution 1 mg/L	0.370
Sample aliquot 3 mL	0.405

SOLUTION:

This exercise presents a situation where the blank obtained from
the reagents has non-negligible absorbance, therefore, it must be
subtracted from the total absorbance signal.

$$\text{Standard solution:} \quad 0.370 - 0.025 = 0.345$$
$$\text{Sample:} \quad 0.405 - 0.025 = 0.380$$

Now, accepting compliance to the Beer's law, a ratio can be
established:

$$\frac{1 \text{ mg/L Mn (standard solution)}}{C \text{ mg/L Mn (sample)}} = \frac{0.345}{0.380}$$

Solving for C, the concentration of Mn in the final test solution
is 1.101 mg/L.

To obtain the concentration of Mn in the solid fertilizer, the
overall analytical procedure, depicted in Figure 5, is considered.

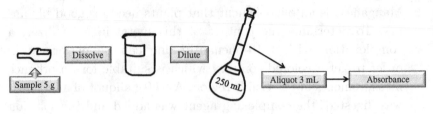

Figure 5.

> Note that the concentration of the analyte in the 250-mL
> solution is the concentration calculated for the final test
> solution because only an aliquot was withdrawn to perform
> the absorbance measurement.

$$C_{Mn} = 1.101 \, \frac{mg}{L} \cdot \frac{1 \, L}{10^3 \, mL} \cdot \frac{250 \, mL}{5.00 \, g}$$

$$= 0.05505 \, \frac{mg}{g} \cdot \frac{10^3 \, \mu g}{1 \, mg} = 55.1 \, \mu g/g$$

3. The composition of a material used in a metallurgical industry
   includes copper concentrations around 3%. To determine the
   content of this metal in an alloy, an aliquot of the solid is weighted
   and dissolved with an acid mixture. Then, a complexation with
   8-hydroxyquinoline is performed and the solution is diluted to
   250.00 mL. The absorbance values of the solutions obtained
   must be between 0.3 and 0.7 in order to keep them within
   the calibration working range. Determine the range of sample
   mass that can be analyzed in order to fulfill this requisite.
   Notes: (1) the molar absorptivity of the complex of copper (II)
   with 8-hydroxyquinoline at the wavelength of the maximum is
   $6.8 \cdot 10^2 \, L/(mol \cdot cm)$; (2) assume 1:1 stoichiometry.

SOLUTION:

This exercise involves the estimation of the mass of sample that
must be analyzed, following the analytical procedure described
in the enunciate, in order to obtain absorbance values within

the calibration working range established for the instrumental determination.

Practical working limits (absorbance): $A_1 = 0.3$ $A_2 = 0.7$

The molar concentration is calculated applying the Beer's law for both absorbance values:

$$0.3 = \varepsilon \cdot b \cdot C_1$$

$$C_1 = \frac{0.3}{6.8 \cdot 10^2 \dfrac{L}{mol \cdot cm} \cdot 1\,cm} = 4.41 \cdot 10^{-4}\,mol/L$$

$$0.7 = \varepsilon \cdot b \cdot C_1$$

$$C_1 = \frac{0.7}{6.8 \cdot 10^2 \dfrac{L}{mol \cdot cm} \cdot 1\,cm} = 1.00 \cdot 10^{-3}\,mol/L$$

The mass of Cu is calculated with the molar mass (63.54 g/mol) of the metal and the volume of the solution (250 mL):

$$\text{Mass}_1 \text{ of Cu} = 4.41 \cdot 10^{-4}\frac{mol}{L} \cdot 63.54\frac{g}{mol} \cdot 250\,mL \cdot \frac{1\,L}{10^3\,mL}$$

$$= 7.00 \cdot 10^{-3}\,g$$

$$\text{Mass}_1 \text{ of Cu} = 1.00 \cdot 10^{-3}\frac{mol}{L} \cdot 63.54\frac{g}{mol} \cdot 250\,mL \cdot \frac{1\,L}{10^3\,mL}$$

$$= 0.0164\,g$$

The amount of sample can be calculated considering the typical content of copper in the samples (3 %):

$$\frac{3\,g\,Cu}{100\,g\,sample} = \frac{7.00 \cdot 10^{-3}\,g\,Cu}{x_1\,g\,sample}$$

$x_1 = 0.2333$ g of sample

$$\frac{3\,g\,Cu}{100\,g\,sample} = \frac{0.0164\,g\,Cu}{x_2\,g\,sample}$$

$x_2 = 0.5467$ g of sample

Therefore, the amount of sample to be analyzed can vary between 0.23 and 0.55 g.

4. Micronized titanium dioxide is a blocking agent that is included in the formulation of some sunscreens in concentrations up to 25 %. This compound acts as a mirror on the skin reflecting the incident UV radiation. The determination of titanium dioxide in sunscreens can be performed, after the adequate pretreatment of the sample, by the formation of a yellow–orange complex $(TiO_2(SO_4)_2)$ that has a maximum of absorbance at 410 nm. For this purpose, a 0.4210 g aliquot of a commercial sunscreen was treated appropriately and diluted to 50.00 mL (solution A). A 10.00 mL aliquot of solution A was used to form the complex and finally diluted to 25.00 mL (solution B). The absorbance provided by 3 mL of B was 0.281. Another 10.00 mL aliquot of solution A was mixed with 7.00 mL of a standard solution containing 5.00 g/L of titanium dioxide, diluted to 25.00 mL as well (solution C), obtaining an absorbance value of 0.451. Determine the concentration of titanium dioxide in the sunscreen, expressed as mass percent.

SOLUTION:

This exercise proposes to determine a concentration using a standard addition point which involves the analysis of two aliquots of sample, one of them spiked with a known concentration of the analyte.

The application of the Beer's law to the experimental absorbance values yields:

$$A_B = 0.281 \quad 0.281 = \varepsilon \cdot b \cdot C_B$$

$$A_C = 0.451 \quad 0.451 = \varepsilon \cdot b \cdot C_C$$

The concentration of TiO_2 in solution C(C_C) is the sum of the concentration in solution B(C_B) and the added concentration:

$$C_C = C_B + \frac{7\,\text{mL} \cdot 5\,\text{g/L}}{25\,\text{mL}}$$

So $C_C = C_B + 1.4\,\text{g/L}$

Note that C_B is referred to 25 mL.

Therefore, Beer's law equations can be expressed as:

$$0.281 = \varepsilon \cdot b \cdot C_B$$

$$0.451 = \varepsilon \cdot b \cdot C_C = \varepsilon \cdot b(C_B + 1.4)$$

Combining these equations:

$$\frac{0.281}{0.451} = \frac{C_B}{C_B + 1.4}$$

Solving the equation for C_B, a value of 2.31 g/L is obtained, which is the concentration of complex in solution B, prepared from the sample.

In order to calculate the concentration of TiO_2, stoichiometry must be taken into account:

Note that concentration values can be estimated easily considering the stoichiometry when molarity is used.

Molar mass of $TiO_2(SO_4)_2 = 272.07$ g/mol.

And therefore, the molarity of $TiO_2(SO_4)_2$ is:

$$M = 2.31 \frac{g}{L} \cdot \frac{1\,mol}{272.07\,g} = 8.49 \cdot 10^{-3}\,mol/L$$

Taking into account the stoichiometry of the complex, the TiO_2 molar concentration coincides with that of the complex. Thus, the concentration of TiO_2 in solution B is calculated using the TiO_2 molar mass (79.9 g/mol), as follows:

$$8.49 \cdot 10^{-3} \frac{mol}{L} \cdot 79.9 \frac{g}{mol} = 0.6784\,g/L \text{ of } TiO_2$$

Finally, the percentage of TiO_2 in the sunscreen sample is calculated easily taking into account the different steps of the sample treatment procedure (see Figure 6):

$$\% \,TiO_2 = 0.6784 \frac{g}{L} \cdot \frac{1\,L}{10^3\,mL} \cdot \frac{25\,mL}{10\,mL} \cdot \frac{50\,mL}{0.4210\,g} \times 100 = 20.14\,\%$$

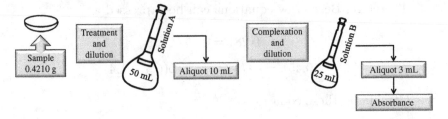

Figure 6.

5. A common practice performed at undergraduate courses consists of determining iron by UV–VIS spectroscopy using an adequate complexing agent. For this purpose, several standard solutions with increasing concentrations of Fe were prepared and sufficient amount of potassium thiocyanate was added to each one to constitute the complex. The absorbance of these solutions was measured at 450 nm and the results are collected in the following table. The procedure was applied to determine Fe in fresh water. For this, 30.00 mL aliquots of water samples were taken, potassium thiocyanate added and, finally, diluted to 100.00 mL. The absorbances obtained for two samples were 0.475 and 0.695. Determine the concentration of Fe in the samples.

C_{Fe} (mg/L)	A_{450}
0.00	0.010
5.00	0.156
10.00	0.307
15.00	0.458
20.00	0.623
25.00	0.772
30.00	0.913

SOLUTION:

A graph of the absorbances versus the concentrations of Fe shows a good linear relationship as can be seen in Figure 7. The

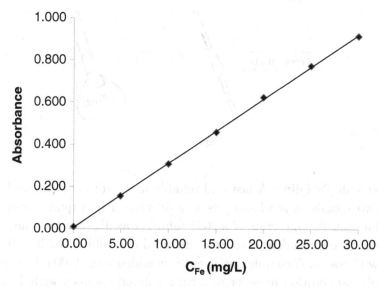

Figure 7.

residuals plot is not presented here but the student is strongly encouraged to make it (as is explained in Chapter 2) and visualize that outliers are not present.

The least-squares fit yields the following equation:

$$A = (0.0066 \pm 0.00413) + (0.0304 \pm 0.00023) \cdot C_{Fe}$$

The concentration of Fe in the solutions obtained from the two samples is calculated straightforwardly by interpolation:

Test solution 1: $C_{Fe} = 15.408\,\text{mg/L}$

Test solution 2: $C_{Fe} = 22.645\,\text{mg/L}$

Taking into account the sample treatment (see Figure 8), the final concentrations are calculated.

Sample 1: $C_{Fe} = 15.408\,\dfrac{\text{mg}}{\text{L}} \cdot \dfrac{100\,\text{mL}}{30\,\text{mL}} = 51.36\,\text{mg/L}$

Sample 2: $C_{Fe} = 22.645\,\dfrac{\text{mg}}{\text{L}} \cdot \dfrac{100\,\text{mL}}{30\,\text{mL}} = 75.48\,\text{mg/L}$

6. Titanium and chromium are frequently added to steels to increase their strength and make them useful for aeronautics

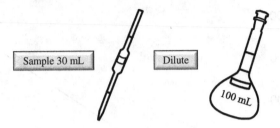

Figure 8.

and shipbuilding. A fast and reliable method to determine these two metals in steel samples was developed in a quality control laboratory. One gram of a steel sample was digested with an acid mixture in a microwave oven, filtrated and diluted to 100.00 mL with water. A complexing agent was added to a 10.00 mL aliquot of that solution in order to form a colored complex with Ti and Cr (suppose 1:1 stoichiometry in both cases). The absorbance of the two complexes at two wavelengths was measured in the visible region, along with standard solutions of both complexes. The results are shown in the following table. Calculate the concentrations of Ti and Cr in the original steel sample.

	Absorbance of the complex	
	410 nm	460 nm
Ti Standard solution 40.0 mmol/L	0.608	0.410
Cr Standard solution 120.0 mmol/L	0.444	0.600
Sample	0.849	0.755

SOLUTION:

This is an example of a multicomponent analysis, in particular to determine two analytes in a mixture. The absorbance measured at each wavelength will be the sum of the individual absorbances of each absorbing species, each of them fulfilling the Beer's law.

First, the molar absorptivity (ε) of each complex at the selected wavelengths must be determined using the absorbances obtained for the respective standard solutions (it is assumed that the metals and the ligand do not absorb by themselves).

Molar absorptivity of the Ti complex at 410 nm:

$$A_{Ti410} = \varepsilon_{Ti410} \cdot b \cdot C_{Ti}$$

$$\varepsilon_{Ti410} = \frac{A_{Ti410}}{b \cdot C_{Ti}} = \frac{0.608}{1\,cm \cdot 40\,\dfrac{mmol}{L}} = 0.0152\,L/mmol \cdot cm$$

Molar absorptivity of the Ti complex at 460 nm:

$$A_{Ti460} = \varepsilon_{Ti460} \cdot b \cdot C_{Ti}$$

$$\varepsilon_{Ti460} = \frac{A_{Ti460}}{b \cdot C_{Ti}} = \frac{0.410}{1\,cm \cdot 40\,\dfrac{mmol}{L}} = 0.0102\,L/mmol \cdot cm$$

Molar absorptivity of the Cr complex at 410 nm:

$$A_{Cr410} = \varepsilon_{Cr410} \cdot b \cdot C_{Cr}$$

$$\varepsilon_{Cr410} = \frac{A_{Cr410}}{b \cdot C_{Cr}} = \frac{0.444}{1\,cm \cdot 120\,\dfrac{mmol}{L}} = 0.0037\,L/mmol \cdot cm$$

Molar absorptivity of the Cr complex at 460 nm:

$$A_{Cr460} = \varepsilon_{Cr460} \cdot b \cdot C_{Cr}$$

$$\varepsilon_{Cr460} = \frac{A_{Cr460}}{b \cdot C_{Cr}} = \frac{0.600}{1\,cm \cdot 120\,\dfrac{mmol}{L}} = 0.0050\,L/mmol \cdot cm$$

C_{Ti} and C_{Cr} being the concentrations of the complexes of Ti and Cr, respectively (consider that complexation was 100 % efficient, and remember stoichiometry!).

Now two equations can be established considering the absorbance of the sample at each target wavelength:

Absorbance signal at 410 nm:

$$A_{410} = \varepsilon_{Ti410} \cdot b \cdot C_{Ti} + \varepsilon_{Cr410} \cdot b \cdot C_{Cr}$$

Absorbance signal at 460 nm:

$$A_{460} = \varepsilon_{Ti460} \cdot b \cdot C_{Ti} + \varepsilon_{Cr460} \cdot b \cdot C_{Cr}$$

Solving this system of two equations provides the following concentrations for the Cr and Ti complexes:

$$C_{Cr} = 73.77 \, \text{mmol/L}$$
$$C_{Ti} = 37.90 \, \text{mmol/L}$$

Since stoichiometry was considered to be 1:1, the molar concentrations of the metals coincide with those of the complexes:

Chromium concentration in the steel sample is 73.77 mmol/L
Titanium concentration in the steel sample is 37.90 mmol/L

7. Coumarin is a natural product that can be present in commercial foodstuff containing vanilla. However, its use as food additive (to mimic vanilla) is prohibited due to its harmful effects on human health. Therefore, maximum permissible limits were set in European legislation for coumarin. As a consequence, it is necessary to determine it in many food products, typically those that claim to contain cinnamon and/or vanilla. UV–VIS derivative spectroscopy can be an adequate analytical technique for this, since the spectra of coumarin and vanilla overlap only partially.

   In this exercise, it is assumed that a binary mixture of vanilla and coumarin is studied and that the concentration of coumarin in an unknown sample has to be calculated from the following data: the spectra of each compound of the binary mixture (Figure 9), their derivative spectra (obtained currently

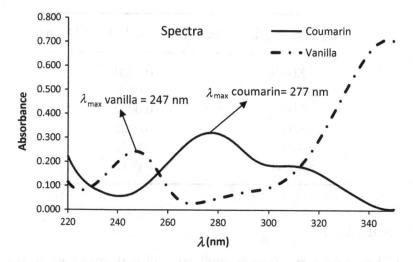

Figure 9.

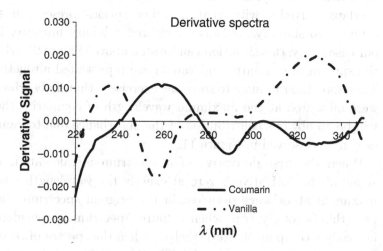

Figure 10.

with the software which controls the instrument) (Figure 10) and the signals of the derivative spectra measured for several coumarin standard solutions at exactly the maximum absorbance wavelength of the vanilla, showed in the table. (This exercise was adapted from Ref. [1], which is a nice book dealing with laboratory practices — in Spanish.)

$C_{coumarin}$ (mg/L)	Derivative signal ($\lambda = 247\,\text{nm}$)
2.00	0.0030
4.00	0.0059
6.00	0.0090
8.00	0.0122
10.00	0.0151
Sample	0.0072

SOLUTION:

This exercise is a typical application of derivative spectroscopy to address the determination of a compound whose spectrum overlaps partially with that of other species present in the sample. To simplify, we only considered a binary mixture. For our case study, this situation can be seen above (Figure 9), where the spectra of coumarin and vanilla are represented altogether. Therefore, the rationale to solve the exercise is that if absorbance were measured at the maximum wavelength of coumarin, there will be an obvious contribution of the absorbance due to vanilla as can be observed in Figure 11.

When the first derivative of a spectrum is calculated, the signal of the derivative is zero at exactly the wavelength where maximum absorbance occurred in the original spectrum. Note that this is totally true when a 'pure' spectrum is considered, i.e., only a component. Nevertheless, when the spectra of several compounds overlap (incompletely), this might not be totally true. Thus, the derivative signal at the wavelength of the maximum for vanilla (interferent), which occurs at 247 nm, will provide the signal corresponding to coumarin (analyte) alone (see Figure 12). This is the wavelength where we should perform our quantitations.

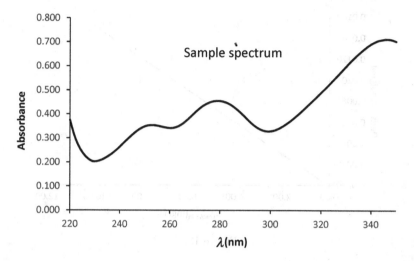

Figure 11.

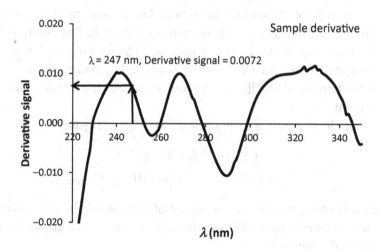

Figure 12.

Recall that derivative spectra are obtained straightforwardly nowadays by means of digital modern instruments (as it was mentioned in the enunciate).

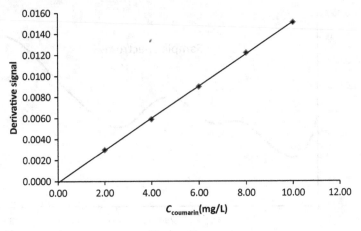

Figure 13.

Once the measurement wavelength has been selected, the derivative signals of the coumarin standard solutions are obtained (see table in p. 164). Figure 13 shows a good straight-line relationship between signals and concentrations. The residuals plot is not presented here but the student is strongly encouraged to make it (as explained in Chapter 2) and visualize that outliers are not present.

The least-squares fit equation (see Figure 13) is:

$$S = (-1.1 \cdot 10^{-4} \pm 9.95 \cdot 10^{-5})$$
$$+ (1.5 \cdot 10^{-4} \pm 1.50 \cdot 10^{-5}) \cdot C_{coumarin}$$

Substituting the derivative signal of the sample in the equation yields the concentration of coumarin we were looking for in the original sample:

$$C_{coumarin} = 4.81 \, \text{mg/L}$$

8. The intense 'cobalt blue' complex is used in the formulation of some paints intended for ceramics. The complex is obtained by mixing Co^{2+} with a ligand (L) containing amino groups. In order to determine the formation constant of the complex, two series of solutions were prepared, where the cation and the ligand

were mixed in different proportions, so that the total amount of moles is constant in all mixtures. The signal of the solutions was measured at the wavelength where the complex absorbs most. The concentration of both species in the solutions and the absorbances are shown in the following table.

$[Co^{2+}]$ (mmol/L)	$[L]$ (mmol/L)	A
0.0	$1.0 \cdot 10^{-2}$	0.000
$1.0 \cdot 10^{-3}$	$9.0 \cdot 10^{-3}$	0.400
$1.5 \cdot 10^{-3}$	$8.5 \cdot 10^{-3}$	0.600
$2.0 \cdot 10^{-3}$	$8.0 \cdot 10^{-3}$	0.780
$2.5 \cdot 10^{-3}$	$7.5 \cdot 10^{-3}$	0.830
$3.0 \cdot 10^{-3}$	$7.0 \cdot 10^{-3}$	0.800
$4.0 \cdot 10^{-3}$	$6.0 \cdot 10^{-3}$	0.750
$6.0 \cdot 10^{-3}$	$4.0 \cdot 10^{-3}$	0.530
$8.0 \cdot 10^{-3}$	$2.0 \cdot 10^{-3}$	0.270
$1.0 \cdot 10^{-2}$	0.0	0.000

SOLUTION:

In this example, the method of continuous variations is utilized to determine the stoichiometry of the complex and also to calculate its formation constant. The absorbances are plotted against the molar fraction of one of the species, here the molar fraction of the metal (X_M) see Figure 14. The molar fraction of the metal is calculated as usual:

$$X_M = \frac{n_M + n_L}{n_{total}}$$

Note: The molar fractions of the metal (and of the ligand) can be calculated using concentration values providing they are referred to the same final volume. In this example, we maintained the traditional terminology (present in many textbooks and reports), instead of the updated 'mole fraction'.

Molar fraction of metal (X_M)	A
0.00	0.000
0.10	0.400
0.15	0.600
0.20	0.780
0.25	0.830
0.30	0.800
0.40	0.750
0.60	0.530
0.80	0.270
1.00	0.000

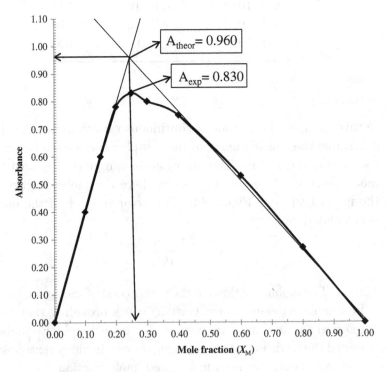

Figure 14.

Figure 14 has two straight portions, which intersect at a point associated to exactly the molar fraction that yields the theoretical total complexation. In this case, the molar fraction is approximately 0.25 and, so, the equation of the complex is CoL_3 (note that $0.25 = 1/4 = 1/(1+3)$).

Therefore, the formation equilibrium is $Co^{2+}+3L \Leftrightarrow CoL_3$, and its constant is:

$$K_f = \frac{[CoL_3]}{[Co^{2+}] \cdot [L]^3}$$

To calculate the formation constant, the following tiered approach is implemented. First, the molar absorptivity of the complex is determined. For this, the information provided by Figure 14 is required. Thus, as it was mentioned in the introductory section, the absorbance value obtained extrapolating the two straight branches of the plot, namely A_{theor}, here 0.960, denotes the theoretical signal corresponding to the quantity of complex formed from the stoichiometric amounts of metal and ligand ($[CoL_3]_{theor}$), without any consideration of dissociation and/or common ion effect. The molar absorptivity of the complex can be calculated using this value (recall that ε depends on the wavelength, not on the concentration).

$$A_{theor} = \varepsilon \cdot b \cdot [CoL_3]_{theor}$$

When $[CoL_3]_{theor}$ is expressed in molar units, it coincides with the concentration of Co^{2+} in the stoichiometric point (stoichiometric considerations). Note that this concentration is read from the table when $A_{experimental} = 0.830$. Thus:

$$\varepsilon = \frac{0.960}{1\,cm \cdot 2.5 \cdot 10^{-3}\,mmol/L} = 384\,L/(mmol \cdot cm)$$

The concentration of the complex in the equilibrium is calculated from the experimental absorbance:

$$A_{experimental} = \varepsilon \cdot b \cdot [CoL_3]_{eq}$$

$$[CoL_3]_{eq} = \frac{A_{experimental}}{\varepsilon}$$

$$= \frac{0.830}{384\,L/(mmol \cdot cm)} = 2.16 \cdot 10^{-3}\,mmol/L$$

The concentration of solubilized metal in the equilibrium will be the difference between the initial concentration at the stoichiometric point (see table) and that which reacted to form the complex:

Take stoichiometry into account!

$$[Co^{2+}]_{eq} = [Co^{2+}]_{in} - [CoL_3]_{eq}$$

$$= 2.5 \cdot 10^{-3}\,\frac{mmol}{L} - 2.16 \cdot 10^{-3}\,\frac{mmol}{L}$$

$$= 3.4 \cdot 10^{-4}\,mmol/L$$

The concentration of solubilized ligand in the equilibrium is calculated similarly, considering the metal–ligand complex stoichiometry (1:3 in this case) and the initial concentration at the stoichiometric point (see table in p. 167):

$$[L]_{eq} = [L]_{in} - 3 \cdot [CoL_3]_{eq}$$

$$= 7.5 \cdot 10^{-3}\,\frac{mmol}{L} - 3 \cdot 2.16 \cdot 10^{-3}\,\frac{mmol}{L}$$

$$= 1.02 \cdot 10^{-3}\,mmol/L$$

And thus, K_f of the CoL_3 complex becomes $6 \cdot 10^9$

9. The determination of copper in aviation fuels can be performed using pyrrolidine dithiocarbamate (PDCA) as complexing agent. The colored complex formed is extracted with methylisobutylketone (IBMK) and the absorbance is measured 30 min later. To determine the stoichiometry of the Cu–APDC complex, a series of solutions are prepared with a constant Cu concentration of 2.5×10^{-4} M and increasing amounts of APDC, which are shown in the following table, along with the transmittance values obtained. Determine the stoichiometry and the formation constant of the complex.

[APDC]	%T
$6.2 \cdot 10^{-5}$	75.16
$1.2 \cdot 10^{-4}$	56.49
$1.9 \cdot 10^{-4}$	42.17
$2.5 \cdot 10^{-4}$	33.96
$3.2 \cdot 10^{-4}$	32.14
$3.8 \cdot 10^{-4}$	31.92
$5.0 \cdot 10^{-4}$	31.77

SOLUTION:

In this exercise, the molar-ratio method is proposed to calculate the stoichiometry of the complex. The absorbances have to be calculated from the transmittance values and plotted against the ligand-to-metal molar ratios (Figure 15).

The ligand-to-metal molar ratios are calculated easily using the concentration values providing they are referred to the same volume.

moles L/moles M	Absorbance
0.25	0.124
0.48	0.248
0.76	0.375
1.00	0.469
1.28	0.493
1.52	0.496
2.00	0.498

Figure 15 shows that a unit molar ratio gives the theoretical total complexation, so the stoichiometry of the complex is 1:1. Thus,

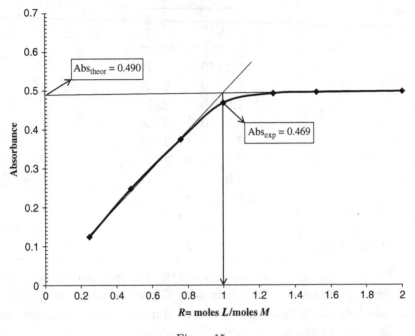

Figure 15.

the formation equilibrium is

$$Cu^{2+} + APDC \Leftrightarrow Cu\text{--}APDC$$

and its corresponding constant is:

$$K_f = \frac{[Cu\text{--}APDC]}{[Cu^{2+}] \cdot [APDC]}$$

As in the previous exercise, the molar absorptivity of the complex is calculated using the absorbance obtained extrapolating the straight portions of the plot, namely A_{theor}:

$$A_{theor} = \varepsilon \cdot b \cdot [Cu\text{--}APDC]_{theor}$$

The 1:1 stoichiometry means that $[Cu\text{--}APDC]_{theor}$ coincides with the initial concentration of Cu^{2+} (molar values). Thus:

$$\varepsilon = \frac{0.490}{2.5 \cdot 10^{-4}} = 1\,960\,L/(mol \cdot cm)$$

The concentration of the complex formed in the equilibrium is calculated from the experimental absorbance:

$$A_{\text{experimental}} = \varepsilon \cdot b \cdot [\text{Cu–APDC}]_{\text{eq}}$$

$$[\text{Cu–APDC}]_{\text{eq}} = \frac{A_{\text{experimental}}}{\varepsilon}$$

$$= \frac{0.469}{1960 \, \text{L}/(\text{mol} \cdot \text{cm})} = 2.39 \cdot 10^{-4} \text{M}$$

The concentration of metal solubilized in the equilibrium will be the difference between the initial concentration (see table in the enunciation) and that which reacted to form the complex:

$$[\text{Cu}^{2+}]_{\text{eq}} = [\text{Cu}^{2+}]_{\text{in}} - [\text{Cu–APDC}]_{\text{eq}}$$

$$= 2.5 \cdot 10^{-4} \, \frac{\text{mol}}{\text{L}} - 2.39 \cdot 10^{-4} \, \frac{\text{mol}}{\text{L}} = 1.1 \cdot 10^{-5} \, \text{M}$$

The concentration of ligand in the equilibrium is calculated similarly, considering the stoichiometry of the complex (1:1 metal–ligand in this case) and the initial concentration (see table in p. 171) and that employed to get the complex:

$$[\text{APDC}]_{\text{eq}} = [\text{APDC}]_{\text{in}} - [\text{APDC}]_{\text{eq}}$$

$$= 2.5 \cdot 10^{-4} \, \frac{\text{mol}}{\text{L}} - 2.39 \cdot 10^{-4} \, \frac{\text{mol}}{\text{L}} = 1.1 \cdot 10^{-5} \, \text{M}$$

And thus, K_{f} of the Cu–APDC complex is $1.98 \cdot 10^6$

10. The next experiment was proposed in a subject on general chemistry to determine the stoichiometry of the Fe-1,10-phenanthroline complex. The slope-ratio method was utilized preparing two series of solutions. One of them has the same amount of ligand while that of the metal increases steadily; in the other, the same concentration of metal is mixed with increasing quantities of ligand. The absorbance of the solutions was measured at 510 nm and the results are shown in the table. Calculate the metal–ligand ratio.

$[L] = 3 \cdot 10^{-4}\,M$		$[Fe] = 3 \cdot 10^{-4}\,M$	
[Fe] (mol/L)	A	[L] (mol/L)	A
$1.0 \cdot 10^{-5}$	0.100	$1.0 \cdot 10^{-5}$	0.031
$2.0 \cdot 10^{-5}$	0.217	$2.0 \cdot 10^{-5}$	0.073
$3.0 \cdot 10^{-5}$	0.359	$3.0 \cdot 10^{-5}$	0.108
$4.0 \cdot 10^{-5}$	0.504	$4.0 \cdot 10^{-5}$	0.145
$5.0 \cdot 10^{-5}$	0.621	$5.0 \cdot 10^{-5}$	0.197

SOLUTION:

In this experiment, the slope-ratio method is utilized to determine the stoichiometry of the complex. The absorbance values are plotted against the concentration of the corresponding species in each series of solutions.

When the concentration of ligand is constant, the concentration of complex formed is proportional to the analytical concentration of metal added, because all the metal is reacting since the ligand is in excess, then for a general complex formula Fe_nL_m:

$$[Fe_nL_m] = \frac{[Fe]}{n}$$

Considering now the expression of the Beer's law:

$$A = \varepsilon \cdot b \cdot [Fe_nL_m] = \varepsilon \cdot b \cdot \frac{[Fe]}{n}$$

Plotting the absorbance of the complex against the concentration of the metal, a straight line is obtained with a slope equal to $\varepsilon b/n$ (Figure 16).

Similarly, when the concentration of the metal is constant, the concentration of the complex formed is proportional to the analytical concentration of the ligand added, because all the

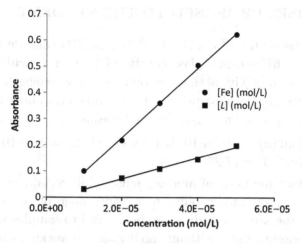

Figure 16.

ligand is reacting since the metal is in excess:

$$[\mathrm{Fe}_n L_m] = \frac{[L]}{m}$$

Considering the Beer's law:

$$A = \varepsilon \cdot b \cdot [\mathrm{Fe}_n L_m] = \varepsilon \cdot b \cdot \frac{[L]}{m}$$

and the corresponding plot of A versus the concentration of the ligand has a slope equal to $\varepsilon b/m$.

From Figure 16, we can establish that

$$\varepsilon b/n = 13\,290 \quad \text{and} \quad \varepsilon b/m = 4\,040$$

Finally, the ratio of the slopes yields the ligand–metal ratio in the complex:

$$\frac{\dfrac{\varepsilon b}{n}}{\dfrac{\varepsilon b}{m}} = \frac{m}{n} = 3$$

As a consequence, the stoichiometry of the Fe-1,10-phenanthroline complex is 1:3 $(\mathrm{Fe}L_3)$.

EXERCISES PROPOSED TO THE STUDENT

11. The absorbance of a $4.75 \cdot 10^{-5}$ M iron (III) chloride solution is 0.112 at 675 nm in a glass cuvette of 1.00 cm. Calculate (a) the molar absorptivity of the compound at this wavelength indicating its units; (b) the absorptivity and its units if the above concentration is expressed in mg/L; (c) the transmittance of the solution.

    Solution: (a) $\varepsilon = 2.36 \cdot 10^3$ L/(mol·cm); (b) $a = 2.38 \cdot 10^{-2}$ L/(mg·cm); (c) $\%T = 77.27\%$.

12. Standard methods of analysis must fulfill certain requirements such as their applicability to a wide range of concentrations. Thus, the same procedure can be applied to samples with different concentrations without modifying the working range of the calibration or the instrumental conditions. The determination of cobalt in outdoor paints is performed by extracting the metal and complexing it with 2,9-dimethyl-1,10-phenanthroline. The complex has a molar absorptivity of $7.95 \cdot 10^3$ at 551 nm. The sample aliquot under study is finally diluted to 100.00 mL. If it is known that outdoor paints of a commercial brand contain 0.25 % of cobalt, which is the mass of sample aliquot that must be weighted in order to obtain an absorbance between 0.2 and 0.8 for the final working solution?

    Note: Assume stoichiometry 1:1

    Solution: The mass of paint sampled can vary between 0.0592 and 0.2380 g.

13. To determine cadmium in a mineral sample, a 3.0000 g aliquot was weighed, grounded and sieved and then dissolved with *aqua regia* in a microwave oven. The solution obtained was diluted to 250.00 mL. A 10.00 mL aliquot of this solution was pipetted, mixed with the appropriate volume of dithizone and buffer solution and diluted to 50.00 mL. The transmittance of this solution was 57.8 %. The molar absorptivity of the complex (stoichiometry 1:1) is 3 000 L/(mol · cm). Calculate the mass percent of cadmium in the mineral.

    (The molar mass of Cd is 112.40.)

    Solution: The mass percent of cadmium in the mineral is 0.37 %.

14. The determination of copper in natural waters can be performed by its complexation with an adequate organic ligand (L). The complex has 1:3 stoichiometry. A 5.00 mL aliquot of a commercial drinking water was mixed with 10.00 mL of a solution of the ligand and diluted to 25.00 mL. An aliquot of 1.00 mL of this solution gives an absorbance of 0.237. Another portion of 5.00 mL of the sample was spiked with 5.00 mL of a 5.00 mg/L copper standard solution and treated in the same way than the first one, giving a 0.545 signal. Determine the copper concentration in the commercial drinking water.

Solution: Copper concentration in the drinking water sample is 0.30 mg/L.

15. An aliquot of 9.7400 g of a lubricating oil sample was ashed in a muffle. When cold, the residue weighed 0.0168 g; it was dissolved with 6 mL of concentrated sulfuric acid and diluted to 250.00 mL. In order to determine vanadium by UV–VIS spectroscopy, 15.00 mL of this solution were mixed with 15.00 mL of an adequate complexing agent and diluted to 50.00 mL. Another 15.00 mL aliquot was treated in the same manner adding 5.00 mL of a 5.00 mg/L vanadium standard solution before its final dilution to 50.00 mL. The absorbances were 0.281 and 0.409, respectively. Calculate the concentration of vanadium in the lubricating oil.

Solution: The concentration of vanadium in the lubricating oil is 1.50 mg/L.

16. The quality control laboratory of a company that manufactures vitamin supplements follows a standardized protocol to determine iron. A 6.0800 g aliquot of a supplement was dried, dissolved in distilled water and diluted to 1 000.00 mL. Then, a 5.00 mL aliquot of this solution was mixed with 10.00 mL of an adequate complexing ligand and diluted to 250.00 mL. The absorbance obtained for this solution was 0.621. A procedural blank was prepared with the same amount of ligand yielding a signal of 0.192. To calculate the iron concentration in the sample, a standard solution was prepared by diluting 15.00 mL of a 10.00 mg/L iron standard solution to 250.00 mL which provided

a 0.571 absorbance. Calculate the mass percent of iron in the vitamin supplement.

Solution: The mass percent of iron in the vitamin supplement sample is 0.56 %.

17. A milk protein can be determined by UV–VIS spectroscopy after its complexation with Fe(III) (stoichiometry 1:1). For this purpose, a 2.1624 g aliquot of a milk powder sample was weighed, dissolved and filtered. Then, an adequate amount of an Fe(III) solution was added and the final volume was set to 250.00 mL in a volumetric flask with distilled water. A 10.00 mL aliquot of this solution provided a 0.380 absorbance. A standard solution was prepared in the same way than the sample starting from 26.40 mg of the protein, this standard led to a 0.473 absorbance. A procedural blank was also prepared, which yielded a 0.025 signal. Calculate the mass percent of protein in the milk sample. (The molar mass of the protein is 281 000.)

Solution: The mass percent of protein in the milk sample is 0.97 %.

18. Manganese and chromium are among the essential elements to get proper stainless steel. The determination of these metals in steel samples can be performed by UV–VIS spectroscopy through their oxidation to MnO_4^- and $Cr_2O_7^{2-}$. Since the spectra of these species overlap partially, the multicomponent analysis has to be applied. To perform it, a 0.9934 g aliquot of a steel sample was dissolved with an acid mixture and diluted to 250.00 mL (solution A). A 50.00 mL aliquot of solution A is treated with potassium persulfate ($K_2S_2O_8$) in the presence of Ag^+ ions (that act as catalysts) and potassium periodate (KIO_4) whereby Mn and Cr were oxidized to MnO_4^- and $Cr_2O_7^{2-}$, and this solution was diluted to 100.00 mL (solution B). The absorbances of solution B at 440 and 545 nm were 0.204 and 0.170, respectively. Calculate the mass percent of Mn and Cr in the steel sample using the data presented in the following table.

λ(nm)	$\varepsilon(MnO_4^-)$	$\varepsilon(Cr_2O_7^{2-})$
440	95.0	369.0
545	2 350.0	11.0

Solution: The mass percents of Mn and Cr in the steel sample are 0.19 % and 2.8 %, respectively.

19. Multielemental analysis can be used to determine two acid–base indicators in a mixture. For this, individual methyl orange and bromocresol green aqueous solutions were prepared and their absorbances measured at 753 and 594 nm. The absorbances of a mixture of these two indicators were also obtained at the same wavelengths. The results are shown in the following table. Calculate the concentrations of the indicators in the mixture.

Solution	753 nm	594 nm
Methyl orange 0.01 M	0.800	0.150
Bromocresol green 0.02 M	0.200	1.000
Mixture	0.550	0.825

Solution: The concentration of methyl orange and bromocresol green in the mixture are $4.59 \cdot 10^{-3}$ M and $1.83 \cdot 10^{-2}$ M, respectively.

20. The determination of cobalt(II) ions in samples containing chromium ions cannot be performed directly as their UV–VIS spectra overlap partially. This problem can be addressed with derivative spectroscopy. The spectrum of each ion and their derivative spectra (obtained with the software of the instrument) are provided in Figures 17 and 18, as well as the derivative signals of several Co(II) standard solutions at the maximum wavelength of the chromium ion and that provided by the sample. Figure 19 shows the spectrum of the sample and Figure 20 its derivative. Using all these data, calculate the concentration of Co(II) in the sample.

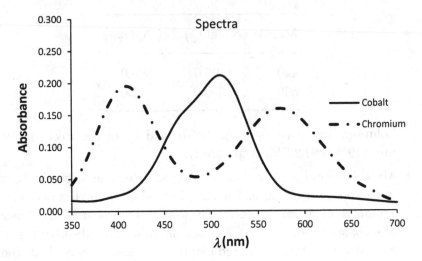

Figure 17.

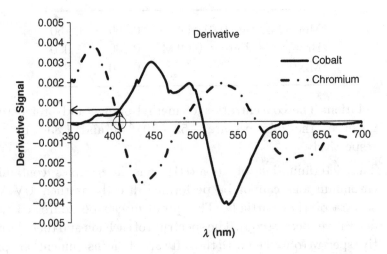

Figure 18.

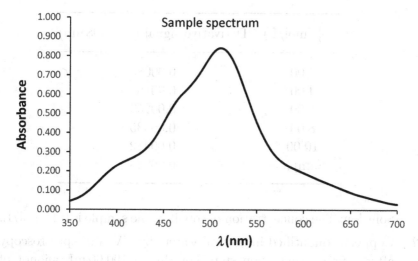

Figure 19.

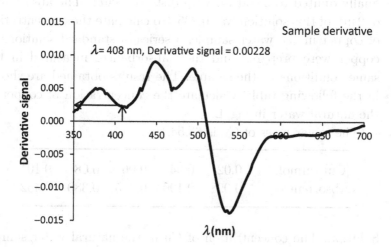

Figure 20.

[Co] (mol/L)	Derivative signal ($\lambda = 408\,\text{nm}$)
2.00	0.00085
4.00	0.00167
6.00	0.00253
8.00	0.00335
10.00	0.00412
Sample	0.00228

Solution: The concentration of cobalt in the sample is 0.11 mol/L.

21. Copper is quantified in natural waters by UV–VIS spectroscopy after a preconcentration step. For this, a 100.00 mL aliquot of water sample was acidified with 2 M sulfuric acid and passed through a column packed with a cation exchange resin. Then, copper was eluted with 20 mL of 2 M sulfuric acid. The eluate was finally diluted to 25.00 mL with distilled water. The absorbance of 2 mL of this solution was 0.235. To calculate the concentration of copper in the water sample, a series of standard solutions of copper were prepared and their absorbances measured in the same conditions as the sample. The results obtained are shown in the following table. Calculate the concentration of copper in the natural water in mg/L.

(The molar mass of Cu is 63.54.)

[Cu] (mmol/L)	0.02	0.04	0.06	0.08	0.10
Absorbance	0.098	0.190	0.275	0.380	0.452

Solution: The concentration of Cu in the natural water sample is 0.80 mg/L.

22. The determination of phosphates in detergents can be done by UV–VIS spectroscopy after the formation of a complex with ammonium molybdate (stoichiometry 1:1) whose absorbance maximum occurs at 715 nm. A disodium hydrogen phosphate standard solution was prepared dissolving 0.0256 g in water and diluting it to 250.00 mL in a volumetric flask. Working standard solutions were prepared pipetting increasing aliquots of this solution, adding the reagents needed to form the complex and diluting to 25.00 mL. A 1.500 g aliquot of detergent powder was ashed, the residue was dissolved in dilute sulfuric acid and made up to 100.00 mL. A 10.00 mL aliquot of this solution was subsequently diluted to 250.00 mL, yielding a 0.350 absorbance. With the data presented in the table, determine the mass percent of phosphorus in the detergent.

Volume (mL)	0.00	0.50	1.00	2.00	3.00	4.00	5.00
Absorbance	0.015	0.078	0.150	0.302	0.436	0.554	0.692

Solution: The mass percent of P in the detergent is 0.19 %.

23. A classical procedure to determine lead in drinking water consists of its complexation with dithizone and the subsequent measurement of the complex by UV–VIS spectroscopy. The stoichiometry and the dissociation constant of the complex were calculated in the laboratory using the method of continuous variations in order to verify the validity of the procedure. Solutions of metal mole fraction varying from 0 to 1 were prepared as shown in the following table and their absorbances measured at the maximum wavelength. Data: The initial concentration of lead and dithizone were $1 \cdot 10^{-3}$ M and $5 \cdot 10^{-3}$ M, respectively.

Mole fraction of Pb (X_{Pb})	Absorbance
0.0	0.000
0.1	0.238
0.2	0.401
0.3	0.698
0.4	0.701
0.5	0.598
0.6	0.482
0.7	0.362
0.8	0.239
0.9	0.122
1.0	0.002

Solution: The stoichiometry is PbL_2 and its dissociation constant is $9.67 \cdot 10^{-7}$.

Note that in this exercise, the ligand is in excess; this should be considered when calculating the concentration of free ligand in the equilibrium.

24. The quality control laboratory of a sewage treatment plant is interested in determining iron in the sludge generated in the plant. A strategy to do so consists of complexing the metal with thiourea to set a very intense colored complex. In order to carry out this application, the stoichiometry and the stability of the complex must be calculated first. With the tabulated data below, determine these two parameters.

Data: The initial concentration was $2 \cdot 10^{-3}$ M for both, iron and thiourea.

X_L	0.9	0.8	0.7	0.6	0.5	0.4	0.3	0.2	0.1
Absorbance	0.254	0.500	0.785	1.100	1.298	1.400	1.150	0.783	0.390

X_L: molar fraction of thiourea

Solution: The stoichiometry of the complex is Fe_3L_2 and its formation constant is $1.09 \cdot 10^{14}$.

25. A metal (M^{2+}) forms a complex with an L whose absorption maximum was visualized at 520 nm. In order to determine the stoichiometry and the formation constant of the complex, the molar-ratio method was applied preparing a series of solutions with a constant metal concentration of $1.35 \cdot 10^{-4}$ M and increasing concentrations of the ligand. The absorbance values obtained for the solutions are shown in the following table.

$[L]$ (mol/L)	Absorbance
$3.0 \cdot 10^{-5}$	0.068
$7.0 \cdot 10^{-5}$	0.163
$1.2 \cdot 10^{-4}$	0.274
$2.0 \cdot 10^{-4}$	0.419
$3.0 \cdot 10^{-4}$	0.514
$4.0 \cdot 10^{-4}$	0.573
$5.0 \cdot 10^{-4}$	0.608
$6.0 \cdot 10^{-4}$	0.626
$7.0 \cdot 10^{-4}$	0.630
$8.0 \cdot 10^{-4}$	0.630

Solution: The stoichiometry of the complex is ML_2 and the formation constant $K_f = 9.72 \cdot 10^8$.

Note that the concentration of ligand was not provided explicitly for the molar ratio $= 2$, hence, it was calculated from the metal concentration considering the stoichiometry as if such a solution had been prepared.

26. The molar-ratio method was utilized to determine the stoichiometry and the dissociation constant of the complex formed by Co(II) with an L. For this, a series of solutions was prepared adding $3.00\,\text{mL}$ of $1 \cdot 10^{-4}\,\text{M}$ Co(II) solution and increasing amounts of the ligand, diluting them all to $50.00\,\text{mL}$. The results obtained are shown in the following table.

L (mmol $\cdot 10^{-4}$)	A
0.00	0.010
4.25	0.195
7.75	0.353
12.00	0.560
15.50	0.707
20.30	0.806
24.00	0.842
27.80	0.857
32.00	0.860
36.30	0.864

Solution: The stoichiometry of the complex is CoL_6 and its dissociation constant is $2.41 \cdot 10^{-35}$.

Note that the concentration of ligand was not provided explicitly for a molar ratio $= 6$, hence, it was calculated from the metal concentration considering the stoichiometry as if such a solution had been prepared.

27. The determination of the stoichiometry of unstable complexes, i.e., complexes with high dissociation constants, must be performed using the slope-ratio method, since the other two methods do not work in these cases. Two series of solutions were prepared for this purpose. In one of them, the amount of ligand was kept constant while the concentrations of the metal increased. In the other series, the opposite was done. With the data presented in the table, calculate the metal-to-ligand ratio.

$[L] = 5.0 \cdot 10^{-4}\,M$		$[M] = 5.0 \cdot 10^{-4}\,M$	
$[M]$ (mol/L)	A	$[L]$ (mol/L)	A
$2.0 \cdot 10^{-5}$	0.169	$2.0 \cdot 10^{-5}$	0.070
$4.0 \cdot 10^{-5}$	0.364	$4.0 \cdot 10^{-5}$	0.124
$6.0 \cdot 10^{-5}$	0.556	$6.0 \cdot 10^{-5}$	0.200
$8.0 \cdot 10^{-5}$	0.762	$8.0 \cdot 10^{-5}$	0.260
$1.0 \cdot 10^{-4}$	0.986	$1.0 \cdot 10^{-4}$	0.335

Solution: The stoichiometry is 1:3.

28. The slope-ratio method was utilized to calculate the stoichiometry of a complex formed by a metal (M^{2+}) and an L. Two series of solutions were prepared, one of them with the same amount of ligand and increasing concentrations of the metal, and the other with the same concentration of metal and increasing concentrations of ligand. The absorbance of the solutions was measured at the wavelength of maximum absorbance, obtaining the values shown in the following table. Calculate the metal-to-ligand ratio.

$[L] = 2.5 \cdot 10^{-3}$ M		$[M] = 2.5 \cdot 10^{-3}$ M	
$[M]$ (mol/L)	A	$[L]$ (mol/L)	A
$0.5 \cdot 10^{-4}$	0.150	$0.5 \cdot 10^{-4}$	0.055
$1.5 \cdot 10^{-4}$	0.267	$1.5 \cdot 10^{-4}$	0.102
$3.0 \cdot 10^{-4}$	0.420	$3.0 \cdot 10^{-4}$	0.173
$4.5 \cdot 10^{-4}$	0.590	$4.5 \cdot 10^{-4}$	0.257
$6.0 \cdot 10^{-4}$	0.725	$6.0 \cdot 10^{-4}$	0.336

Solution: The stoichiometry of the complex is 1:2.

EXERCISE REFERENCES

[1] Maurí, A.; Llobat, M.; Herráez, R. (2010). *Laboratorio de Análisis Instrumental*. Ed. Reverté, Spain.

CHAPTER 4

INFRARED SPECTROMETRY

José Manuel Andrade-Garda and
María Paz Gómez-Carracedo

OBJECTIVES AND SCOPE

The objective of this chapter is to present students the very basics of the chemical interpretation of an infrared (IR) spectrum. Explanations will be restricted to typical functional groups and relatively simple structures. Also, only the medium IR region and organic molecules will be considered. Quantitation issues are commented briefly since the multivariate statistical treatment is out of the scope of this book and classical calibration will be treated deeply in the other chapters. Therefore, the chapter focuses on the basics of IR spectra interpretation, something that somehow is not considered in many textbooks in analytical chemistry.

1. INTRODUCTION

1.1. Basic concepts

Infrared (IR) spectrometry (or spectrometry in the infrared region, IR) is a molecular analysis technique of paramount importance in chemistry. Not in vain it is a powerful, ubiquitous workhorse to study and ascertain the main characteristics of molecular structures.

Further, and in addition to this classical application, its use has fostered in the last two decades, thanks to its combination with multivariate chemometric techniques (mainly full-spectrum regression techniques like partial least squares, PLS). This conferred on IR spectrometry an unprecedent power to quantitate one or several species of interest in the analyzed aliquots. In conjunction with nuclear magnetic resonance (NMR) and mass spectrometry (MS), it constitutes a basic tool for most studies on organic and inorganic materials performed nowadays.

Frequently, references to the 'infrared region' assume implicitly that work is performed at the medium IR range (MIR or mid-IR, 4000–$600\,cm^{-1}$), which is considered to offer a spectral 'fingerprint' of a substance [1, 2]. However, nowadays the near infrared range (NIR, 13300–$4000\,cm^{-1}$) has become a very powerful and popular tool. The NIR range is used to study, mainly, harmonic vibrations of the functional groups. Practical applications using the far infrared region (FIR, 400–$20\,cm^{-1}$) are still pretty scarce, although they are increasing, thanks to the so-called terahertz spectroscopy [3], a term which somehow supersedes the FIR one. A relevant advantage of this region is that plastics, cloth and semiconducting materials are transparent, opening up ways to *in situ* analyses. The region deals usually with skeletal vibrations of the molecule, although the signals are unspecific. Thus, the spectrometric measurements in both the MIR and NIR ranges constitute a common workhorse in many laboratories and can be applied routinely even in complex fields, such as food, beverages and pharmaceutical analyses, to mention some relevant industrial arenas. For these applications, the use of multivariate chemometric techniques is of most importance, if not mandatory.

Some great advantages of mid-IR spectrometry are: (i) samples can be analyzed in gaseous, liquid and solid forms; (ii) it is applicable to both organic and inorganic substances; and (iii) it does not require large amounts of original sample or complex sample treatment processes. Not surprisingly, many methodologies were developed thinking on measuring the original samples directly, something for which there is a plethora of commercial accessories that can be acquired according to the needs of each laboratory.

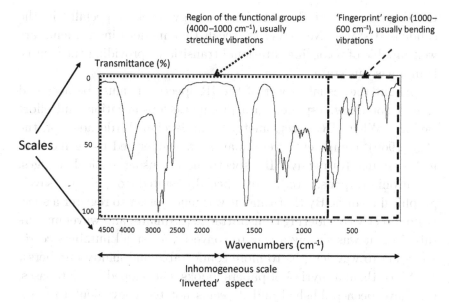

Figure 1.

Considering the student's point of view, the large number of spectral peaks observed in the IR spectra (mainly the MIR range) may be surprising and 'misleading'. Further, and due to historical reasons and routine common practice, the visual appearance of the spectra (see Figure 1) is 'opposed to normal spectra' they are more used to (e.g., UV–VIS spectra) and the abscissa units are not, directly, wavelengths; most times they are the wavenumbers; i.e., the inverse of the wavelength, when measured in cm). All this can confuse the student.

Hence, the student has to remember, first, that the origin of all peaks (when they are narrow) or spectral bands (otherwise) in this region is in the vibrational and/or rotational energy transitions that the molecule suffers when it is exposed to a beam of polychromatic IR radiation. Sometimes, a bond involving two atoms experiences a transition, sometimes a whole functional group (group of atoms) does. From this, it obviously follows that when the complexity of the molecule is high, the IR spectral complexity will be (could be) high. In general, only vibrational bands are studied due to the limited resolution offered by most instrumental equipment and to the degeneration of the rotational energy levels, especially in the condensed states. Nevertheless, recall that modern instruments are well capable of recording rotational transitions, providing the sample is in a gaseous state.

About the visual aspect of the IR spectra, it must be accepted that it somehow responds to a pragmatic 'optimization and effort saving'. When IR spectrometry was developed (decades of the 1950–1960), equipments were analogic and operated using mechanical systems (dispersive IR spectrometry, using optical wedges, synchronized pens, etc.), not specially suited to get (or save!) graphical records. By that time, it was much easier to register a spectrum directly considering transmittance (ordinate) and wavenumbers (abscissa, it was simply a way to convert very small numbers corresponding to wavelengths to more comfortable and handy numbers).

Since then, a myriad of papers, books, encyclopedias, databases, etc. have been published and it seems not too convenient to make a drastic change to reformat all the historical information. In the end, nowadays, computers spend just a few microseconds to convert absorbance to transmittance (or *vice versa*) and to display the abscissa units you may desire.

1.2. Spectral interpretation

The qualitative interpretation of an IR spectrum (in this text, limited to simple organic substances and to the MIR range) relies on comparing the position (and sometimes, the shape) of the peaks/ spectral bands to tabulated values for different groups of atoms. That is, the molecular structure is not ascertained just 'in one step'. On the contrary, the constituent parts of the molecule (functional groups) are deduced and, then, you assemble them in a most probable structure (you will need some additional information), much like a crossword.

It is worth noting that although the use of tabulated values (some of them are listed below) is essential, many students tend to assume that they are 'universally accurate' in all situations. Unfortunately, this is far from being true. As you may already have studied, the particular position of a spectral band in the IR region (i.e., the rotational and/or vibrational transition) changes (even a lot) depending on factors such as the analyte concentration, the concentration of other species that can modify the atomic environment of the functional group, the state in which the measurement is made (the bands become situated at slightly different positions depending on whether the gas phase or other condensed states are used), the temperature at which the measurement is made, etc. Hence, it is important to maintain a general idea of what you are trying to identify and be flexible when using the tables.

Some general guidelines can be given, somewhat restricted to the cases considered in this chapter (recall Figure 1 also):

a. The spectral peaks at higher wavenumbers (4000–1600 cm^{-1}) correspond to bond stretching. It is possible that overtones of bending vibrations appear in this region, but they are usually weak.

b. The spectral peaks situated at low wavenumbers (1600–600 cm^{-1}) correspond usually to bendings of atomic bonds.

c. The spectral region around 2000 cm^{-1} has usually very few characteristic spectral peaks (this explains a typical change on the abscissa scale that can be seen in all IR spectra, see Figure 1). If peaks appear here, they can be assigned straightforwardly to typical functional groups.

d. The 1300–910 cm^{-1} area is the 'fingerprint' region, or characteristic region. It is common to see many peaks in this zone, although not all of them can be ascribed easily to specific functional groups or vibration modes.

e. In general, saturated and unsaturated hydrocarbons can be differentiated rather simply. Double C–C bonds are visualized at 1600 cm^{-1} as they yield a strong peak which is attributed to the stretching of their bonds. Triple C–C bonds are observed at 2100 cm^{-1}, with an intense and very narrow peak. Branched hydrocarbons, with either isopropyl or butyl groups, are recognized because they cause doublets in the common stretching and bending bands of the CH groups.

f. The substituted aromatic compounds can be identified easily because they show three or four narrow and intense peaks at the 1600–1500 cm^{-1} range, due to the C=C stretching vibrations (the exact positions depend on the substituents). The 910–650 cm^{-1} region is important to study aromatic structures because the type of ring substitution can sometimes be ascertained as the graphical pattern observed there would inform on the type of substitution. A very important help for this is found around 2000 cm^{-1}. Here, the graphical pattern is usually enough to evaluate the ring substitution (see Section 1.3). The peaks correspond to overtones and combinations of out-of-the-plane and in-the-plane aromatic C–H bending.

g. Functional groups containing the C=O bond are identified quite fast. This is possible because this functional group presents a fundamental – intense – vibration mode in the 1700–1750 cm^{-1} interval. Besides, if the molecule contains carboxylic acid groups, they can be identified by two intense bands. One, around 1250 cm^{-1} (1320–1210 cm^{-1} interval) due to the combination of the stretching of the C=O bond and the bending of the O–H bond. The other (broad, quite unspecific) around 3400 cm^{-1}, associated to the O–H stretching (likely, coupled to other OH groups, due to intermolecular hydrogen bonds).

A final note is in order here. In this chapter, the term 'aromatic' refers only to structures derived from benzene. This decision was taken because this book is intended for initial courses of analytical chemistry. Should this simplification not be done, the chapter would be overwhelmingly extensive, and probably, not very useful from a pedagogic viewpoint.

1.3. Characteristic typical vibrations in the mid-IR region

It is always difficult to select a title for a scientific work or, simply, for a section of a chapter. This is also the case here because although the title of this section seems quite reasonable, it immediately rises doubts about the term 'typical vibrations'. What vibrations can be considered as 'typical'? How many are they?

In literature (for instance, the references given in this chapter), you can find books (some of them really thick!) plenty of tables of vibrations and with detailed studies of many functional groups and/or structures. Since this book is intended as a first approach for the (self-)training of undergraduate students, we felt that it would not be reasonable to get onboard such a complex journey. The student should be aware that only some (very common) functional groups and relatively simple examples were considered for this introductory chapter. The examples deal with hydrocarbons because they are of general use although, of course, some inorganic compounds (or, at least, functional groups) can be studied by IR spectrometry as well. Those who need more technical details, review the spectroscopic characteristics of other functional groups or, simply, interested readers are kindly forwarded to any of the references given at the end of this chapter [4–13] as a quite good starting point.

In the next lines, you will find a combination of 'tables' of fundamental vibrations and graphical outlooks of their appearance. Such a presentation was preferred instead of the typical list or sets of tables of vibrations because it appears more pedagogic and student-friendly. We tried to select typical good shapes for this first presentation but the student must be definitely aware that the shapes of the bands can change depending on the other parts of the structure of the molecule, the solvent, the concentration of the analyte and concomitants, the instrumental setup, etc. This should have been learnt in the 'theoretical classes' and will not be detailed here.

(a) Hydrocarbons

Figures 2–6 resume the most typical vibration bands obtained when C–H and C–C bonds are present in a molecule.

(b) Alcohols

Mid-IR spectra can easily denote the presence of alcohols in molecules; the very wide band around $3400 \, cm^{-1}$ is a clear indication of OH groups (Figure 7).

(c) Carboxylic group

The C=O group yields a very characteristic band around $1700 \, cm^{-1}$. Its 'exact' position will denote the type of functional group (Figure 8).

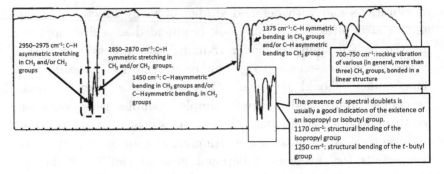

Figure 2.

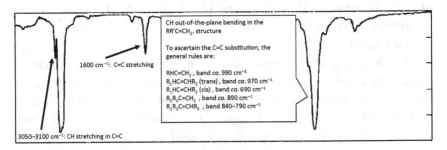

Figure 3.

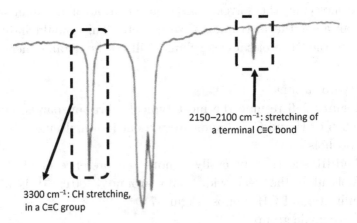

Figure 4.

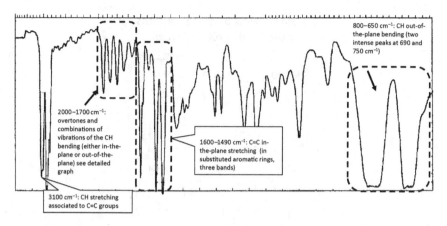

800–650 cm⁻¹: CH out-of-the-plane bending (two intense peaks at 690 and 750 cm⁻¹)

2000–1700 cm⁻¹: overtones and combinations of vibrations of the CH bending (either in-the-plane or out-of-the-plane) see detailed graph

1600–1490 cm⁻¹: C=C in-the-plane stretching (in substituted aromatic rings, three bands)

3100 cm⁻¹: CH stretching associated to C=C groups

Figure 5.

(d) Amines and nitriles

The presence of N–H bonds can be identified most times by the presence of several bands in the mid-IR spectrum. The nitrile group is ascertained fast by the existence of a very sharp peak around $2250\,\text{cm}^{-1}$ (Figures 9 and 10).

1.4. Prediction of the position of the bands in IR spectrometry

The reference values presented in the figures correspond to mean values obtained from many experiences and interpretation of associated spectra. This means that such values, although trustworthy, cannot be taken strictly for any situation. As it was explained above, a reasonable, flexible mindset will accept that a $10\,\text{cm}^{-1}$ displacement (or more, even up to $50\,\text{cm}^{-1}$) can be perfectly normal in some situations.

Although the tables are based on experimental information, this does not impede the existence of a sound physico-chemical support behind them. A simple approach, presented in many textbooks, is the so-called model of the harmonic oscillator. A brief reminder will be given here in order to complete this overview on IR spectrometry.

First, bear in mind that the recorded spectra will only show the processes associated to the absorption of IR radiations whose

Detailed view of typical patterns in the 2000–1700 cm⁻¹ region

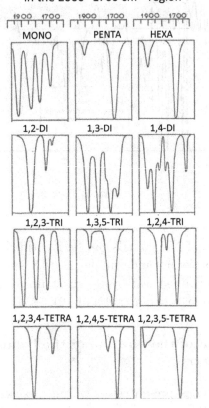

Figure 6.

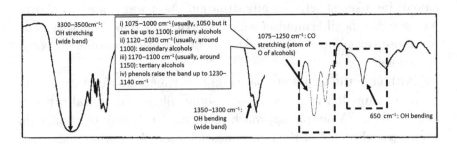

Figure 7.

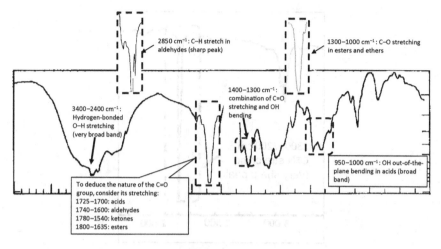

Figure 8.

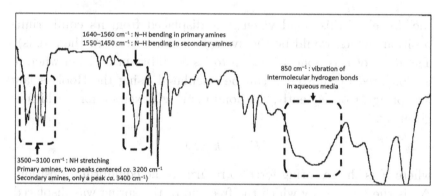

Figure 9.

energies correspond exactly to those of the transitions between two vibrational modes, providing the vibration itself changes the electric dipole of the molecule. This limitation is due to the selection rules governing the quantum mechanical studies (which are not addressed here). Following, homonuclear diatomic molecules cannot be studied by this technique. Let us see now how we can predict (approximately) the position of the IR spectral bands.

Considering that a bond between two atoms can be modeled roughly as when a spring connects two masses (this is a preliminary model, but acceptable for our purposes here), it is possible to evaluate

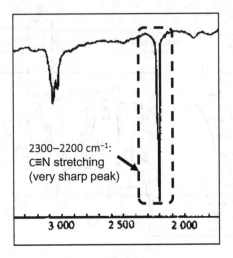

Figure 10.

the 'force' of this bond when it is displaced from its equilibrium position (which would be the average distance between the atoms). The force of this bond to return to its equilibrium position when it is elongated or shortened can be calculated using the Hooke's law for springs. Consequently, the bond ('spring') exerts a force which is given by:

$$F = -k \cdot \Delta x$$

where F is the restoring force to return to the equilibrium position, Δx is the amount by which the free end of the spring was displaced from its equilibrium position (cm^{-1}) and k is a positive real number characteristic of the spring (force constant of the spring, dyne*cm^{-1}). The negative sign indicates that a force to return to the equilibrium position is needed to compress or to stretch the bond.

The force constant, in turn, is related to the masses situated at the ends of the spring and with the frequency of vibration. When two masses connected by a spring are considered, this relationship is given by:

$$\nu = \frac{1}{2\Pi} \cdot \sqrt{\frac{k}{\mu}}$$

where ν is the vibration frequency (s^{-1}) and μ is the reduced mass, that can be defined as $\mu = [m_1 \cdot m_2/(m_1 + m_2)]$ where m is the mass of each atom (in grams).

It is possible to gather approximate values for the force constant for different masses and also for double and triple bonds, employing physico-chemical studies. Then, you can immediately calculate the approximate frequency of vibration for that bond.

Obviously, the situation is more complex for molecules with more than two atoms. This is due to the existence of both stretching and bending vibrations. In this respect, remember that for a molecule with Z atoms, $3Z-6$ normal vibration modes will be observed ($3Z-5$ if it is lineal). Out of them, $Z-1$ are stretching vibrations (i.e., vibrations where the bonds are elongated or shortened or — in other words — the atoms move around their relative positions), and $2Z-5$ are bending vibrations (i.e., the angles formed between two bonds are enlarged or shortened).

In many books and vibration frequency tables, ν and δ are used to denote the stretching and bending vibrations, respectively. Furthermore, the stretching vibrations can be 'symmetric' or 'asymmetric' depending on whether the symmetry of the molecule changes with the vibration mode. If the stretching vibration is 'symmetric' an 's' subscript can be used, otherwise an 'as' or 'a' subscript can be employed. In some textbooks and scientific papers, the geometry of the vibration is specified further using other additional Greek letters:

ν_γ	Out-of-plane bending
ν_β	In-plane bending
δ'	Deformation (stated in general terms, bends, rocking, etc.)
ω	Shacking (rotation)
r	Rocking
r_β	In-plane rocking
r_γ	Out-of-plane rocking
τ	Torsions

π The vibration is performed in parallel (to a bond, for example)

σ The vibration is performed in perpendicular (to a bond, for example)

1.5.   Quantitation in IR spectrometry

It had been traditionally stated that (mid-)IR spectroscopy was not a suitable technique to quantify an analyte due to a low signal-to-noise ratio and because the work had to be done on printed graphical records and also because of the variability in sample preparation. This was true many years ago, when dispersive IR instruments were in use but it is not nowadays. Indeed, this analytical technique is used in almost any chemical industry to quantify analytes, thanks to the Fourier transform instruments and the use of multivariate regression models (whose implementation is studied within a branch of analytical chemistry called chemometrics). However, these models are out of the scope of the introductory level of this chapter and will not be considered further.

To quantify an analyte using the classical Lambert–Beer–Bouguer law, we have to resource to the basics presented in other chapters in this book. Here, we will only explain a relevant issue, which in turn is general for any spectrometry and, even, chromatography.

Obviously, the spectral band selected must be caused only by the analyte of interest. Therefore, it is important to consider the solvent that can be used and the concomitants. In general, it is possible to find out a band which is specific for the compound of interest. Then, we can use it to make a calibration at such a wavenumber. It is quite typical to consider the height (transmittance or absorbance) of the peak and to perform a calibration using different standards. For this to succeed, you must always be aware of the absorbance that can be unequivocally attributed to the analyte. A satisfactory means to ascertain this is to use the maximum, or height, of the spectral peak after some 'baseline correction' is made. That is, the maximum of the absorbance peak should be defined carefully, taking care to establish

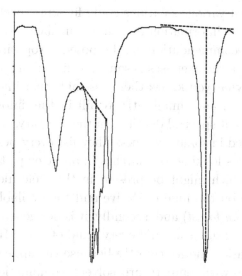

Figure 11.

where it begins and where it finishes and, also, what the background is. The two extreme points of the peak can be joined with a straight line (the baseline) and the height is measured vertically from the maximum until the intersection with the baseline (see Figure 11). Nowadays, the computers associated with all IR spectrophotometers can make these calculations straightforwardly. Remember to keep the same baseline parameters throughout the calibration process (Figure 11).

1.6. First approach to the exercises

The exercises presented below do present a situation which may not be the most favorable one for the students (or for a chemist!). In effect, they are intended to elucidate the structure of a (pure) molecule from a spectrum, with only few additional data and without the student being inserted in an experimental set up. However, in this chapter, only a limited number of problems will be presented and also only simple cases will be discussed. In our opinion, the student will be able to get the correct solutions to the proposed exercises once the solved ones are studied. In some occasions, additional help is given.

It is possible that various solutions could fit some spectra. Whenever this is known, the information is included in the exercise.

Finally, a recommendation can be posed. Proposing a molecular structure to solve the exercises is somehow like filling in a crossword. First, identify what you know; then, assemble that information with the data provided and, finally, try to fill in the 'information gaps' with other relevant spectral details. On the contrary, we discourage a strategy observed in some students: They list every peak (band) and inspect the tables looking for matches in an attempt to assign functional groups which might be present in the molecule. Firstly, this avoids thinking on the final objective and the available information (*the trees hide the forest*) and, secondly, it is not generally feasible or time-efficient to assign each and every band of a mid-IR spectrum to particular vibration modes (mostly because the appearance of mid-IR spectra varies with temperature, solvent, combination of vibration modes, etc.). Finally, and as a small anecdote, our students said that it was much easier to solve the exercises where the empirical equation was provided. They felt that those where only the elemental composition was indicated were a bit more complex.

REFERENCES

[1] Bellanato, J.; Hidalgo. A. (1971). *Infrared Analysis of Essential Oils.* Heyden & Sons Ltd, London.
[2] Conley, R. T. (1979). *Espectroscopia Infrarroja.* Alhambra, Madrid.
[3] Davies, A. N. (2014). Terahertz spectroscopy. *Spectroscopy Europe,* 26(1), 23–24.
[4] Günzler, H.; Gremlich. H-U. (2002). *IR Spectroscopy, An Introduction.* John Wiley-VCH, New York.
[5] Forrest, T.; Rabine, J-P.; Rouillard, M. (2011). *Organic Spectroscopy Workbook.* John Wiley and Sons, Chichester.
[6] Larkin, P. (2011). *Infrared and Raman Spectroscopy; Principles and Spectral Interpretation.* Elsevier.
[7] Mayo, D. W.; Miller, F. A.; Hannah, R. W. (2004). *Course Notes on the Interpretation of Infrared and Raman Spectra.* Wiley-Interscience, Hoboken (New Jersey).
[8] Nyquist, R. (2001). *Interpreting Infrared, Raman, and Nuclear Magnetic Resonance Spectra.* Elsevier.

[9] Pretsch, E.; Clerc, T.; Seibl, J.; Simon, W. (2001). *Tablas para la determinación estructural por métodos espectroscópicos.* Springer-Verlag, Barcelona.

[10] Roeges, N. P. G. (1994). *A Guide to the Complete Interpretation of Infrared Spectra of Organic Structures.* John Wiley & Sons, Chichester.

[11] Rubinson, K. A.; Rubinson, J. F. (2000). *Análisis Instrumental.* Prentice Hall, Mexico.

[12] Smith, B. C. (1988). *Infrared Spectral Interpretation: A Systematic Approach.* CRC Press, Boca Raton.

[13] Socrates, G. (2005). *Infrared and Raman Characteristic Group Frequencies Tables and Charts.* John Wiley & Sons, Chichester.

[14] Morcillo, J.; Madroñero, R. (1962). *Aplicaciones Prácticas De La Espectroscopia Infrarroja.* Facultad de Ciencias, Universidad de Madrid, Madrid.

[15] Silverstein, R. M.; Bassler, G. C.; Morrill, T. C. (1991). *Spectrometric Identification of Organic Compounds.* John Wiley and Sons, New York.

[16] Skoog, D. A.; West, D. M. (1984). *Análisis Instrumental*, 2nd ed. Interamericana, México.

[17] Szymanski, H. A. (1963). *Infrared Band Handbook*, Vol. 1 and Appendices 1, 2, 3 and 4. Plenum Press, New York.

[18] Szymanski, H. A. (1964). *Interpreted Infrared Spectra*, Vols. 1–3. Plenum Press, New York.

Note: The mid-IR spectra employed in the exercises in this chapter were adapted and modified from their original sources. In particular, spectra for exercises 1, 6, 8, 16, 18, 19, 21, 23, 31, 34, 35, 42, 45, 48, 52, 54, and 55 were taken from Morcillo and Madroñero [14]; exercises 2, 5, 7, 9, 10, 12, 17, 20, 22, 24, 25, 29, 32, 36, 41, 43, 49, 50, 53, 56, 57 and 58 proceed from the Conley's book [2]; exercises 3, 4, 11, 26, 28, 30, 33, 38 and 39 utilized the comprehensive encyclopedia from Szymanski [17, 18]; exercises 13, 14, 37, 40 and 44 were developed from Bellanato and Hidalgo's book [1]; exercises 15 and 51 are based on examples discussed in Skoog and West's textbook [16], whose original source was 'Catalog of selected ultraviolet spectral data', Thermodynamics Research Center Data Project, Thermodynamic Research Center, Texas, A&M University, College station, Texas. The spectra they used corresponded to a collection of unpublished data from 1975. And exercises 27, 46 and 47 were adapted from Silverstein, Bassler and Morrill [15].

WORKED EXERCISES

1. A gaseous exhaust from a chimney was analyzed using mid-IR spectrometry. The figure shows the IR spectrum obtained after removing impurities by spectral subtraction. Such impurities were due to nitrogen oxides and sulfur. The spectrum corresponds to a compound formed only by C and H. Propose its molecular structure. (*Note*: in order to improve the visualization of the spectral details, a region was measured twice using different pathlengths) (Figure 12).

SOLUTION:

The enunciation itself and the general appearance of the bands (mostly, the relative simplicity of the spectrum) indicates that the compound is in gas phase. This example was selected as the first one because it shows (at least, somehow) some rotational energy transitions 'embedded' within the vibrational ones. They can be seen only for the spectrum recorded with a larger pathlength (the full spectrum plot). In addition, the spectrum is very simple and seems to be of a linear hydrocarbon chain (there are no characteristics related to double bonds and/or aromatics, as evidenced by the absence of bands at 1600 and 700–900 cm^{-1}). We are told that it only contains C and H. The main spectral characteristics are:

- 2950–2975 cm^{-1}: It is the typical band of the asymmetric stretching vibration of a CH bond in CH$_3$ groups.

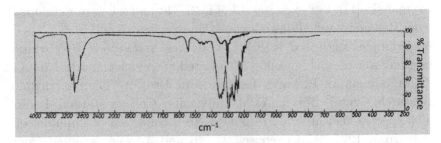

Figure 12.

- $2850–2870\,\mathrm{cm}^{-1}$: It is the typical band of the symmetric stretching vibration of a CH bond in CH_3 groups.
- $1300\,\mathrm{cm}^{-1}$: The peak observed with the smaller pathlength (less intense peak) suggests the symmetric bending vibration of a CH bond, when included in a CH_3 group. When, the optical path of the cell was increased, a second band appears around $1350–1375\,\mathrm{cm}^{-1}$. This can also be associated to the asymmetric bending $(1400\,\mathrm{cm}^{-1})$ of a CH bond in a CH_3 group.

Now, considering that the sample comes from a chimney, that it is a result of a combustion process and that no complex structures are detected, it can be deduced that the compound is methane, as indeed is the correct solution.

2. The spectrum of a gaseous sample, which is composed only of C and H, is shown in Figure 13. With the additional information that its relative molecular mass is 72, indicate its molecule structure.

SOLUTION:

- This is again a quite simple spectrum that corresponds to a hydrocarbon structure which must be short (because of its low molecular mass) and is in the gas phase (according to the initial definition). It does not reveal bands that can reasonably be attributed to C=C bonds, aromatics or triple bonds (peaks around 1600, 700–900 or $2000\,\mathrm{cm}^{-1}$). Then, a saturated molecule is expected. The spectral characteristics

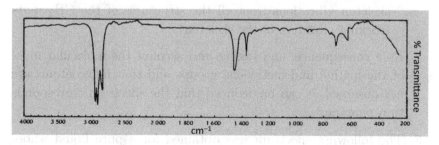

Figure 13.

that give us important clues to reach a probable structure are:

- 2950–2975 cm^{-1}: It is the typical band of the asymmetric stretching vibration of the CH bond included in CH$_3$ groups, overlapped with the asymmetric stretching vibration band of the CH bond in CH$_2$ groups (these two stretchings almost always result in a single band).

- 2850–2870 cm^{-1}: It is the typical band of the symmetric stretching vibration of the CH bond in CH$_3$ groups, overlapped with the symmetric stretching vibration band of the CH bond in CH$_2$ units (these two stretchings almost always result in a single band).

- 1450 cm^{-1}: The band can be attributed to the spectral overlapping of two bands associated to: (i) the asymmetric bending vibration of the CH bond present in CH$_3$ groups and (ii) the symmetric bending vibration band of the CH bond in CH$_2$ groups.

- 1375 cm^{-1}: This band can be attributed to the spectral overlapping of two bands: (i) the symmetric bending of the CH bond (in CH$_3$ groups) and (ii) the asymmetric bending of CH (in CH$_2$ units).

In the latter two cases, and as a mnemonic rule, note that the match is symmetrical with asymmetrical.

- 750 cm^{-1}: This typical peak (small intensity) is characteristic of saturated hydrocarbon chains in which three or more CH$_2$ groups are connected linearly. The more CH$_2$ groups bonded linearly, the more intense the peak will be. It is attributed to skeleton vibrations from all the structures of the CH$_2$ units (rocking vibration of the skeleton).

As a consequence, and taking into account the molecular mass of the methyl and methylene groups and that heteroatoms are not observed, it can be deduced that the spectrum corresponds to n-pentane.

3. The following spectrum was obtained for a pure liquid whose empirical formula was calculated as C$_7$H$_{16}$ (Figure 14). Propose

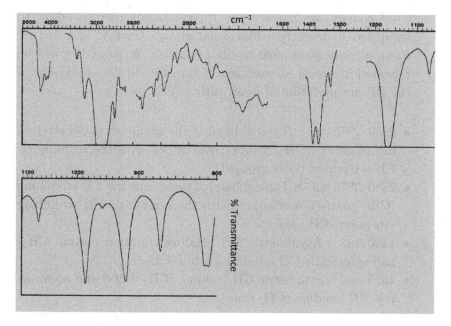

Figure 14.

its most probable structure. Observe that the overall spectrum displayed in the figure is a composition obtained after measuring different regions at different pathlengths (using a typical variable pathlength cell). This was done to best visualize the structure of the sample.

SOLUTION:

Most students prefer the exercises where the empirical equation is given because they feel more comfortable when looking for the solution. This is a typical example.

First, it can be seen quite easily that the spectrum corresponds to a fully saturated hydrocarbon. As in the previous example, the presence of heteroatoms, aromatic structures or double/triple bonds cannot be expected. Nevertheless, this example is not trivial at all because the spectrum was selected to show how the presence of branched functional groups can be ascertained from some particular details.

Of course, we can see the typical bands corresponding to saturated hydrocarbon functional groups (methyl and methylene), already mentioned in the previous exercises (they will be repeated in almost all exercises as they constitute a backbone of the IR interpretation of hydrocarbons):

- 2950–2975 cm^{-1}: Typical band of the asymmetric CH stretching vibration (CH$_3$ groups), overlapped with the asymmetric CH stretching (CH$_2$ groups).
- 2850–2870 cm^{-1}: Typical band of the symmetric CH stretching (CH$_3$ groups), overlapped with the symmetric CH stretching vibration (CH$_2$ units).
- 1450 cm^{-1}: Asymmetric CH bending vibration (within CH$_3$) and symmetric CH bending (within CH$_2$).
- 1375 cm^{-1}: Symmetric CH bending (CH$_3$ units) and asymmetric CH bending (CH$_2$ ones).

However, unlike the previous exercises, two new spectral peaks around 1380 (in fact, a double peak, or 'doublet') and 1170 cm^{-1} are observed (they are not always visualized clearly). It is worth noting that the existence of isopropyl and/or *t*-butyl groups in the hydrocarbon molecules almost always causes splitting ('doublets') of some CH characteristic bands. In this case, a clear splitting of the symmetrical CH bending (included in CH$_3$ groups, 1380 cm^{-1}) can be observed. To decide whether we have an isopropyl or *t*-butyl group, we have to concentrate now in the region around 1150–1250 cm^{-1}.

In this example, you can see a very nice band at 1170 cm^{-1}, which corresponds to a general skeleton deformation vibration of the CH(CH$_3$)$_2$ structure. The *t*-butyl group should present its corresponding structural bending band displaced toward *ca.* 1250 cm^{-1}. This is not the case here and, so, we select the isopropyl alternative.

So far, with the information we obtained and considering the empirical equation, the 2,4-dimethyl pentane structure can be proposed: (CH$_3$)$_2$CH–CH$_2$–CH(CH$_3$)$_2$.

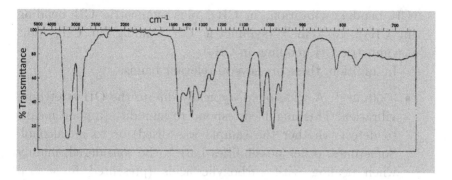

Figure 15.

Final note: Another relevant information that would help us to easily reject possible linear structures (compatible with the empirical equation at first glance; e.g., $(CH_3)_2CH(CH_2)_3CH_3$) would be the lack of a peak at $750\,cm^{-1}$ (rocketing of several linear CH_2 groups), although unfortunately that region was not considered in the plot displayed here.

4. The spectrum above (Figure 15) corresponds to a quite common liquid solvent in laboratories. It consists only of C, H and O and its relative molecular mass is 74. The record has been made using a liquid cell with a 0.01 cm pathlength. What was the solvent?

SOLUTION:

In this example, we have information on elemental composition and molecular mass. Hence, it is obvious that the molecule will have a hydrocarbon structure to which a group with oxygen should be connected. The key to solve this type of exercises is usually to discover the functional group that contains oxygen and, then, consider the molecular mass.

In the spectrum, we can observe again the typical bands associated to CH bonds. Just to summarize: $2950–2975\,cm^{-1}$ (asymmetric CH stretching, in CH_3, and asymmetric CH stretching, in CH_2); $2850–2870\,cm^{-1}$ (symmetric CH stretching, in CH_3, and symmetric CH stretching, in CH_2); $1450\,cm^{-1}$ (asymmetric

CH bending vibration, in CH_3, plus symmetric CH bending in CH_2); $1375\,cm^{-1}$ (symmetric CH bending, in CH_3, plus asymmetric CH bending, in CH_2).

In addition, there are several relevant bands:

- $3500\,cm^{-1}$: A wide band, typically due to the OH stretching vibration. This might correspond to humidity (a good means to detect whether the sample was dried) or to an alcohol. Sometimes, other possibilities have to be considered, mainly when dealing with carboxylic acids (presented in several examples below).

- 1350–$1300\,cm^{-1}$: A wide band is observed. This is characteristic of the OH bending. In general, it is a strong signal which may hide or overlap with other modes of vibration (i.e., CH groups).

- $1100\,cm^{-1}$: this intense spectral peak is characteristic of the alcohols, as it corresponds to the stretching of the CO bond (the oxygen giving rise to the alcohol). It is very important to locate the position of this peak because it will give us indications on the type of alcohol:

  o 1075–$1000\,cm^{-1}$ (usually, at 1050 but it can be up to 1100): primary alcohols.
  o 1120–$1030\,cm^{-1}$ (usually, around 1100): secondary alcohol.
  o 1170–$1100\,cm^{-1}$ (usually, around 1150): tertiary alcohol.
  o phenols raise the band until 1230–$1140\,cm^{-1}$.

With all this information in mind, the molecule studied in this exercise is 2-butanol: CH_3–CH_2–CH $OHCH_3$

5. The following mid-IR spectrum (Figure 16) was obtained in a forensic study after analyzing a solvent found in a garage. Its relative molecular mass (the only data that could be saved from a burnt receipt) was 84. Propose its molecular structure.

SOLUTION:

As in the previous examples, the presence of double and triple bonds, as well as aromatic rings can be discarded. Therefore, it

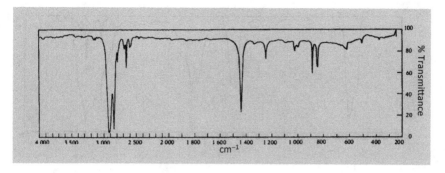

Figure 16.

is expected that the sample corresponds to a light hydrocarbon, likely a saturated structure. In addition, you can visualize a relevant issue: the splitting of some peaks, most notably those associated to the CH stretching and bending at 2950 and $850\,cm^{-1}$, respectively. This suggests (as a first approximation) the existence of t-butyl and/or isopropyl groups. However, we cannot see the characteristic skeleton vibration peaks of these structures (in exercise 3, they were noted as $1390\,cm^{-1}$ for isopropyl and $1250\,cm^{-1}$ for t-butyl groups). Following, we can reject the presence of these groups in this sample.

As a consequence, we must face the possibility that the spectrum corresponds to a cycloalkane (whose structures are a bit difficult to perceive). Indeed, in this case, we reached such a conclusion after excluding other possibilities. As a rule of thumb, when many doublets are seen in a mid-IR spectrum, you can think of a cycloalkane (unless other possibility appears more clear for you).

Then, when you review the molecular masses of the simplest cycloalkanes, the molecular mass of 84 shows immediately that the compound under study is the typical solvent for oil and grease in garages: Cyclohexane.

6. In a process set up in petroleum refineries, propane (sold as retail fuel gas) and propylene (used to produce polymers) must be separated. An automated mid-IR spectrophotometer to monitor the separation was installed at the exit of the production

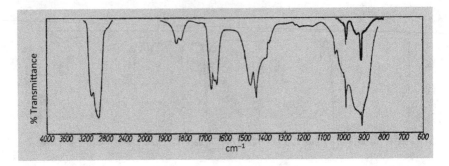

Figure 17.

unit. Decide if the spectrum above (Figure 17) corresponds to propane or to propylene (some bands were registered at several pathlengths).

SOLUTION:

This is a quite simple example. The main key point is to determine whether the C=C bond shows its spectral characteristics. This is characterized by a strong stretching vibration band around $1600 \, cm^{-1}$. Further, you can also see the vibrations of the aliphatic CH bonds which were discussed in previous exercises, in summary:

- $2950–2975 \, cm^{-1}$: Asymmetric CH stretching (in methyl) plus asymmetric CH stretching (in methylene).
- $2850–2870 \, cm^{-1}$: Symmetric CH (in methyl) and symmetric CH stretching (in methylene).
- $1450 \, cm^{-1}$: Asymmetric CH bending (CH_3) plus symmetric CH bending (CH_2).
- $1375 \, cm^{-1}$: Symmetric CH bending (CH_3) and asymmetric CH bending (CH_2).

In addition, in this example, you can see other relevant spectral characteristics. The use of various optical paths can aid in the interpretation (this is not always done):

- $900 \, cm^{-1}$: Out-of-the-plane bending (considering the plane formed by the C=C bond) of the CH_2 group.

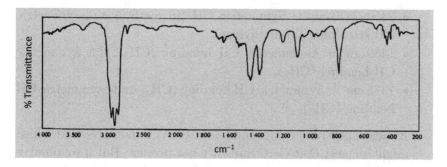

Figure 18.

- 995 cm^{-1}: Out-of-the-plane bending of the CH bond (when it is included in a C=C functional group).
- 1850 cm^{-1}: First harmonic peak (or overtone) associated to the out-of-the-plane CH bending.

Another important, typical characteristic is the strong signal at 3100 cm^{-1}, which is due to the CH stretching when this bond is involved in a C=C unit. This characteristic is almost always clearly visible.

Consequently, the spectrum corresponds to propylene.

7. A gaseous compound obtained in a refinery presents the mid-IR spectrum depicted above (Figure 18). The compound is constituted by C and H and it was found that its most probable relative molecular mass is 70. What is the molecular structure of the compound?

SOLUTION:

This is a bit difficult case study because it does not show too typical peaks. However, the example is very interesting. First, we can see that the spectrum does not reveal a complex structure and that it displays the fine characteristic bands of the CH$_3$/CH$_2$ groups commented previously:

- 2950–2975 cm^{-1}: Asymmetric CH stretching (in methyl) plus asymmetric CH stretching (in methylene).

- 2850–2870 cm^{-1}: Symmetric CH (in methyl) and symmetric CH stretching (in methylene).
- 1450 cm^{-1}: Asymmetric CH bending (CH$_3$) plus symmetric CH bending (CH$_2$).
- 1375 cm^{-1}: Symmetric CH bending (CH$_3$) and asymmetric CH bending (CH$_2$).

A clear spectral peak around 1600 cm^{-1} is not observed which might suggest that C=C units are not present. But if we observe closely the various peaks around 3000 cm^{-1}, we can see that one of them is *ca.* 3100 cm^{-1} and, therefore, it indicates that double bonds are indeed present in the molecule. The problem is that the stretching vibration of the C=C bond in simple molecules (like that presented here, with a very low molecular mass) can be very weak (on the contrary, conjugated systems potentiate this band) and we must look carefully at other weak bands to draw conclusions.

Such is the case of the peak appearing around 800 cm^{-1}. This band should be attributed to the out-of-the-plane CH bending, when this bond is associated to C=C structures. In particular, of the type R$_1$R$_2$C=C–HR$_3$.

As the spectrum does not present relevant additional bands, the molecular structure is that of methyl butene: (CH$_3$)$_2$–C=CHCH$_3$.

8. In an archaeological excavation, a closed vessel with a spicy gas was discovered. Using elemental analysis techniques, they concluded that is constituted by C and H. The relative molecular mass was estimated as 26. Deduce the gas structure from the IR spectrum. To reveal details, an addition spectral band obtained with another optical path was collected (Figure 19).

SOLUTION:

The compound under study must be very simple, according with the stated data. The appearance of the IR spectrum shows that it has not aromatic structures (a distinctive peak at 1600 cm^{-1}

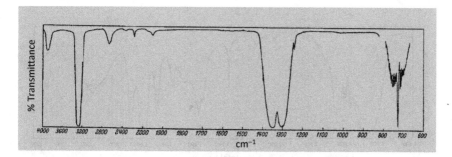

Figure 19.

is not present). The typical spectral peaks of the CH_3 and/or CH_2 groups can be visualized (they will not be repeated here).

The most significant feature is a very strong spectral peak around $3300\,\text{cm}^{-1}$. This must be identified as the CH stretching when this bond belongs to a $C{\equiv}C$ bond. If the presence of a $C{\equiv}C$ is suspected, it is usually confirmed by a sharp peak around $2100\,\text{cm}^{-1}$ ($C{\equiv}C$ stretching). In this example, such a peak is observed although it is weak (recall that the sample is in the gas phase).

The somehow poorly defined band at $700\,\text{cm}^{-1}$ corresponds to the CH out-of-the-plane bending. The appearance of this band is due to the rotational transitions because the sample is in the gas phase. The overtone of this vibration occurs (not very resolved) around $1400\,\text{cm}^{-1}$.

Combining all this information with the molecular weight, the spicy gas was acetylene.

9. Deduce the structure of a compound whose relative molecular mass is 58 and its spectrum is shown in Figure 20. Additional information: it is a volatile and sweet-smelling compound.

SOLUTION:

We have information on the molecular mass of the compound but not on its composition. Therefore, we must proceed steadily 'dissecting' the figure.

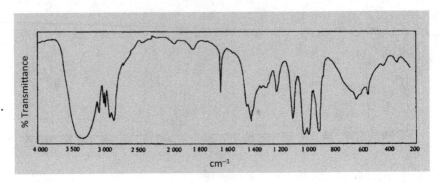

Figure 20.

First, you will see the, now, very well-known bands corresponding to the vibrations of the methyl and methylene groups:

- 2950–$2975\,\text{cm}^{-1}$: CH asymmetric stretching band in methyl groups, and/or the CH asymmetric stretching vibration in methylene groups.
- 2850–$2870\,\text{cm}^{-1}$: CH symmetric stretching in methyl groups, and/or CH symmetric stretching in methylene groups.
- $1450\,\text{cm}^{-1}$: CH asymmetric bending vibration band (in CH_3) and/or CH symmetric bending (in CH_2). In this case, the peak is displaced slightly, likely due to a major unresolved band (we will consider it in next lines).
- $1375\,\text{cm}^{-1}$: CH symmetric bending vibration (in CH_3) and/or CH asymmetric bending (in CH_2). It can hardly be seen because of a wide, unresolved band.

Note: Recall that the existence of a given band does not necessarily guarantee that the two types of functional groups are present in the molecule; it would be only one, hence, the use of the term 'and/or'.

- A distinctive, sharp peak is seen at $1600\,\text{cm}^{-1}$. This probably corresponds to the characteristic C=C stretching, as detailed in other previous exercises. This guess is confirmed with the $3100\,\text{cm}^{-1}$ absorption peak (although not excellently resolved), associated to the C–H stretching vibration (when involved in a C=C bond).

With regard to the broadest peak, its appearance immediately suggests an alcohol (see exercise 4). Of course, the broad band at 3300–3500 cm^{-1} corresponds to the OH stretching. The other most relevant bands are:

- 1350–1300 cm^{-1}: A rather broad, ill-defined peak that can be attributed to the OH bending vibration. The band *ca.* 650 cm^{-1} is also due to the OH bending vibration.
- 1025–1000 cm^{-1}: This intense peak is characteristic of primary alcohols, and is caused by the CO(–H) stretching vibration.
- 915 and 990 cm^{-1}: They hold the key to decide the type of substituents in the double bond. These peaks are due to the out-of-the-plane CH bending of the CH$_2$ groups (the plane defined by the double bond). These two peaks are typical of CH$_2$=CH–R structures.

Therefore, all the information suggests that the structure under study corresponds to the allyl alcohol (or 1-propen-3-ol): CH$_2$=CHCH$_2$OH

10. The mid-IR spectrum presented in Figure 21 corresponds to a compound used in the paint and solvent industries. Its empirical formula is C$_{10}$H$_{18}$. Suggest its most probable structure.

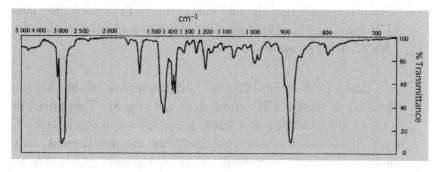

Figure 21.

SOLUTION:

Although the spectrum does not appear highly complex, the correct solution is not immediate. First, the empirical equation suggests a fairly saturated compound, with only one or two unsaturated carbons (remember the C_nH_{2n+2} rule). Analyzing the spectrum, you will surely recognize some spectral peaks (as in most previous examples and, so, not many details will be repeated here):

- 3050–$3100\,cm^{-1}$: Although this peak is weak, it is clearly defined. Its appearance is typical of the CH stretching vibration (when the C is involved in a C=C functional group).
- 2950–$2975\,cm^{-1}$: CH asymmetric stretching (CH_3) and/or CH asymmetric stretching (CH_2).
- 2850–$2870\,cm^{-1}$: CH symmetric stretching (CH_3) and/or CH symmetric stretching (CH_2).
- $1450\,cm^{-1}$: CH asymmetric bending (CH_3) and/or CH symmetric bending (CH_2). The peak is a bit displaced due to the adjacent broad band.
- $1375\,cm^{-1}$: CH symmetric bending (CH_3) and/or CH asymmetric bending (CH_2). The peak is overlapped with the adjacent band.
- $1600\,cm^{-1}$: Our initial guess about unsaturated C=C groups is confirmed here, this is the typical C=C stretching. It seems clear that, at least, the molecular structure has a C=C bond.
- 890–$900\,cm^{-1}$: The intense absorption peak is attributable (mostly considering the previous band) to the out-of-the-plane CH bending. In addition, this position is characteristic of a RR'C=CH$_2$ structure.

Note that a peak around $750\,cm^{-1}$, characteristic of the skeleton vibration of several CH_2 units, does not appear. This forces us to reject the existence of a linear structure with a unique C=C bond. Since the aromatic characteristics are not present, only one option remains: a cycloalkane (note that as in example 5, we detected such a structure by excluding other possibilities).

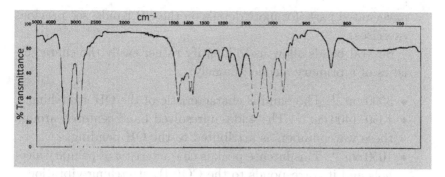

Figure 22.

With this idea in mind, the doublets at 1400 and 1000 cm^{-1} tell us about the possible presence of isopropyl or *t*-butyl groups. If we assume an isopropyl group, the peaks at 800 and 1175 cm^{-1} can be linked to its skeleton vibrations.

With all the information collected (and after trying different options because the solutions are not trivial at all) we can conclude that the final structure of the compound is 4-isopropyl-1-methylene cyclohexane: CH_2 =(cyclohexane)–$CH(CH_3)_2$

Final note: The para substitution (which is correct) is not likely to be ascertained from the IR spectrum alone. Any other similar structure would be compatible with the information we have. The important points in this example are: (i) to recognize that a C=C bond is present, (ii) to realize that a cycloalkane is there, and (iii) to visualize the isopropyl group.

11. The spectrum above (Figure 22) shows the mid-IR characteristics of a sweet-smelling liquid which is used in cosmetics and in some perfumes. Identify its structure considering that its empirical equation is $C_5H_{12}O$.

SOLUTION:

Once more, you can easily recognize the typical spectral bands associated to the CH bonds of the hydrocarbon (2950–2975, 2850–2870, 1450 and 1375 cm^{-1}). Therefore, let us focus on the oxygenated part (for more details on the CH bands, see previous exercises).

Several bands allow us to identify rather easily the characteristics of a primary alcohol; namely:

- $3500\,cm^{-1}$: The band is characteristic of the OH stretching.
- $1350\text{--}1400\,cm^{-1}$: The wide, unresolved band centered around these wavenumbers is attributed to the OH bending.
- $1050\,cm^{-1}$: This intense peak is characteristic of primary alcohols and it corresponds to the CO(–H) stretching vibration.

Notice the doublet at $1400\,cm^{-1}$, the CH bending, which may be due to the presence of isopropyl or *t*-butyl groups. We cannot see a rather strong band around $1250\,cm^{-1}$, which characterizes the butyl groups, and therefore we can think of an isopropyl group. This can be confirmed observing the medium-intense bands at $840\text{--}850$ and $1140\,cm^{-1}$ (in this case, the latter locates a bit low, when compared with the reference value at $1170\,cm^{-1}$).

Hence, we can conclude that the component of the perfume is isopentyl alcohol:

$$(CH_3)_2CHCH_2CH_2OH$$

12. Propose a molecular structure for a compound whose mid-IR spectrum is displayed in Figure 23 and it is known that its empirical equation is $C_8H_{14}O_2$.

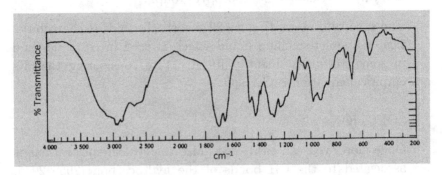

Figure 23.

SOLUTION:

This is a spectrum wherefrom two solutions can be obtained easily. The correct option will be studied here, but the other alternative would also be acceptable (we could get more experimental information by simply comparing the spectrum of the unknown to the spectra of the two candidate pure substances).

First, observe that despite the 'undefined' appearance of the spectrum, two or three major spectral bands are clear and, somehow, they 'hide' most of the other spectral peaks. This is a very common situation for carboxylic acids (although, maybe not so dramatically as in this example). We can interpret some relevant bands:

- $3500\,cm^{-1}$: This is a very broad band and poorly defined. It can be assigned to the typical OH stretching from the acidic group. However, the band is much wider than in the previous example (for instance) because of the hydrogen bridges that occur among the carboxylic groups of the different molecules of the analyte which, indeed, make several OH units vibrate 'together'. Such associations modify the free vibration of the OH group and, hence, this band is much broader than for the alcohols.
- $1700\,cm^{-1}$: This band corresponds typically to the C=O stretching although, in general, it appears much more defined and visible. Here, it is partially overlapped with the $1600\,cm^{-1}$ band.
- 1400–$1300\,cm^{-1}$: The two broad bands centered around those wavenumbers correspond to a combination of the C=O stretching and the OH bending (the second signal is usually more intense).
- 950–$1000\,cm^{-1}$: This band is associated to the out-of-the-plane OH bending (plane defined by the acid group).
- $1650\,cm^{-1}$: This peak (here, displaced slightly from its typical value, around $1600\,cm^{-1}$) indicates that there is a C=C bond somewhere in the molecule.

- 2950–2975 cm^{-1}: CH asymmetric stretching (in CH$_3$ groups) and/or CH asymmetric stretching (in CH$_2$).
- 2850–2870 cm^{-1}: CH symmetric stretching (CH$_3$) and/or CH symmetric stretching (CH$_2$).
- 1450 cm^{-1}: CH asymmetric bending (CH$_3$) and/or CH symmetric bending (CH$_2$).
- 1375 cm^{-1}: It is difficult to see the CH symmetric bending (CH$_3$) and/or CH asymmetric bending (CH$_2$) in this spectrum.
- 700 cm^{-1}: It is the distinctive skeleton vibration of several, linear, consecutive CH$_2$ groups.

To deduce the type of substitution of the C=C bond, we have to resort to the CH out-of-the-plane typical bands. Unfortunately, in this example, they are not very clear. The general rule is:

- RHC=CH$_2$ would yield a band *ca.* 990 cm^{-1}
- R$_1$HC=CHR$_2$ (trans) would yield a band *ca.* 970 cm^{-1}
- R$_1$HC=CHR$_2$ (cis) would yield a band *ca.* 690 cm^{-1}
- R$_1$R$_2$C=CH$_2$ would yield a band *ca.* 890 cm^{-1}
- R$_1$R$_2$C=CHR$_3$ would yield a band 840–790 cm^{-1}

The appearance of this spectrum does not permit a sound decision on this issue.

The correct solution (according to the spectral database) is CH$_3$(CH$_2$)$_4$CH=CHCOOH, although other possibilities are also compatible with the graphical information; for instance, CH$_2$=CH(CH$_2$)$_5$COOH. This case study reflects the statement given in the introduction about the need of additional information to draw sound conclusions from an IR spectrum. In many occasions, it is not possible to get a unique answer from the spectrum alone.

13. Figure 24 shows the mid-IR spectrum of an industrial liquid product, whose purity is monitored routinely using this instrumental technique. The product is composed only of C, H and O and its relative molecular mass is 72. Propose the most probable structure for the product.

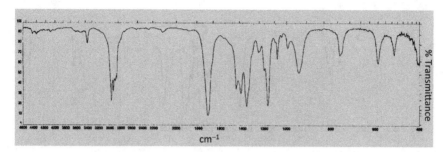

Figure 24.

SOLUTION:

The molecular mass suggests that the structure should not be complicated. This is confirmed observing that there are not evidences for the existence of aromatic structures and/or double bonds. In short, you will recognize (see previous exercises) the typical vibrations at 2950–2975 cm^{-1}; 2850–2870 cm^{-1}; 1450 cm^{-1}; and 1375 cm^{-1}.

The intense band at *ca.* 1720 cm^{-1} stands out. It can be assigned to the C=O stretching vibration. This value is quite high for a carboxylic acid and, mostly, note that the spectrum has no broad unresolved bands (see example 12). It seems that we have a ketone (we can reject, at least by now, the ester because of the low molecular weight).

The ketone is confirmed by the intense peak at 1200–1100 cm^{-1}, which is the skeleton bending of the three carbons from the ketone (C–CO–C). It is worth noting that it is not always easy to differentiate between a ketone and an ester.

Finally, considering the molecular mass, we conclude that the compound is butanone: $CH_3CH_2COCH_3$.

14. A new component of a perfume is constituted only by C, H and O. Its relative molecular mass was found to be 88. Predict the structure of this compound using the mid-IR spectrum in Figure 25 considering that the concomitants in the solution do not modify its spectral profile significantly.

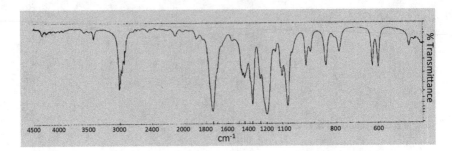

Figure 25.

SOLUTION:

It is worth comparing the spectral characteristics of this spectrum to the previous exercise. The relative molecular mass and the appearance of the spectrum does not suggest the existence of phenyl groups or, even, C=C bonds (there is no CH stretching band around $3100\,\mathrm{cm^{-1}}$). However, some other typical characteristics can be observed:

- 2950–$2975\,\mathrm{cm^{-1}}$: Asymmetric stretching of the CH bond in CH_3 groups and/or asymmetric stretching of the CH bond in CH_2 groups.
- 2850–$2870\,\mathrm{cm^{-1}}$: Symmetric stretching of the CH bond in CH_3 groups and/or symmetric stretching of the CH bond in CH_2.

Unfortunately, any of these, usually, characteristic bands can be seen too clearly in this example.

- $1450\,\mathrm{cm^{-1}}$: CH asymmetric bending when this bond pertains to CH_3 groups and/or CH bending, in CH_2 groups.
- $1375\,\mathrm{cm^{-1}}$: Symmetric bending of the CH bond in CH_3 groups and/or CH bending associated to CH_2 groups.

The distinctive peak around $1750\,\mathrm{cm^{-1}}$ denotes the typical C=O stretching vibration. Since broad bands are not obvious in the spectrum (they would point toward alcohols and/or carboxylic acids), we should suspect on the existence of either a ketone or an ester. We can see two intense peaks around 1250 and $1050\,\mathrm{cm^{-1}}$. The former is as intense as that (sometimes, even more) of

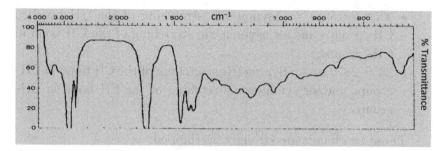

Figure 26.

the carboxylic group itself and it is linked to the asymmetric stretching of the C–O–C group, whereas the latter corresponds to its symmetric stretching. Both bands are highly characteristic of an ester group.

The correct solution is $CH_3CH_2COOCH_3$ (ethyl acetate or ethyl ethanoate) although the $CH_3COOCH_2CH_3$ option is another reasonable alternative.

15. In an environmental forensic study conducted after a massive death of fishes in a local river, a sweet odor was detected in the area. The extraction of a sample of contaminated water yielded a colorless liquid substance (after the solvent was eliminated and the residue was dried with sodium sulfate). Its corresponding empirical equation was $C_6H_{12}O$ and its boiling point was 130 °C. The mid-IR spectrum of that residue is presented in Figure 26 above. Propose a structure for the pollutant.

SOLUTION:

The appearance of the spectrum is not too 'clean', likely due to concomitants causing a high background from the sample preparation steps. Despite this, we can still visualize key spectral bands to ascertain the most relevant characteristics of the substance causing the death of the fishes. This example was included here to show that the sample preparation step can influence the appearance of the IR spectrum we are interested in.

- 2950–2975 cm^{-1}: Asymmetric stretching of the CH bond in CH$_3$ groups and/or asymmetric stretching of the CH bond in CH$_2$ groups.
- 2850–2870 cm^{-1}: Symmetric stretching of the CH bond in CH$_3$ groups and/or symmetric stretching of the CH bond in CH$_2$ groups.

These two bands are strongly overlapped.

- 1450 cm^{-1}: CH asymmetric bending when this bond pertains to CH$_3$ groups and/or CH bending, in CH$_2$ groups.
- 1375 cm^{-1}: Symmetric bending of the CH bond in CH$_3$ groups and/or CH bending associated to CH$_2$ groups.

These latter bands are visualized poorly in this spectrum and there are many more best examples throughout this chapter.

- A strong peak can be seen around 1700 cm^{-1}, a clear indication for a C=O group (that would be the C=O stretching vibration). In order to assign a functional group to this bond, we find out a very clear indication in the sharp spectral peak around 2850 cm^{-1}. This shape is quite typical of the CH stretching vibration of the aldehydes: it appears at the slope of the peaks associated to aliphatic CH stretching vibrations.
- 700–750 cm^{-1}: This peak is typical of the rocking vibration of various (in general, more than three) CH$_2$ groups, bonded in a linear structure.

Hence, the pollutant happen to be CH$_3$(CH$_2$)$_4$COH (hexanal or hexanalaldehyde, typically used in varnish- and paint-related industries although, curiously, in small quantities it can be used in the flavor industry).

16. The quality control of a polymer of daily use worldwide is done routinely using mid-IR spectrometry. The spectrum (see Figure 27) reflects essentially the monomer characteristics (whose relative molecular mass is 104). Propose a structure for the monomer.

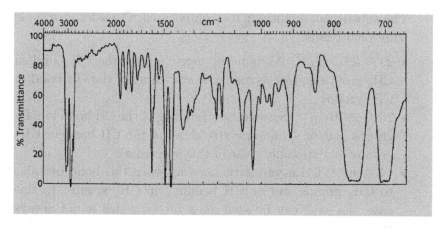

Figure 27.

SOLUTION:

This case study is a marvelous and typical example of a mono-substituted aromatic ring. We will review some characteristic bands (in next exercises, they will merely be referred to).

- $3100 \, cm^{-1}$: CH stretching associated to C=C groups. When this is visualized in conjunction with the next information, the aromatic ring appears very clearly.
- $1600-1490 \, cm^{-1}$: Three narrow, intense peaks corresponding to the in-the-plane stretching vibration of the C=C bonds of the aromatic ring.
- $2000-1700 \, cm^{-1}$: This excellent visualization of the overtones and combinations of vibrations of the CH bending (either in-the-plane or out-of-the-plane) is a clear indication of a monosubstituted ring (the decreasing heights, to the right).
- $800-650 \, cm^{-1}$ region: The two strong, symmetric peaks at 690 and $750 \, cm^{-1}$ evidence the monosubstitution (out-of-the-plane bending of the aromatic CH bonds).

With this information available, we can simply calculate $104 - 77$ (total mass − phenyl), which yields an 'unassigned mass' of 27.

The substituent seems not too complicated. Consider now some additional bands:

- 2950–2975 cm^{-1}: Asymmetric stretching of the CH bond in CH$_3$ groups and/or asymmetric stretching of the CH bond in CH$_2$ groups.
- 2850–2870 cm^{-1}: Symmetric stretching of the CH bond in CH$_3$ groups and/or symmetric stretching of the CH bond in CH$_2$ groups (not specially clear in this spectrum).
- 1450 cm^{-1}: CH asymmetric bending when this bond pertains to CH$_3$ groups and/or CH bending, in CH$_2$ groups. In this case, the peak can be seen quite nicely, which is not always the case.
- 1375 cm^{-1}: Symmetric bending of the CH bond in CH$_3$ groups and/or CH bending associated to CH$_2$ groups.

All these pieces of information suggest a group with two atoms of carbon. Which group? A relevant hint is to consider the molecular mass. Then, a C=C group appears as the most probable option. In effect, this can be confirmed by the intense peak at 910 cm^{-1}, characteristic of the CH out-of-the-plane vibration (the plane of the C=C group).

To conclude, the monomer we are studying is styrene (the polymer is the very well-known and broadly used polystyrene: CH$_2$ =CH–Ph).

17. A pure, dense liquid, whose empirical equation is C$_{10}$H$_{14}$, yields the following mid-IR spectrum (Figure 28). Propose its most probable structure.

SOLUTION:

This example contains many characteristics of the hydrocarbons exemplified so far.

- 2950–2975 cm^{-1}: Asymmetric stretching of the CH bond in CH$_3$ groups and/or asymmetric stretching of the CH bond in CH$_2$ groups.
- 2850–2870 cm^{-1}: Symmetric stretching of the CH bond in CH$_3$ groups and/or symmetric stretching of the CH bond in CH$_2$ groups (not specially clear in this spectrum).

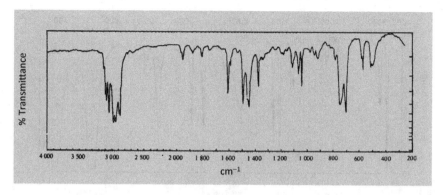

Figure 28.

- 1450 cm^{-1}: CH asymmetric bending when this bond pertains to CH$_3$ groups and/or CH bending, in CH$_2$ groups.
- 1375 cm^{-1}: Symmetric bending of the CH bond in CH$_3$ groups and/or CH bending associated to CH$_2$ groups.

In addition, we can also observe:

- 3100 cm^{-1}: CH stretching associated to C=C groups. When this is visualized in conjunction with the next information, the aromatic ring appears very clear.
- 1600–1450 cm^{-1}: Three narrow, intense peaks corresponding to the in-the-plane stretching vibration of the C=C bonds of the aromatic ring.
- 2000–1700 cm^{-1}: This fairly good visualization of the overtones and combinations of vibrations of the CH bending (either in-the-plane or out-of-the-plane) is a clear indication of a monosubstituted ring.
- 800–650 cm^{-1}: We can see two intense peaks, approximately at 650 and 750 cm^{-1}, that although not too symmetrical and slightly overlapped, suggest that there is a monosubstituted ring (out-of-the-plane bending of the aromatic CH bonds).

Hence, all the information we have indicates that the most probable structure is that associated to N-butylbenzene: CH$_3$CH$_2$CH$_2$CH$_2$–Ph.

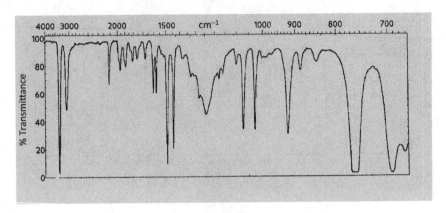

Figure 29.

18. A group of students synthesized an organic compound as a practical exercise to get their grades. They obtained a colorless, viscous liquid. The compound should be constituted only by C and H, with a relative molecular mass of 102. The spectrum obtained from the final product is recorded in Figure 29. What compound do you think they obtained?

SOLUTION:

First, the absence of typical functional groups containing O can be discarded. There are neither broad bands around $3500\,\mathrm{cm}^{-1}$ nor an intense characteristic peak around $1700\,\mathrm{cm}^{-1}$. Nitrogen also seems to be absent because neither the characteristic broad bands of the amines nor a sharp peak around $2200\,\mathrm{cm}^{-1}$ (nitrile group) can be observed (these groups will be studied in the last examples of this chapter). Hence, the compound synthesized by the students likely, in effect, contains only C and H.

To start with, note that the spectrum is quite typical of an aromatic compound. In this sense, the student should not be disturbed by several doublets that appear due to the presence of a conjugated system of C–C bonds (as it will be clear when the solution is presented). In other spectra that can be found in literature for this compound, the doublets are not so clear and, instead, the corresponding bands are broader. Maybe the students got the spectrum in a very diluted CS_2 or CCl_4 solution (a similar spectrum can be found in the public NIST website).

Major spectral characteristics are:

- $3100 \, \text{cm}^{-1}$: CH stretching vibration when the C is associated to a C=C group. The two intense peaks between 690 and $790 \, \text{cm}^{-1}$ and the pattern between 1700 and $2000 \, \text{cm}^{-1}$ yield very good clues on the aromaticity of this compound.
- $1600–1490 \, \text{cm}^{-1}$: This region is not visualized fine in the spectrum due to the presence of a doublet. It contains three narrow peaks that, frequently, correspond to the in-the-plane stretching vibration of the C=C bonds of the aromatic ring. When the aromatic ring is (at least) monosubstituted, the energy of the three C=C bonds is no more degenerated and, so, they vibrate at slightly different wavenumbers.
- $2000–1700 \, \text{cm}^{-1}$: This is a typical region to be studied when the substitution of the aromatic groups is ascertained. The peaks here correspond to overtones and combinations of vibrations of the CH bending (either in the plane or out of the plane). Here, the pattern is a bit distorted because of the doublets. Nevertheless, it still clearly points toward a monosubstituted structure.
- $800–650 \, \text{cm}^{-1}$: We can see two intense, fairly symmetrical peaks around 690 and $790 \, \text{cm}^{-1}$. They correspond to out-of-the-plane bending of the aromatic CH bonds and suggest that there is a monosubstituted ring. The band *ca.* $690 \, \text{cm}^{-1}$ was distorted by the bending of the CH bond of the C≡C group (see next explanations).

We can review now the weight of the molecular 'fraction' that we have not still assigned: $102 - 77$ (total − phenyl) = 25. This clearly points toward a C≡CH group. We can fully confirm this by observing two sharp, intense peaks which are very relevant for this exercise:

- $2150–2100 \, \text{cm}^{-1}$: The sharp (not very intense) peak at this wavenumber indicates the stretching of a terminal C≡C bond.

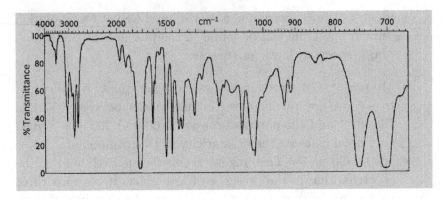

Figure 30.

- 3300 cm^{-1}: CH stretching, when the C is involved in a C≡C bond. The signal is not always so intense as in this example. Note that you will not see this band in the acetylene spectrum (example 8) because of its symmetry.
- The CH bending (when the C is involved in a C≡C group) was explained above as it cannot be seen clearly in this example. It overlaps with the CH bendings of the phenyl group.

Therefore, all the information we have indicates that the compound synthesized by the students was phenylacetylene (a colorless, viscous liquid): CH≡C–Ph.

19. A toxic compound released accidentally by a varnish industry has a C_8H_8O empirical equation. Its mid-IR spectrum was registered in a 0.05 cm pathlength liquids cell after its extraction from water. Propose a structure for this compound (Figure 30).

SOLUTION:

As for the previous examples, the aliphatic and aromatic characteristics are quite clear and we will not insist on them. If we assume that a phenyl group is in the molecule, two atoms of carbon have to be assigned further. So, let us focus on the substituent(s).

- Around 1700 cm^{-1}: The intense and sharp peak suggests that a C=O functional group is there (this peak will be attributed to the characteristic C=O stretching). The carboxylic group can be discarded because there is no characteristic broad, undefined band around $3300\,\text{cm}^{-1}$.

- Around $2850\,\text{cm}^{-1}$: There is a sharp peak which should correspond to an aldehyde group (CH stretching in aldehydes).

Therefore, the pollutant is phenylacetaldehyde: $COHCH_2$–Ph.

20. A substance whose relative molecular mass is 120 can be employed as a monitoring compound to evaluate environmental pollution. Its mid-IR spectrum was obtained in a liquid flow-through microcell (Figure 31). Can you assign a molecular structure to this compound?

SOLUTION:

Once more, we can see very clearly the existence of a monosubstituted aromatic ring (review previous exercises) and, so, we have to discuss about the substituent.

It is clear that many bands are 'splitted' (spectral doublets) and this is usually a good indication of the existence of either an isopropyl or a butyl group. We can subtract the mass of the phenyl group to the total amount ($120 - 77 = 43$) and, so, the butyl group can be discarded. Additional information which can

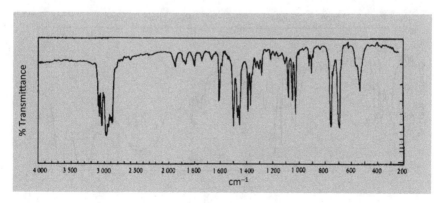

Figure 31.

be interpreted quite simply:

- 2950–2975 cm^{-1}: Asymmetric stretching of the CH bond in CH$_3$ groups and/or asymmetric stretching of the CH bond in CH$_2$ groups.
- 2850–2870 cm^{-1}: Symmetric stretching of the CH bond in CH$_3$ groups and/or symmetric stretching of the CH bond in CH$_2$ groups.
- 1450 cm^{-1}: Asymmetric bending of the CH bond in CH$_3$ groups and/or bending of the CH bond in CH$_2$ groups.
- 1375 cm^{-1}: Symmetric bending of the CH bond in CH$_3$ groups and/or symmetric bending of the CH bond in CH$_2$ groups.

Following, this is the spectrum of the pollutant known as cumene: Ph–CH(CH$_3$)$_2$.

21. The next mid-IR spectrum corresponds to a colorless liquid whose elemental composition was found to be C$_7$H$_9$N (Figure 32). Determine its structure.

SOLUTION:

The empirical equation suggests a phenyl group (the degree of insaturation is very high), whose spectral characteristics are quite clear. Ascertaining the functional group where the atom of

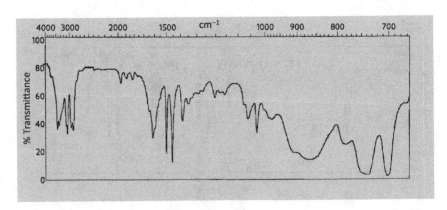

Figure 32.

N is located seems the key to determine the substituent. Indeed, this is a quite typical case study. A primary amine appears as a probable solution because of the two characteristic spectral bands centered around $3200\,\text{cm}^{-1}$, which correspond to the NH stretching. A secondary amine would yield only a band around $3300–3500\,\text{cm}^{-1}$.

The broad band centered *ca.* $850\,\text{cm}^{-1}$ is also very characteristic and it has to be attributed to the intermolecular hydrogen bonds the amines establish in an aqueous medium. All of them generate a broad unspecific vibration band. Primary amines also have a characteristic NH bending vibration at $1625\,\text{cm}^{-1}$ which, in this case, overlaps with the bands caused by the aromatic ring that can be visualized considering the next bands:

- $3100\,\text{cm}^{-1}$: CH stretching when the bond is linked to a C=C structure.
- $1600–1490\,\text{cm}^{-1}$: Three sharp peaks (one of them overlaps with a peak from the amine) due to the in-the-plane stretching of the C=C aromatic bonds.
- $2000–1700\,\text{cm}^{-1}$: Although the pattern is not totally clear (remember, this region corresponds to overtones of the bending modes), a monosubstitution seems to be present.
- $800–650\,\text{cm}^{-1}$: Although the spectrum appears a bit saturated in this region (for instance, because the solution of the compound was too concentrated, because the pathlength was too thick, or because of the hydrogen bonds mentioned above), two intense peaks are observed (out-of-the-plane CH bending of the aromatic CH bonds) and, so, the hypothesis of the monosubstituted ring is reinforced.

Additional bands are:

- $2950–2975\,\text{cm}^{-1}$: Asymmetric bending of the CH bond in CH_3 groups and/or asymmetric bending of the CH bond in CH_2 groups.
- $2850–2870\,\text{cm}^{-1}$: Symmetric bending of the CH bond in CH_3 groups and/or symmetric bending of the CH bond in CH_2 groups (hardly visible here).

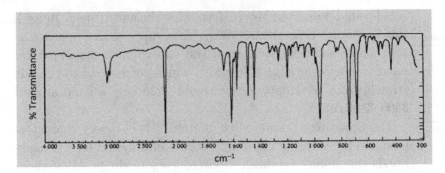

Figure 33.

- 1450 cm^{-1}: Asymmetric bending of the CH bond in CH$_3$ groups and/or CH bending in methylene groups. In this case, the band appears a bit displaced from its typical position, maybe because of the so large background (or baseline).
- 1375 cm^{-1} (hardly visible): Symmetric bending of the CH bond in methyl groups and/or CH bending (in methylene groups).

Following, all the information points toward benzylamine: NH$_2$CH$_2$–Ph.

22. A compound constituted only by C, H and N, relative molecular mass 129, contains a phenyl group (Figure 33). Propose its structure.

SOLUTION:

The indication about the phenyl group is very useful because it avoids us studying a series of characteristic bands carefully (see previous examples where the aromatic ring was discussed). We can also observe quite clearly that the ring should be monosubstituted (see the 690–750 cm^{-1} region).

A critical peak (for this example) is the narrow, intense one at 2200 cm^{-1}. Although the student might think of a triple C–C bond, the type of N would be, then, very difficult to ascertain. Therefore, let us think first about typical N structures ('typical' here should be restricted to the low number of possibilities we consider at this undergraduate level). That sharp peak and the absence of broad bands in the spectrum (which might otherwise suggest the presence of amines) unequivocally points toward the stretching of the C≡N bond (nitrile group).

Simple calculations with the information we already have on the 'assigned' molecular/atomic mass indicate that $129 - (77 + 12 + 14) = 26$. This suggests that a C=C bond might be present in the structure, as well. The peak at $990\,\mathrm{cm}^{-1}$ is, then, considered. This corresponds typically to the out-of-the-plane bending of the CH bond in 'trans' RCH=CHR' structures.

The unknown compound was, then, cinamonitrile: N≡CCH= CH–Ph.

EXERCISES PROPOSED TO THE STUDENT

Remember that the acronym Ph is used throughout to denote a phenyl group.

23. The spectrum corresponds to a valuable light gaseous hydro-carbon obtained in petrochemical refineries worldwide which, after passing through the refinery purification systems, is formed only by C and H. Identify the structure of the hydrocarbon (Figure 34). Additional information: routine quality control analyses suggest that the structure should contain eight atoms of hydrogen, at most.

Solution: $CH_3CH_2CH_3$

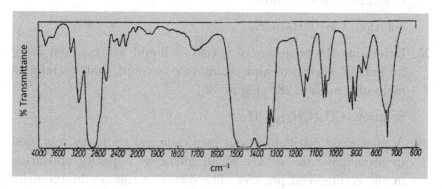

Figure 34.

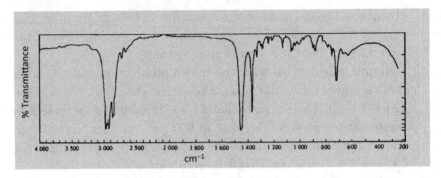

Figure 35.

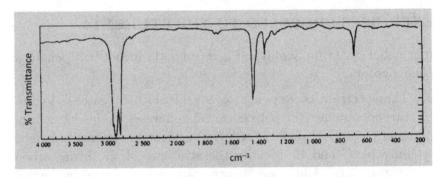

Figure 36.

24. The graphical record in Figure 35 presents the mid-IR spectrum of a laboratory solvent whose empirical equation is C_6H_{14}. Determine its structure.

   Solution: $CH_3(CH_2)_4CH_3$

25. Determine the structure of a viscous liquid hydrocarbon using its mid-IR spectrum and taking into account that its relative molecular mass is 282 (Figure 36).

   Solution: $CH_3(CH_2)_{18}CH_3$

26. A pilot-scale distillation unit in a chemical company was used to investigate the nature of an unknown impurity appearing in the commercial product. The distillate (likely, the impurity alone) was measured by mid-IR spectrometry using different

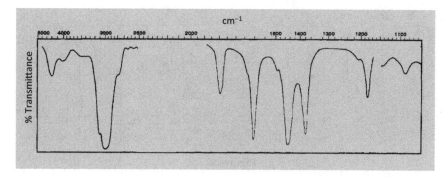

Figure 37.

pathlengths (this is why the spectrum looks like a sum of several parts, each registered with a pathlength). Propose a structure for the impurity (Figure 37).

Additional information: elemental analyses indicate that the empirical equation is C_6H_{10} and it does not correspond to a cycle.

Solution: $CH_2=C(CH_3)-C(CH_3)=CH_2$ (alternative possibilities are also acceptable, like: $CH_2=CHCH_2CH_2CH=CH_2$)

Hint: This is a difficult example. The empirical equation indicates a moderate degree of insaturation, probably two C=C bonds. The unresolved peak around $3000\,cm^{-1}$ indicates the presence of C–H groups linked to the C=C groups. It is not simple at all to decide between the RHC=CHR′ (cis or trans) and RR′C=CH$_2$ structures. Likely, the best clue is found when evaluating the position of the unresolved peak. The former structure tends to yield a band around $3050\,cm^{-1}$, whereas the RR′C=CH$_2$ one locates the CH stretching peak toward $3100\,cm^{-1}$, which is the case here.

However, the visualization is complicated, maybe because they had to use several pathlengths to avoid excessive absorbance (too low transmittance).

27. A mid-IR measurement was made on a yellow liquid obtained from the distillation of an extract of flowers. Deduce the molecular structure of the distilled compound (Figure 38). Its

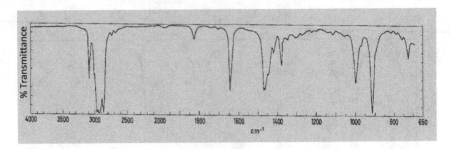

Figure 38.

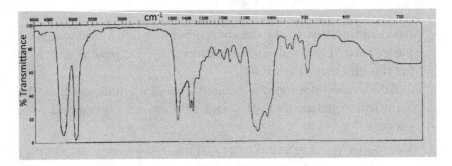

Figure 39.

composition contains only C and H and its relative molecular mass was calculated to be 112.

Solution: $CH_2{=}CH(CH_2)_5CH_3$

28. The next mid-IR spectrum corresponds to a substance that is liquid at room temperature (*ca.* 20 °C). Its composition is $C_6H_{14}O$. Propose the most probable structure for this compound (Figure 39).

Solution: $CH_3{-}CH(CH_3){-}(CH_2)_2CH_2OH$

Hint: The most clear suggestion about the existence of either an isopropyl or a *t*-butyl group is the doublet of the symmetrical CH stretching at 1350 cm^{-1}. The structural bending band of the isopropyl group can be seen here at somewhat low wavenumbers, *ca.* 1100 cm^{-1}. A good indication on the lack of a long linear chain is the absence of a clear peak around 750 cm^{-1}.

29. The substance considered in this example is very common in the cosmetics industry. Its empirical equation is $C_4H_{10}O$. Propose the structure of this compound (Figure 40).

Solution: $CH_3(CH_2)_2CH_2OH$

Hint: It is instructive to compare this spectrum with that in exercise 4. There, the C–O stretching band at $1100\,cm^{-1}$ was very clear (and that suggested a secondary alcohol).

30. A pure compound was said to contain only C and H and yielded the mid-IR spectrum shown in Figures 41a and 41b in a 0.01 cm pathlength cell. As an additional information, the spectroscopist was informed that the relative molecular mass of the compound was 110. Propose its molecular structure.

Solution: $CH{\equiv}C(CH_2)_5CH_3$

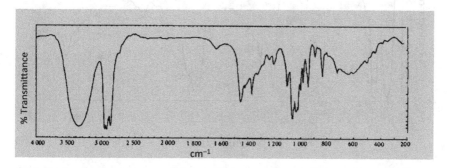

Figure 40.

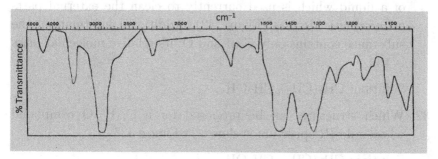

Figure 41(a)

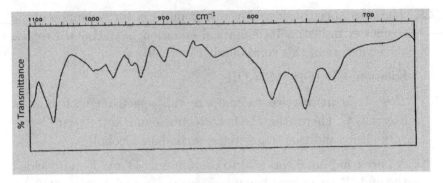

Figure 41(b)

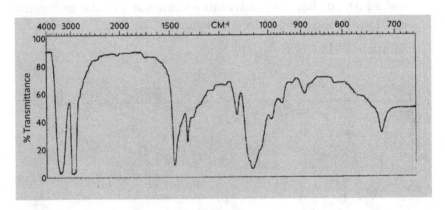

Figure 42.

31. Using the mid-IR spectrum given above (0.03 mm pathlength) and the additional information, propose the molecular structure of a liquid which is used currently to clean the external parts of informatic devices (Figure 42). Additional information: the substance contains only C, H and O; its relative molecular mass is 130.

    Solution: $CH_3(CH_2)_6CH_2OH$

32. Which structure can be proposed for a $C_{10}H_{22}O$ compound whose mid-IR spectrum is shown in Figure 43?

    Solution: $CH_3(CH_2)_8CH_2OH$

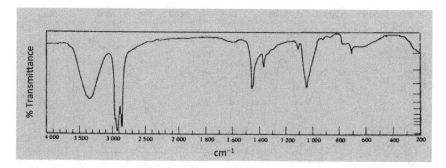

Figure 43.

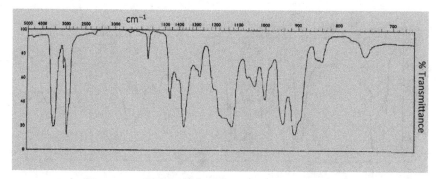

Figure 44.

33. A non-cyclic compound, whose relative molecular mass is 114, yields the mid-IR spectrum in Figure 44. Considering that it is not a primary alcohol, propose its structure.

Solution: $CH_2=CH(CH_2)_3-CHOH-CH_3$

Hint: The spectrum is a bit difficult to visualize but two relevant characteristics can be of help. First, observe the C=C stretching ($1600\,cm^{-1}$) and other characteristics associated to this functional group and, then, the C–O stretching band around $1150\,cm^{-1}$ which suggests that it should be a secondary alcohol (recall the enunciate of the exercise).

34. A compound used for cosmetics is extracted from animal grease. It is solid at room temperature, its empirical equation is $C_6H_{10}O_4$

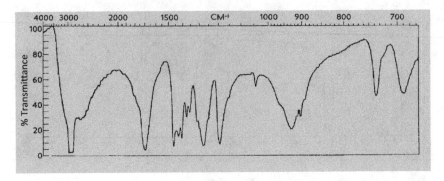

Figure 45.

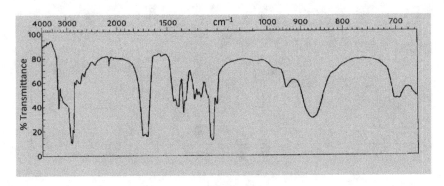

Figure 46.

and is measured by mid-IR spectrometry after mixing it with nujol. Determine its structure (Figure 45).

Solution: $COOH(CH_2)_4COOH$

35. A researcher is trying to synthesize a new compound. To confirm the final result, the product is analyzed by mid-IR spectrometry using a slurry prepared with nujol (Figure 46). To which of the following structures does the spectrum correspond to?

(a) $CH_3C \equiv CCOOH$ (b) $CH \equiv CCH_2COOH$
(c) $CH_2 = CHCH_2COOH$ (d) $CH_3COCHCH_2$.

Solution: Alternative b

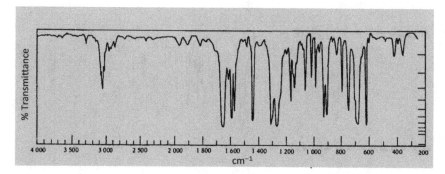

Figure 47.

36. A $C_{13}H_{10}O$ compound yields the mid-IR spectrum measured in liquid phase that is depicted in Figure 47. Propose its molecular structure.

Solution: Ph–CO–Ph

Hint: The Ph–C=O structure can be seen quite clearly. However, the other substituent is not easy to determine. There are not special bands that may be attributed to additional functional groups. The high degree of insaturation, the subtraction of the number of atoms of the Ph–C=O structure, as well as the empirical equation, lead rather straightforwardly to the final decision.

37. Propose the most probable structure for the $C_6H_{10}O$ molecule whose IR spectrum is in Figure 48.

Solution: Cyclohexanone

Hint: As it was mentioned in the resolved exercises, it is far from trivial to determine cyclic compounds from their spectra. In this example, the empirical equation might suggest an alkane but we cannot see the typical $750 \, \text{cm}^{-1}$ peak of the linear chains. On the contrary, we can see a lot of spectral doublets and this is the best signal pointing in the direction of a cyclic structure.

38. Assuming that a given $C_6H_{14}O$ compound can be obtained free from concomitants (or at least, that their concentrations are so

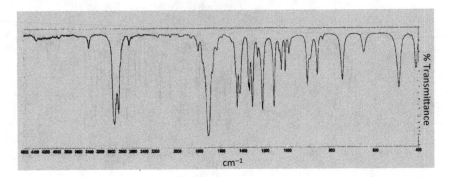

Figure 48.

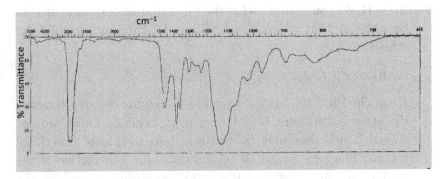

Figure 49.

low that cannot be visualized by mid-IR spectrometry), propose a structure for the spectrum in Figure 49.

Solution: $CH_3(CH_2)_3-O-CH_2CH_3$

Hint: This spectrum is compatible with other options that can be proposed for the ether substituents.

39. Ascertain the structure for a molecule containing only C, H and O, with a relative molecular mass of 100 and whose spectrum is shown in Figure 50.

Solution: $CH_2=CH-O-(CH_2)_3CH_3$

Hint: This spectrum is compatible with other options that can be proposed for the substituents of the ether. The C=C group should always be considered at a terminal position.

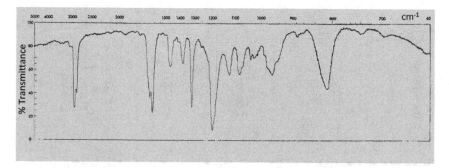

Figure 50.

40. Associate (and justify) each spectrum with its corresponding structure (Figures 51a–51c).

(1) $CH_3(CH_2)_{11}CH_2COH$ (2) $CH_3COO–(CH_2)_3CH_3$
(3) $(CH_3)_2CHCH_2NH_2$.

Solution: $1 \to C$; $2 \to A$; $3 \to B$

41. A molecule constituted by C, H and O, relative molecular mass 106, leads to the mid-IR spectrum depicted in Figure 52. Determine its structure and explain which band you would prefer (if not interfered) to monitor its amount in a given commercial drug.

Solution: Ph–COH; monitored using the typical C=O stretching band

42. A solvent is used typically in garages to clean drops of car painting. A drop of the solvent between two KBr windows yielded the spectrum in Figure 53. Identify the structure of the solvent considering that its relative molecular mass is 92.

Solution: Ph–CH$_3$

43. A hydrocarbon found in seawater after a shipwreck contains only C and H. Its relative molecular mass was estimated as 78, using elemental chemical analysis. Which molecular structure does it possess (Figure 54)?

Solution: Benzene

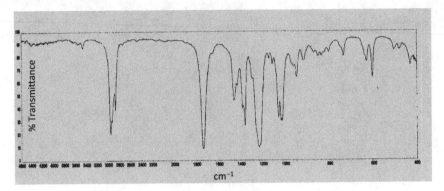

Figure 51(a)

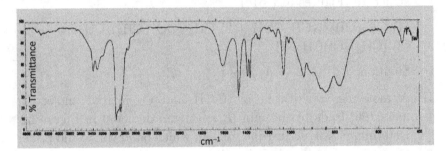

Figure 51(b)

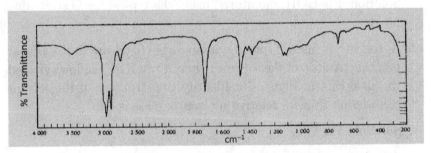

Figure 51(c)

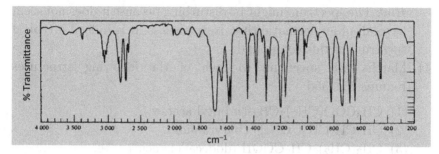

Figure 52.

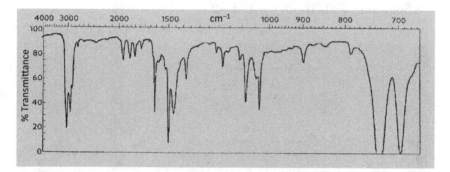

Figure 53.

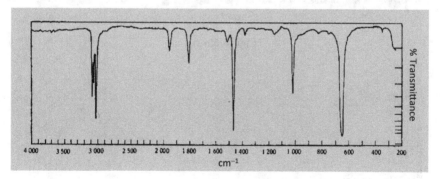

Figure 54.

Hint: The spectrum might be disturbing because it does not seem 'typical' of an aromatic ring. Nevertheless, it is indeed a typical spectrum of this compound.

44. Match each spectrum to each of the following structures Figures 55a–55d:

    (1) $CH_3COO(CH_2)_3CH_3$ (*n*-butyl acetate),

    (2) $CH_3CH_2CH_2CONH_2$ (butyramide),

    (3) $CH_3(CH_2)_9CH_2COOH$ (dodecanoic acid),

    (4) CH_3CH_2COONa (sodium propionate).

Solution: a→1, b→3, c→4, d→2

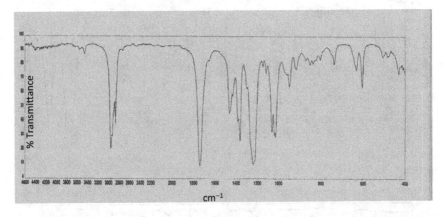

Figure 55(a)

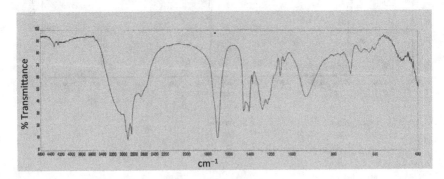

Figure 55(b)

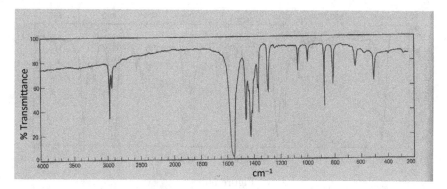

Figure 55(c)

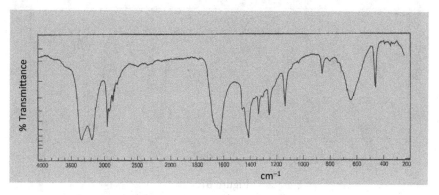

Figure 55(d)

45. The IR spectrum of this exercise pertains to an impurity obtained when producing benzene industrially. If you are told that its relative molecular mass is 120, propose a structure for the impurity (Figure 56).

Solution: 1,2,4-trimethylbenzene

Hint: It is not easy to determine whether the ring is di-(1-methyl, 2-ethyl benzene) or tri-substituted. Both possibilities are compatible with the spectrum. However, the tri-substitution is the correct answer.

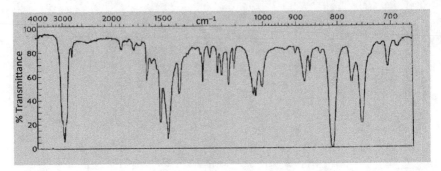

Figure 56.

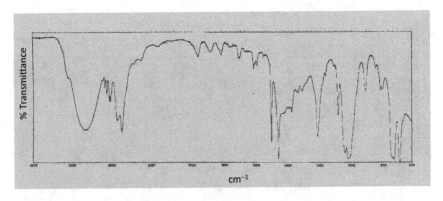

Figure 57.

46. An odorous compound present in various fruits has the mid-IR spectrum shown in Figure 57. Its relative molecular mass is 108 and it is formed only by C, H and O. Reason what structure it possesses.

Solution: Ph–CH$_2$OH

47. The spectrum corresponds to an essential oil that can be extracted from many fragrant plants. It is formed only by C, H and O and its relative molecular mass is 122 (Figure 58). Ascertain its structure.

Solution: Ph–CH$_2$ CH$_2$OH

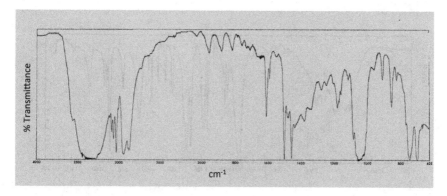

Figure 58.

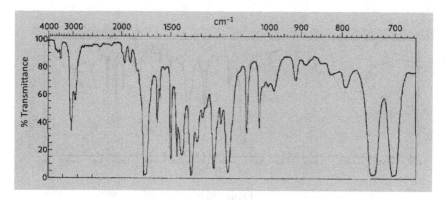

Figure 59.

48. Deduce the structure of a liquid compound (at room temperature) whose relative molecular mass is 134, formed by C, H and O, and whose IR spectrum is shown in Figure 59.

Solution: Ph–CH$_2$–CO–CH$_3$

Hint: The aromatic monosubstitution appears well defined (700–750 cm^{-1}). Ester groups can be discarded because different possibilities do not match the molecular mass. A possibility is Ph–CO–CH$_2$CH$_3$ but this transforms fast and spontaneously to the correct solution.

49. Determine the structure of the C$_7$H$_6$O$_2$ molecule by considering its mid-IR spectrum (Figure 60).

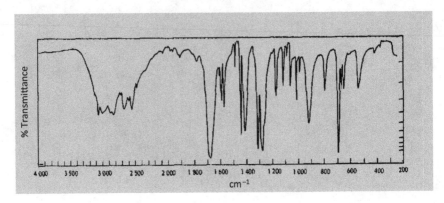

Figure 60.

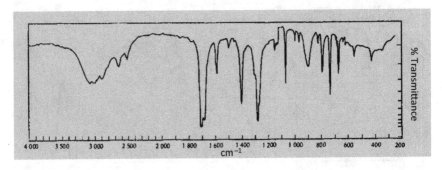

Figure 61.

Solution: Ph–COOH

50. Use the mid-IR spectrum of a liquid $C_8H_6O_4$ substance to propose its most probable structure (Figure 61).

Solution: Two possibilities can be given: 1,2- or 1,3-disubstituted benzene (COOH substituents).

51. A compound formed by C, H and N leads to the following mid-IR spectrum (Figure 62) in a continuous liquid flow-through cell. Propose its structure.

Solution: Ph–CN

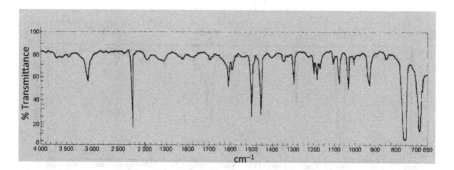

Figure 62.

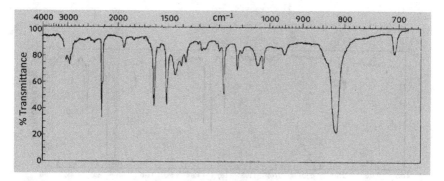

Figure 63.

52. The spectrum in Figure 63 corresponds to a low-melting-point liquid, relative molecular mass 117, constituted by C, H and N. What structure would you suggest for it?

Solution: CH_3–Ph–CN (1,4 disubstitution)

53. An industrial process to obtain phenylacetonitrile is monitored routinely by mid-IR spectrometry. During the start-up process, the product is not obtained pure and, so, a spectrum is measured each 15 min until the pure compound is obtained. Which of the three spectra recorded below corresponds to phenylacetonitrile (Figures 64a–64c)?

Solution: Spectrum C

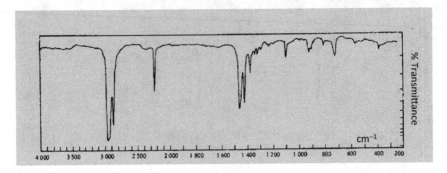

Figure 64(a)

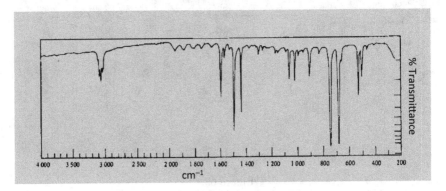

Figure 64(b)

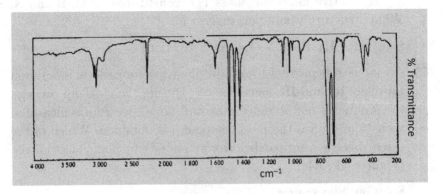

Figure 64(c)

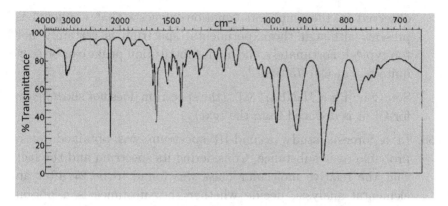

Figure 65.

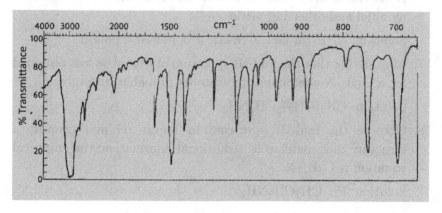

Figure 66.

54. A substance whose composition is C_9H_7N presents the appended mid-IR spectrum. Deduce its molecular structure (Figure 65).

   Solution: $CN-Ph-CH=CH_2$ (1,4-disubstitution)

55. Identify the structure of the $C_7H_{10}NCl$ compound whose mid-IR spectrum is shown in Figure 66.

   Hint: A tertiary amine is the first, intuitive choice because the text states that we have a salt. The high degree of insaturation observed in the empirical equation suggests that a phenyl is present, although the experimental spectrum is certainly not too typical. Fortunately, the monosubstitution peaks can be seen quite fine at 690–750 cm^{-1}.

Solution: $[Ph–CH_2NH_3]^+$ Cl^- (the spectrum does not show peaks for Cl, it is deduced from the text)

56. In a forensic study, a mid-IR spectrum was obtained for a probable toxic substance. Considering its spectrum and the fact that the relative molecular mass was found to be 27 after an elemental analysis, decide whether the substance is a poison (Figure 67).

   Note: the asterisks correspond to bands which may be affected by nujol and other impurities.

   Solution: CNH (in fact, CNNa), it is indeed a poison

57. Considering that the mid-IR spectrum of Figure 68 was obtained for a $C_4H_{11}N$ compound, indicate its probable structure.

   Solution: $CH_3(CH_2)_2CH_2NH_2$

58. Observe the mid-IR spectrum in Figure 69 and propose a structure that matches it. Additional information: the empirical equation is $C_8H_{11}N$.

   Solution: $Ph–CH_2CH_2NH_2$

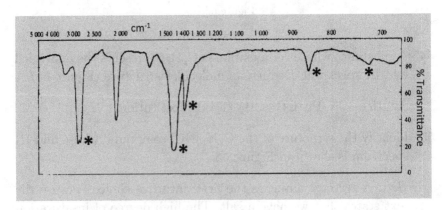

Figure 67.

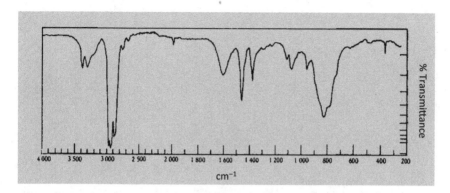

Figure 68.

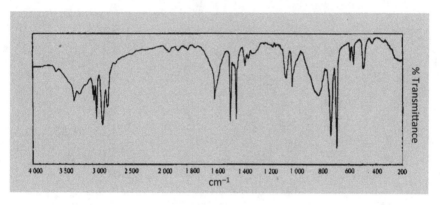

Figure 69.

CHAPTER 5

ATOMIC SPECTROMETRY

Rosa María Soto-Ferreiro and Alatzne Carlosena-Zubieta

OBJECTIVES AND SCOPE

This chapter deals with typical calculations performed in atomic spectrometry. First, a conceptual review is presented to classify the different techniques, discuss their advantages and disadvantages, as well as some of their typical applications. The different quantitation methods will be reviewed (calibration, internal standard, and standard addition method) and examples will be deployed to decide whether a sample matrix interferes the measurements. Efforts will be devoted to get the student used to calculating analyte concentrations after multistage sample treatments and deciding whether a result is reliable enough (this issue is strongly linked to Chapter 2).

1. INTRODUCTION

Atomic spectrometry is a wide concept that includes the techniques that are based on two main principles: (i) atoms are obtained in their fundamental electronic state, from the sample, and they interact with a radiation beam, yielding the so-called optical atomic spectroscopy; and (ii) atomic ions are produced that are subsequently detected by mass spectrometry, leading to atomic mass spectrometry. This gross classification is based not only on the different fundamentals of the techniques but also on the instrumentation they use.

Optical atomic spectrometry includes the following techniques:

— atomic absorption spectrometry (AAS),
— atomic emission spectrometry (AES),
— atomic fluorescence spectrometry (AFS).

With regards to atomic mass spectrometry, several techniques arise according to the different sources of ions utilized, the most popular one being the inductively coupled plasma mass spectrometry (ICP-MS).

A brief, conceptual description of the main atomic spectrometry techniques will be made in this chapter, including their fundamentals, their instrumentation and some of their most typical applications. For further information, readers are kindly forwarded to any of the textbooks extensively describing these instrumental techniques of analysis [1–5].

2. OPTICAL ATOMIC SPECTROMETRY

All instruments used in optical atomic spectrometry include an atomizer to create an atomic vapor from the sample. Here, atomic means that the atoms are in their fundamental electronic state, hence the name of the technique. Following, these atoms can either be excited and subsequently emit radiation or they can absorb radiation from a source. In the former case, the emission of radiation is recorded (emission-based techniques), whereas in the latter, the attenuation of a radiation beam is studied (absorption-based techniques). Modern instruments incorporate different atomizers and offer the possibility of exchanging them automatically. AFS will not be included in this brief introduction due to its little development.

2.1. Atomic absorption spectrometry (AAS)

AAS is based on free atoms absorbing visible or UV radiation at specific wavelengths. The ratio between the incident and the exiting radiation over a very narrow wavelength range is measured and the absorbance values (see Chapter 3 for its definition) so obtained are related to the concentration of the atoms in the sample in a calibration function (see Chapter 2).

2.1.1. Instrumentation

An atomic absorption spectrometer basically consists of the following components: a radiation source, which provides the external radiation beam; an atomizer, in which gaseous atoms are produced; a monochromator, where an appropriate wavelength is selected; and the detector, where the radiation is transformed into an electrical measurable signal. Figure 1 shows a diagram of such an instrument.

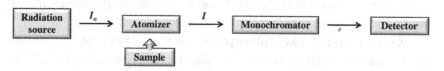

Figure 1. Schematic diagram of an atomic absorption spectrometer.

Source

The radiation sources used in AAS are usually element-specific. That is to say, they emit the characteristic spectrum of the element to be determined. Two types are used commonly, the hollow cathode lamps (HDLs) and the electrodeless discharge lamps (EDLs). Continuum sources are steadily being introduced nowadays.

Hollow cathode lamps can be designed for single-element or multielement determination. Single-element HDLs have a higher energy output and supply adequate radiation intensity for most of the metals determined commonly by AAS. Multielement HDLs provide less energy but they are useful in routine operation since changing the lamp after each elemental determination is avoided.

Electrodeless discharge lamps are utilized for the determination of volatile elements such as mercury, selenium, arsenic, tin, etc., for which the HDLs give low energy outputs and short life times.

Atomizer

Main atomizers in AAS are the flame and the graphite furnace devices. In addition, a closed quartz tube can be used when a vapor is generated from the sample.

Flame atomic absorption spectrometry (FAAS) is used for liquid solutions. The sample is introduced through a nebulizer and converted into an aerosol which is injected into the atomizer by the gases that give rise to the flame. The gaseous atoms of the target element absorb the characteristic radiation emitted by the source and this is registered finally. The sensitivity of this technique is limited by the very low efficiency of the nebulization process, the dilution of the analyte in the gases and the short residence time of the fundamental atoms in the flame pathlength.

Electrothermal atomic absorption spectrometry (ETAAS) is used for liquids although it is also suitable for slurries and solids. The atomizer consists of a graphite cylinder whose extremes are open, and with a central hole. It is approximately 1–3 cm in length and 3–8 mm in diameter. A small quantity of sample is introduced into the tube with an autosampler and is completely atomized. A significant improvement in sensitivity, versus flame atomization, is achieved due to the efficient introduction of the sample, the absence of a dilution effect and the higher residence time of the atoms in the atomizer. Atomization in ETAAS is performed following a temperature program, established by the operator, including four fundamental steps: drying, pyrolysis, atomization and clean up. The objective of this program, is to achieve the atomization of the analyte in the absence of matrix interferences.

Vapor generation atomic absorption spectrometry (VGAAS) is used for elements that form volatile species after a chemical reaction as, for example, the cold vapor of mercury and the hydrides of selenium, arsenic, tin, lead, etc. This technique, using either a quartz tube or a graphite furnace as atomizer, provides the highest sensitivity when determining these elements by AAS.

Monochromator

The absorbance in AAS is measured at a characteristic wavelength which is separated at the monochromator from other emission lines and background emission. Diffraction gratings are most commonly used to discriminate line sources. Double-beam instruments are frequently used with flame atomizers, to take into account any variations in the intensity of the source and sensitivity of the detector, since the absorbance signal is registered continuously. However, single-beam spectrometers are employed with graphite furnace as the baseline is adjusted before each run.

Detector

A photomultiplier tube is the detector of choice in most AAS instruments since it is more sensitive than other photon detectors available. Besides, it has a large wavelength coverage, large dynamic range, high amplification gain and low noise.

2.1.2. Interferences

Interferences that may be encountered in AAS can be classified as physical, chemical and spectral. The impact of each type of interference on the measurements depends on the atomization process. They will be described briefly below.

Physical interferences are caused by the physical properties of the solutions analyzed, affecting the sample introduction step. When flame atomization is utilized, the nebulization efficiency can be affected by density, viscosity, amount of dissolved salts, etc. The autosampler used in graphite furnace atomization can also suffer their effects when pipetting the sample. Vapor generation techniques are not significantly affected by this type of interferences. Matching the physical properties of the solutions analyzed, standards and test samples, will compensate for these effects.

The two most common chemical interferences in FAAS are the formation of non-volatile compounds containing the analyte and the ionization of the analyte. The former can be minimized by changing the fuel-to-oxidant ratio or by increasing the temperature of the flame (using different fuels and oxidants). Another approach is to add a releasing agent (which reacts with the interferent) or a protecting agent (which reacts with the analyte) to the samples. The ionization of the target element can be reduced adding a high concentration of an ionization suppressor (species that ionizes more easily than the analyte) to the sample. Thus, the increased concentration of electrons in the flame prevents the ionization of the analyte.

Chemical interferences in ETAAS are due to the formation of analyte compounds during the atomization process having different volatility than the analyte itself, modifying its atomization temperature. This kind of interferences can be reduced adding an adequate chemical modifier to the sample before measuring it, which can react with the analyte or with the matrix components.

Spectral interferences may arise when an absorption line or band of an interferent overlaps with an absorption line of the analyte. Overlapping of atomic absorbance lines is seldomly a problem because they are very narrow. However, broad absorption bands from molecules or the scattering of source radiation can bring important spectral interferences, which is known as background absorption. This phenomenon is not very serious in flame atomization, but may be of relevance in ETAAS. Therefore, graphite furnace instruments currently include the so-called *background correctors*. These devices perform generally two measurements: the total absorbance and the background absorption, and then the electronics of the spectrometer provides the corrected absorbance of the analyte. The continuum source and the Zeeman background correctors are the most common systems used to correct for the background in ETAAS.

2.1.3. Applications

Atomic absorption techniques have been traditionally applied to single-element determination of metals and metalloids. FAAS applications are limited by their sensitivity to the analysis of samples with moderate concentrations of target elements; i.e., mg/L. In return, results are obtained very fast and instrumental handling is very simple. Limits of detection (LODs, calculated in classical terms, as explained in Chapter 2) are in the 1–$20\,\mu g/L$ range. VGAAS is the technique of choice when elements that form volatile species are determined, achieving LOD values between 0.01 and $0.1\,\mu g/L$; meanwhile, ETAAS provides adequate sensitivity for trace analysis for the remaining elements (classical LOD: 0.02–$0.5\,\mu g/L$), but the sample throughput is smaller than that with FAAS and the operation is a bit more complicated.

2.2. Atomic emission spectrometry (AES)

AES is based on the emission of ultraviolet and/or visible radiation following the thermal or electrical excitation of atoms. The atomizer in AES acts also as an excitation source. Thus, the atoms are excited by the energy provided by the source and after that, they emit their characteristic spectra. The intensity of the radiation emitted can be related to the concentration of the atoms in the sample.

2.2.1. Instrumentation

An atomic emission spectrometer basically consists of the following components: an excitation source, where the sample is atomized and the gaseous atoms are subsequently excited; a monochromator, where the appropriate wavelength is selected; and a detector, where the radiation is transformed into an electrical measurable signal. Figure 2 shows a schematic diagram of an instrument to perform atomic emission measurements.

Excitation sources

Among the excitation sources that can be utilized, arcs and sparks have limited applications to the qualitative or semiquantitative

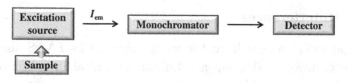

Figure 2.   Schematic diagram of an atomic emission spectrometer.

analysis of solid samples. A flame (like that used for FAAS) yields adequate sensitivity only for the alkaline elements. Finally, a plasma is the excitation source most widely used, in particular, the inductively coupled plasma (ICP).

A plasma is a hot, highly ionized gas, and therefore very energetic, that contains an abundant concentration of cations and electrons. The plasmas used in atomic emission are based on argon. The ICP is produced in a device known as torch, which consists of three concentric quartz tubes through which argon flows. The sample is introduced into the instrument using a nebulizer and carried out to the plasma by an argon flow. Because plasmas operate at much higher temperatures than flames, they provide better atomization and a higher population of excited states.

Monochromator

A monochromator or a polychromator is used to separate the polychromatic radiation exiting the plasma into individual wavelengths, so that the characteristic emission lines from each element can be identified and quantitated. Both devices allow multielement analysis but a monochromator performs a sequential measurement of each wavelength, while a polychromator monitors all of them simultaneously.

Detector

A photomultiplier tube is a characteristic detector and is commonly included in sequential instruments. However, for simultaneous spectrometers (which include polychromators), multichannel detectors must be utilized, such as diode arrays and charge transfer devices, which are able to detect several analytical wavelengths simultaneously.

2.2.2. Interferences

Physical interferences have the same effect as in FAAS since the sample is introduced using a nebulizer. Chemical interferences are reduced due to the high temperature and the inert atmosphere of the plasma compared with the flame. Ionization of the analyte is inhibited by the high concentration of electrons provided by the ionization of the argon.

Severe spectral interferences are relatively rare in inductively coupled plasma atomic emission spectrometry (ICP-AES) since each element has many specific wavelengths that may be used to circumvent any interference. Therefore, the choice of the adequate wavelength will be made, not only in terms of sensitivity but also taking into account how to avoid interferences from the emission wavelengths of other constituents in the sample. On the other hand, some molecular interferences can occur from OH, SO and PO, leading to background signals at some regions of the electromagnetic spectrum.

2.2.3. Applications

As it was mentioned above, multielemental analysis can be performed with ICP-AES, which is the main advantage over traditional atomic absorption techniques. Moreover, most elements of the periodic table can be determined by this technique, with large linear calibration ranges, up to four or five orders of magnitude. The classical LOD values obtained with this technique are intermediate between FAAS and ETAAS, ranging from 0.01 to 5 μg/L.

3. INDUCTIVELY COUPLED PLASMA MASS SPECTROMETRY (ICP-MS)

ICP-MS is an analytical technique used for elemental determination which combines a source capable of producing atomic ions from the sample, the ICP, with a mass spectrometer where the ions are separated according to their mass-to-charge ratio (the m/z ratio) and the number of ions of each class detected.

3.1. Instrumentation

Figure 3 shows an ICP-MS instrumental diagram, which consists mainly of an ion source, the ICP, and a mass spectrometer.

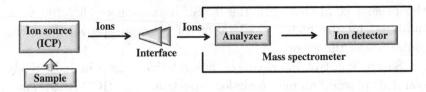

Figure 3. Schematic diagram of an ICP-MS.

3.1.1. Ion source

The ICP and the nebulizer used for sample introduction are similar to those utilized in ICP-AES. The atoms of the elements in the sample are converted into ions and, then, brought into the mass spectrometer via an interface constituted by metal cones. Such an interface transfers the ions from the argon stream at an atmospheric pressure (1–2 torr) into the low pressure region of the mass spectrometer ($<1 \cdot 10^{-5}$ torr), while the temperature decreases as well.

3.1.2. Mass spectrometer

The mass spectrometer consists of an analyzer where the ions are separated by their m/z ratio, and a detector, where the amount of each type of ions is detected. The mass spectrometer most commonly used employs a quadrupole mass filter, which consists of four parallel rods. A combination of direct- and alternate-current voltages is applied to each pair of opposed rods resulting in an electrostatic filter that allows ions of only a given m/z ratio to pass through the channel in the middle of the rods. Changing the voltages applied to the quadrupole allows for a sequential scanning of the mass range (more details on mass spectrometers are given in Chapter 7).

Some commercially available instruments include a magnetic sector analyzer, also called sector field mass spectrometer, which provides high resolution to separate analyte peaks from polyatomic overlaps. Improvements in detection limits and measurement accuracy and precision are also achieved with these systems. Nevertheless, this is a very expensive instrumentation, complex to operate and maintain.

An electron multiplier detector is used in almost all commercial ICP-MS instruments. It provides high sensitivity because the ions colliding with the active surface of the detector release a large number of electrons; these are attracted by the next surface of the detector, releasing further electrons and so forth, amplifying the signal. Detector electronics accounts for and stores the total signal corresponding to each mass, creating a mass spectrum, which provides information about the elemental composition of the sample and allows calculating concentrations from the magnitude of each peak.

3.2. Interferences

Physical interferences associated with the sample introduction step can be compensated for by matching the physical properties of all the solutions analyzed, as it was mentioned above for the techniques using a nebulizer.

Main interferences in ICP-MS are spectral interferences due to the overlap of ions that have the same m/z ratio than the isotope of interest of the analyte. There are two types of spectral interferences:

a. *Isobaric interferences*: There is a direct overlap between some isotope from an element other than the analyte and the isotope of interest at the same nominal m/z ratio, e.g., $^{54}Fe^+$ overlaps with $^{54}Cr^+$.

b. *Polyatomic interferences*: Due to polyatomic ions with the same m/z ratio as the nominal one of the analyte, these ions can proceed from species of the plasma (mainly Ar), some components of the sample and/or the reagents employed during sample pretreatment; e.g., $^{40}Ar^{35}Cl^+$ overlaps with $^{75}As^+$.

Isobaric overlaps cannot be resolved using spectrometric methods, not even with high resolution spectrometers. In order to avoid them, either an alternative isotope of the analyte can be measured or mathematical corrections may be applied. The effect of polyatomic ions can be managed using high-resolution analyzers that are very expensive, or including a collision/reaction cell in quadrupole instruments. These latter devices are located before the quadrupole and are pressurized with a gas that reacts or collides with polyatomic ions.

3.3. Applications

Most of the elements of the periodic table can be determined by ICP-MS and multielemental analysis can also be performed. The main advantages of ICP-MS over AES are the much simpler (mass) spectra (containing less peaks and lower background) and the satisfactory LOD values which can be achieved for a large number of elements, of the order of ng/L.

REFERENCES

[1] Skoog, D. A.; Holler, F. J.; Crouch, S. R.; (2007). *Principles of Instrumental Analysis*. 6th ed. Thomson Brooks/Cole, Belmont (California).

[2] Robinson, J. W.; Frame, E. M. S.; Frame II, G. M. (2014). *Undergraduate Instrumental Analysis*. 7th ed. CRC Press, Boca Raton (Florida).

[3] Thomas, R. (2013). *Practical Guide to ICP-MS. A Tutorial for Beginners*. 3rd ed. CRC Press, Boca Raton (Florida).

[4] Cullen, M. (ed.) (2004). *Atomic Spectroscopy in Elemental Analysis*. Blackwell Publishing, CRC Press, Oxford.

[5] Sanz-Medel, A.; Pereiro, R.; Costa-Fernández, J. M. (2013). An overview of atomic spectrometric techniques, in, J. M. Andrade-Garda (ed.), Basic Chemometric Techniques in Atomic spectroscopy, RSC, Cambridge.

WORKED EXERCISES

1. The determination of magnesium in sewage water samples can be performed by FAAS. For this, one sample was filtered and diluted six-fold. A 10 mL aliquot of this diluted sample was used to measure its absorbance. Several Mg calibration standard solutions were prepared in order to obtain the calibration line. Their absorbances are presented in the following table. Calculate the concentration of Mg in the sewage water sample, expressed as mg/L and mass/volume percent, considering that its signal was 0.241.

C_{Mg} (mg/L)	0.00	0.20	0.40	0.60	0.80	1.00	1.20
Absorbance	0.000	0.062	0.115	0.168	0.212	0.290	0.343

SOLUTION:

The calibration plot (see Figure 4a) shows a good straight-line relationship but the residuals reveal that the calibration point corresponding to 0.80 mg/L of Mg behaves as an outlier (Figure 4b).

That standard is therefore discarded and the calibration straight line calculated again:

$$A = (0.0011 \pm 0.00234) + (0.2857 \pm 0.00331) \cdot C_{Mg}$$

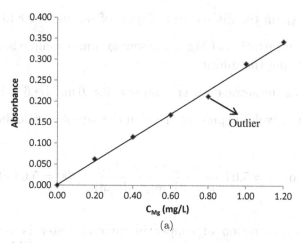

(a)

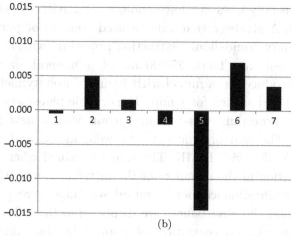

(b)

Figure 4.

Note that the intercept of the calibration equation provides information about the measurement step and should not be forced to zero. Note also that popular spreadsheets offer values for the coefficient of determination, which is the squared correlation coefficient.

Substituting the absorbance of the unknown in the equation yields the concentration of Mg in the test aliquot of the diluted sample:

C_{Mg} in the diluted test aliquot of sample $= 0.840 \, \text{mg/L}$

The concentration of Mg in the sewage water sample is calculated considering the dilution step:

C_{Mg} in the sewage water sample $= 0.840 \, \text{mg/L} \cdot 6 = 5.04 \, \text{mg/L}$

The mass/volume percent of Mg in the sample is calculated easily as:

$$\%(\text{m/v}) = 5.04 \, \frac{\text{mg}}{\text{L}} \cdot \frac{1 \, \text{g}}{10^3 \, \text{mg}} \cdot \frac{1 \, \text{L}}{10^3 \, \text{mL}} \cdot 100 = 5.04 \cdot 10^{-4} \, \%$$

2. The concentration of copper in mineral water is very low so FAAS usually does not have enough sensitivity to determine it directly. A strategy that can be used consists of performing a preliminary liquid–liquid extraction procedure with an adequate organic solvent. Thus, 250.00 mL of a mineral water sample were introduced in a funnel with 10 mL of isobutylmethylketone (IBMK) and shaken for 1 min. The organic phase was separated and the procedure was repeated again with a new portion of IBMK. The two organic extracts were mixed and diluted to 25.00 mL also with IBMK. The signal obtained when aspirating this solution in the nebulizer of the instrument was 0.223.

   The calibration copper standard solutions were prepared in IBMK and the absorbances are displayed in the following table. Calculate the concentration of copper in the mineral water sample.

C_{Cu} (mg/L)	0.00	1.00	2.00	3.00	4.00	5.00	6.00
Absorbance	0.007	0.075	0.140	0.200	0.267	0.332	0.400

SOLUTION:

A plot of the absorbance values versus the concentration of Cu shows a good straight-line relationship (see Figure 5a). The residuals plot reveals that outliers are not present (Figure 5b).

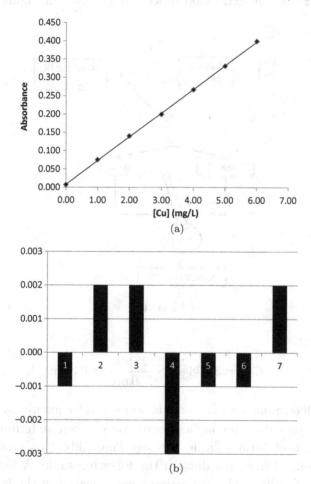

(a)

(b)

Figure 5.

The calibration least-squares fit obtained is:

$$A = (0.0080 \pm 0.00149) + (0.0650 \pm 0.00041) \cdot C_{\text{cu}}$$

The concentration of Cu in the IBMK extract is obtained by substituting the absorbance of the test aliquot in the least-squares fit equation:

$$C_{\text{Cu}} \text{ in the IBMK extract} = 3.308 \,\text{mg/L}$$

The concentration of Cu in the mineral water sample is calculated considering the extraction procedure depicted in Figure 6:

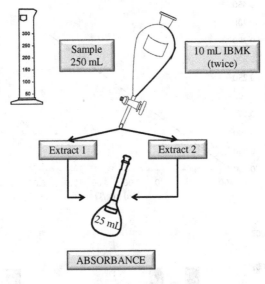

Figure 6.

Thus:

$$C_{\text{Cu}} = 3.308 \,\frac{\text{mg}}{\text{L}} \cdot \frac{25 \,\text{mL}}{250 \,\text{mL}} = 0.33 \,\text{mg/L}$$

3. The determination of arsenic is performed frequently by VGAAS as it provides the highest sensitivity when determining elements that form volatile species. The calibration straight line is obtained from the data in the following table. A bottle with 100 g of pulverized muscle tissue was received in the laboratory for determining As. A 5.0000 g aliquot of the sample was weighted

and submitted to acid digestion in a microwave oven. The solution obtained is filtered and diluted to 25.00 mL. A 3.00 mL aliquot of this solution is subsequently diluted to 10.00 mL wherefrom 5 mL were withdrawn to measure their absorbance. A procedural blank is prepared in the same way obtaining an absorbance value of 0.008.

Determine: (a) the linear range; (b) the sensitivity; (c) the concentration of As in the mussel sample considering that the signal obtained is 0.182 for sample 1 and 0.060 for sample 2; (d) evaluate if the calculated concentrations can be quantitated reliably.

C_{As} (mg/L)	0.00	0.15	0.30	0.40	0.60	0.80	1.00
Absorbance	0.0009	0.026	0.052	0.073	0.104	0.125	0.150

SOLUTION:

(a) A graph of the absorbance versus the concentration of As shows that there is no good fit between the experimental signals and the concentration (Figure 7a). Meanwhile, the residuals plot exhibits a parabolic profile which indicates that the relationship is not linear in the overall working range (Figure 7b).

So, first of all, the linear range must be established. For this, the highest calibration points are discarded until a straight line is obtained. In this case, the linear range covers till 0.6 mg/L of As, and the least-squares fit equation obtained is:

$$A = (0.0007 \pm 0.00141) + (0.1740 \pm 0.00396) \cdot C_{As}$$

(b) The sensitivity is the slope of the calibration line, i.e., 0.1740 Au/(mg/L).

(c) The concentration of As in the samples was calculated as follows:

As it was explained in Chapter 2, before using the calibration equation, it is necessary to check whether the intercept is zero or not, since an absorbance signal was measured for the procedural blank. This can be done by means of a t-test, as

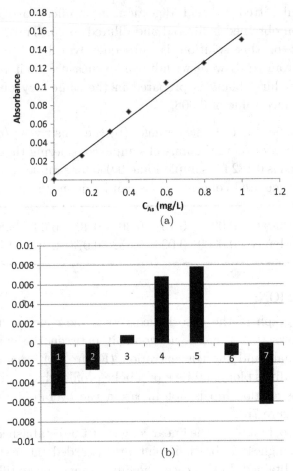

Figure 7.

detailed in Chapter 2. Hence:

$$t_{\text{experimental}}$$
$$= \frac{|0 - a|}{s_a} = \frac{0.0007}{0.00141} = 0.50 \quad t_{\text{tabulated}(95\%,3)} = 3.182$$

Since $t_{\text{experimental}}$ is lower than

$$t_{\text{tabulated}(95\% \text{ confidence, } n-2 \text{ degrees of freedom})},$$

it can be concluded that the intercept is statistically zero and therefore the calibration fit does not need further correction.

The absorbance of the procedural blank is subtracted from the absorbance of the samples and the corrected values interpolated in the calibration. Thereby, the concentration of As in the final test solutions is calculated.

Sample 1: $A_{\text{corrected}} = 0.182 - 0.008 = 0.174$
Sample 2: $A_{\text{corrected}} = 0.060 - 0.008 = 0.052$

The absorbance value of sample 1 is above the linear range of the calibration so the concentration of As in this sample should not be calculated. An additional dilution step might be included in the experimental procedure to obtain a lower signal. The concentration of As in the final test solution obtained for sample 2 is:

$$C_{\text{As}} = 0.295 \, \frac{\text{mg}}{\text{L}} \cdot \frac{10^3 \, \mu\text{g}}{1 \, \text{mg}} \cdot \frac{1 \, \text{L}}{10^3 \, \text{mL}} = 0.295 \, \mu\text{g/mL}$$

To calculate the concentration of As in the mussel sample, the overall analytical procedure must be considered (Figure 8). Thus,

$$C_{\text{As}} = 0.295 \, \frac{\mu\text{g}}{\text{mL}} \cdot \frac{10 \, \text{mL}}{3 \, \text{mL}} \cdot \frac{25 \, \text{mL}}{5.0000 \, \text{g}} = 4.91 \, \mu\text{g/g}$$

Sample 2: $C_{\text{As}} = 4.91 \, \mu\text{g/g}$

(d) To justify if the calculated concentration can be quantified reliably, the limit of quantitation (LOQ) of the procedure has to be calculated. In classical terms (see Chapter 2, Equation (29)) this is:

$$\text{LOQ}_{\text{inst}} = \frac{0.0007 + 10 \cdot 0.00141}{0.1740}$$

$$= 0.0850 \, \text{mg/L} = 0.0850 \, \mu\text{g/mL}$$

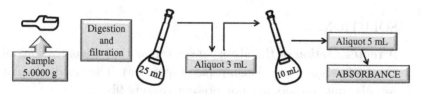

Figure 8.

And the procedural LOQ is obtained considering the steps of the analytical procedural:

$$LOQ_{proc} = 0.0850 \, \frac{\mu g}{mL} \cdot \frac{10 \, mL}{3 \, mL} \cdot \frac{25 \, mL}{5.0000 \, g} = 1.42 \, \mu g/g$$

The updated LOQ definition (Chapter 2, Equation (32)) leads to 2.11 $\mu g/g$.

Then, as the concentration of As in the sample is larger than the procedural LOQ, it can be said that the measurement of the concentration was reliable.

4. An environmental laboratory determined the content of chromium in a sample of ash from a power plant by ETAAS. From a bottle with 100 g of sample, replicate aliquots of 0.2527 g, 0.2485 g and 0.2509 g were weighted in Teflon vessels. Concentrated nitric acid was added (4 mL) and a microwave digestion performed. The solutions thus obtained were filtered and diluted to 25.00 mL with ultrapure water. Aliquots of 5.00 mL were taken and evaporated to dryness in order to eliminate the excess of acid, the residues were dissolved with ultrapure water and diluted to 10.00 mL. The signals obtained were 0.055, 0.048 and 0.052, for each aliquot of ash sample. A procedural blank was prepared in the same way obtaining a signal value of 0.007. The signals obtained for the standard solutions of the calibration curve are shown in the following table. Calculate the concentration of Cr in the sample and assess whether it can be quantified reliably.

C_{Cr} ($\mu g/L$)	0.00	2.00	4.00	6.00	8.00	10.00	12.00
Absorbance	0.006	0.057	0.085	0.130	0.175	0.208	0.240

SOLUTION:

A representation of the absorbances versus the concentrations of Cr shows a good relationship (see Figure 9a). The residuals plot reveals that outliers are not present (Figure 9b).

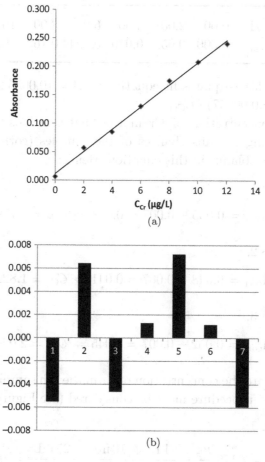

Figure 9.

The least-squares fit equation is:

$$A = (0.0115 \pm 0.00412) + (0.0195 \pm 0.00057) \cdot C_{\mathrm{Cr}}$$

To check whether the intercept is zero, a t-test was conducted, obtaining $t_{\text{experimental}} = 2.79$, which is higher than $t_{\text{tabulated}(95\%, n-2=5)} = 2.57$. Therefore, the intercept is not statistically zero and a new least-squares fit must be obtained subtracting the signal of the calibration blank (A_{Bcal}) from the signals of the standard solutions (A).

C_{Cr} (μg/L)	0.00	2.00	4.00	6.00	8.00	10.00	12.00
$A - A_{Bcal}$	0.000	0.051	0.079	0.124	0.169	0.202	0.234

The new least-squares fit equation is: $A = (0.0055 \pm 0.00412) + (0.0195 \pm 0.00057) \cdot C_{Cr}$.

The concentration of Cr in the final solutions is calculated substituting the absorbances of the samples (corrected by the procedural blank) in this equation. Hence:

Replicate 1:

$$A_{corr} = 0.055 - 0.007 = 0.048 \Rightarrow C_{Cr} = 2.179\,\mu g/L$$

Replicate 2:

$$A_{corr} = 0.048 - 0.007 = 0.041 \Rightarrow C_{Cr} = 1.821\,\mu g/L$$

Replicate 3:

$$A_{corr} = 0.052 - 0.007 = 0.045 \Rightarrow C_{Cr} = 2.026\,\mu g/L$$

To calculate the concentration of Cr in the ash sample, the overall treatment procedure must be considered (see Figure 10).

Replicate 1:

$$C_{Cr} = 2.179\,\frac{\mu g}{L} \cdot \frac{1\,L}{10^3\,mL} \cdot \frac{10\,mL}{5\,mL} \cdot \frac{25\,mL}{0.2527\,g} = 0.431\,\mu g/g$$

Replicate 2:

$$C_{Cr} = 1.821\,\frac{\mu g}{L} \cdot \frac{1\,L}{10^3\,mL} \cdot \frac{10\,mL}{5\,mL} \cdot \frac{25\,mL}{0.2485\,g} = 0.366\,\mu g/g$$

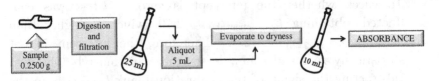

Figure 10.

Replicate 3:

$$C_{Cr} = 2.026 \frac{\mu g}{L} \cdot \frac{1\,L}{10^3\,mL} \cdot \frac{10\,mL}{5\,mL} \cdot \frac{25\,mL}{0.2509\,g} = 0.404\,\mu g/g$$

Now the average concentration of Cr in the sample can be expressed as: $\bar{x} \pm ts/\sqrt{n}$.

The concentration of Cr in the ash sample is $(0.40 \pm 1.73)\,\mu g/g$, which yields a relative standard deviation (RSD) = 7.5 %.

To check whether the concentration of Cr in the sample can be quantitated reliably, the LOQ of the procedure must be calculated. In classical terms (see Chapter 2), this is calculated as:

$$LOQ_{inst} = \frac{0.0055 + 10 \cdot 0.00412}{0.0195} = 2.40\,\mu g/L$$

wherefrom

$$LOQ_{proc} = 2.40 \frac{\mu g}{L} \cdot \frac{1\,L}{10^3\,mL} \cdot \frac{10\,mL}{5\,mL} \cdot \frac{25\,mL}{0.2500\,g} = 0.48\,\mu g/g$$

The updated LOQ definition (Chapter 2, Equation (32)) leads to $0.69\,\mu g/g$.

Note that the concentration of Cr in the sample, $0.40\,\mu g/g$, is lower than the LOQ_{proc}, $0.48\,\mu g/g$, therefore, the concentration of Cr calculated in the sample should not be quantitated. In such a situation, the method must be modified to enhance its sensitivity. If this quantitative result is delivered, it should be accompanied by the LOQ so that the user is conscious on the limitation of such a value.

5. The determination of cadmium in a liquid industrial effluent can be performed by ICP-AES using indium as internal standard. For this, a series of cadmium standard solutions containing 0.1 % nitric acid were prepared adding the same amount of indium to all of them. Two samples were analyzed following the procedures described below:

Sample 1: A 2.00 mL aliquot of sample was diluted to 10 mL with 1 % nitric acid and evaporated to dryness; the residue was dissolved with 5.00 mL of ultrapure water to perform the measurement.

Sample 2: 3.50 mL of sample were treated with 1 mL of 20 % nitric acid and subsequently diluted to 10.00 mL with ultrapure water. A 5.00 mL aliquot of this solution was measured.

The emission intensities are shown in the following table. Calculate the concentration of Cd in the industrial effluent samples expressed as mg/L.

C_{Cd} (mg/L)	0.00	0.10	0.20	0.30	0.40	0.50	Sample 1	Sample 2
Cd signal	4.52	11.38	18.20	25.72	31.95	43.63	27.29	20.74
In signal	19.84	22.74	25.25	25.80	27.03	30.21	35.82	28.20

SOLUTION:

This example is about the internal standard calibration method. The ratio between the analyte and the internal standard signals is calculated and plotted against the concentration of analyte.

C_{Cd} (mg/L)	0.00	0.10	0.20	0.30	0.40	0.50	Sample 1	Sample 2
Relative signal	0.23	0.50	0.72	1.00	1.18	1.44	0.76	0.74

A graph of the relative signals versus the concentration of Cd indicated a good straight-line relationship (Figure 11a) and the residuals plot reveals that outliers are not present (Figure 11b).

So the concentration of cadmium in the test aliquot is calculated by interpolation in the calibration equation.

The least-squares fit equation is: $S = (0.245 \pm 0.0166) + (2.401 \pm 0.0547) \cdot C_{Cd}$

Sample 1: Relative signal $= 0.76 \Rightarrow C_{Cd} = 0.216\,\text{mg/L}$

Sample 1 was treated as shown in Figure 12.

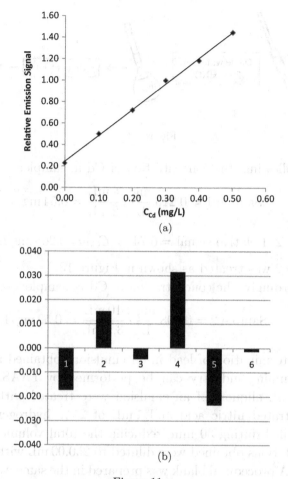

(a)

(b)

Figure 11.

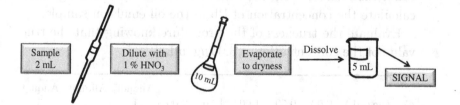

Figure 12.

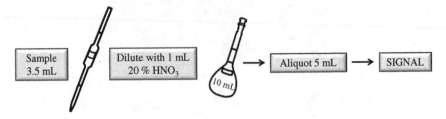

Figure 13.

And, following, the concentration of Cd in sample:

$$\text{Sample 1} = 0.216 \, \frac{\text{mg}}{\text{L}} \cdot \frac{5 \, \text{mL}}{2 \, \text{mL}} = 0.54 \, \text{mg/L}$$

Sample 2: Relative signal $= 0.74 \Rightarrow C_{Cd} = 0.208 \, \text{mg/L}$

Sample 2 was treated as shown in Figure 13.

Accordingly, the concentration of Cd in sample:

$$\text{Sample 2} = 0.208 \, \frac{\text{mg}}{\text{L}} \cdot \frac{10 \, \text{mL}}{3.5 \, \text{mL}} = 0.59 \, \text{mg/L}$$

6. The determination of lead in oil emulsions obtained as residues in a canning industry can be performed by FAAS. For this, 100.00 mL aliquots of an emulsion were treated with 15 mL of concentrated nitric acid and 1 mL of 25 % hydrogen peroxide and boiled during 30 min, reducing the total volume to 50 mL. The solutions obtained were diluted to 250.00 mL with ultrapure water. A procedural blank was prepared in the same way. A series of calibration Pb standard solutions were prepared to obtain the calibration curve. With the results shown in the following table, calculate the concentration of Pb in the oil emulsion sample.

Evaluate the trueness of the procedure knowing that the true value of the concentration is $3.00 \, \mu\text{g/mL}$.

C_{Pb} (μg/mL)	0.00	0.50	1.00	1.50	2.00	Aliquot 1	Aliquot 2	Aliquot 3
Absorbance	0.002	0.040	0.073	0.110	0.147	0.090	0.092	0.087

Procedural blank: $A = 0.007$

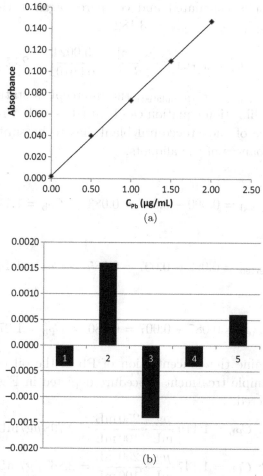

(a)

(b)

Figure 14.

SOLUTION:

The regression plot shows a good relationship between absorbance and concentration (Figure 14a) and outliers are not detected in the residuals plot (Figure 14b). The least-squares fit resulted:

$$A = (0.0024 \pm 0.00102) + (0.0720 \pm 0.00083) \cdot C_{Pb}$$

As a signal is reported for the procedural blank, it is necessary to check whether the intercept of the calibration is zero.

$t_{\text{experimental}}$ is calculated and compared against the tabulated value $t_{\text{tabulated}(95\%,n-2=3)} = 3.182$.

$$t_{\text{experimental}} = \frac{|0 - a|}{S_a} = \frac{0.0024}{0.00102} = 2.35$$

Since $t_{\text{experimental}} < t_{\text{tabulated}}$, the intercept is statistically zero and the calibration equation can be used as such. Previously, the absorbance of the procedural blank has to be subtracted from the absorbances of the aliquots.

Aliquot 1:

$$A_{\text{corrected}} = 0.090 - 0.007 = 0.083 \Rightarrow C_{\text{Pb}} = 1.119\,\mu g/mL$$

Aliquot 2:

$$A_{\text{corrected}} = 0.092 - 0.007 = 0.085 \Rightarrow C_{\text{Pb}} = 1.147\,\mu g/mL$$

Aliquot 3:

$$A_{\text{corrected}} = 0.087 - 0.007 = 0.080 \Rightarrow C_{\text{Pb}} = 1.078\,\mu g/mL$$

To determine the concentration of Pb in the oil emulsion, the overall sample treatment procedure depicted in Figure 15 must be considered.

Aliquot 1: $C_{\text{Pb}} = 1.119\,\dfrac{\mu g}{mL} \cdot \dfrac{250\,mL}{100\,mL} = 2.798\,\mu g/mL$

Aliquot 2: $C_{\text{Pb}} = 1.147\,\dfrac{\mu g}{mL} \cdot \dfrac{250\,mL}{100\,mL} = 2.867\,\mu g/mL$

Aliquot 3: $C_{\text{Pb}} = 1.078\,\dfrac{\mu g}{mL} \cdot \dfrac{250\,mL}{100\,mL} = 2.695\,\mu g/mL$

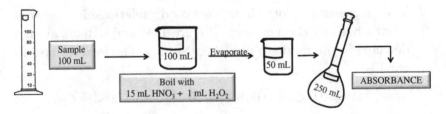

Figure 15.

Therefore, the average concentration of Pb in the oil emulsion is $(2.79 \pm 0.22)\,\mu\text{g/mL}$, with an RSD $= 3.1\,\%$.

The evaluation of trueness is performed with the Student's t-test, as detailed in Chapter 2:

$$t_{\text{experimental}} = \frac{\mu - \bar{x}}{s}\sqrt{n} = \frac{3.00 - 2.79}{0.09}\sqrt{3} = 4.09$$

$t_{\text{tabulated}(95\,\%, n-1=2)} = 4.303$

Since $t_{\text{experimental}} < t_{\text{tabulated}}$, we can conclude that the procedure is true (it has no bias).

7. The determination of calcium in yoghurt can be done using FAAS. The calibration is performed with aqueous standard solutions containing 0, 0.2, 0.4, 0.6 and 0.8 $\mu\text{g/mL}$ of this element. The absorbance values were 0.003, 0.050, 0.103, 0.157 and 0.200, respectively. Three 1.0000 g aliquots were withdrawn from a 250 g yoghurt pot, analyzing them independently. Each aliquot was mixed with 10 mL of concentrated nitric acid and digested in a microwave oven. The solutions were filtered and diluted to 25.00 mL with ultrapure water. Aliquots of 2.00 mL of these solutions were diluted to 10.00 mL with ultrapure water. The absorbance signals obtained are shown in the following table:

	Mass (g)	Signal
Aliquot 1	1.1231	0.120
Aliquot 2	1.0587	0.118
Aliquot 3	1.1023	0.123

Another portion of 1.0952 g was weighted and spiked with 10 μL of a 2000 $\mu\text{g/mL}$ Ca stock standard solution, obtaining a signal of 0.160. The procedural blank gave a signal of 0.005.

Calculate the concentration of Ca in the yoghurt, and the recovery of the procedure.

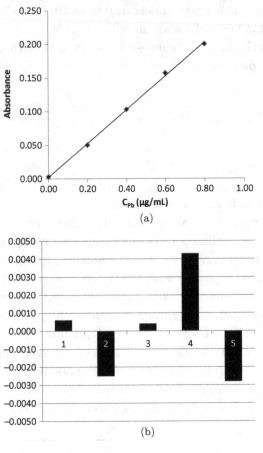

Figure 16.

SOLUTION:

The calibration plot (Figure 16a) shows a good straight-line relationship between signals and concentrations. The residuals plot (Figure 16b) reveals that outliers are not present. The calibration equation is:

$$A = (0.0024 \pm 0.00257) + (0.2505 \pm 0.00525) \cdot C_{\text{Ca}}$$

In order to take account of the procedural blank, the intercept of the calibration curve has to be checked against zero. Here,

$t_{\text{experimental}} = 0.93$; which is lower than $t_{\text{tabulated}(95\%,3)} = 3.18$. Therefore, the intercept is statistically zero and the calibrate equation can be used as calculated.

Recall that the absorbance of the procedural blank has to be subtracted from the absorbance of the aliquots of sample.

$$\text{Aliquot 1: } A_{\text{corrected}} = 0.120 - 0.005 = 0.115$$

$$\Rightarrow C_{\text{Ca}} = 0.449\,\mu\text{g/mL}$$

$$\text{Aliquot 2: } A_{\text{corrected}} = 0.118 - 0.005 = 0.113$$

$$\Rightarrow C_{\text{Ca}} = 0.442\,\mu\text{g/mL}$$

$$\text{Aliquot 3: } A_{\text{corrected}} = 0.123 - 0.005 = 0.118$$

$$\Rightarrow C_{\text{Ca}} = 0.462\,\mu\text{g/mL}$$

The concentration of Ca in the yoghurt is calculated considering the overall sample treatment procedure shown in Figure 17:

$$\text{Aliquot 1: } C_{\text{Ca}} = 0.449\,\frac{\mu\text{g}}{\text{mL}} \cdot \frac{10\,\text{mL}}{2\,\text{mL}} \cdot \frac{25\,\text{mL}}{1.1231\,\text{g}} = 49.973\,\mu\text{g/g}$$

$$\text{Aliquot 2: } C_{\text{Ca}} = 0.442\,\frac{\mu\text{g}}{\text{mL}} \cdot \frac{10\,\text{mL}}{2\,\text{mL}} \cdot \frac{25\,\text{mL}}{1.0587\,\text{g}} = 52.187\,\mu\text{g/g}$$

$$\text{Aliquot 3: } C_{\text{Ca}} = 0.462\,\frac{\mu\text{g}}{\text{mL}} \cdot \frac{10\,\text{mL}}{2\,\text{mL}} \cdot \frac{25\,\text{mL}}{1.1023\,\text{g}} = 52.390\,\mu\text{g/g}$$

Finally, the concentration of Ca in the yoghourt is $(51.52 \pm 3.33)\,\mu\text{g/g}$; with an RSD $= 2.6\%$.

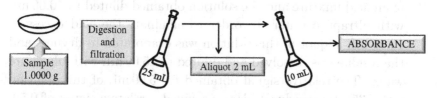

Figure 17.

To assess the recovery of the procedure, another sample aliquot was analyzed, which had been spiked with a known amount of calcium before performing the treatment procedure. The spiked concentration was:

$$C_{Ca_{spiked}} = \frac{10\,\mu L \cdot \dfrac{1\,mL}{10^3\,\mu L} \cdot 2000\,\dfrac{\mu g}{mL}}{1.0952\,g} = 18.26\,\mu g/g$$

Observe that the volume of the spike ($10\,\mu L$) was considered negligible compared to the mass of the sample aliquot analyzed.

The total concentration of calcium in the spiked aliquot was calculated as before, and using the same calibration:

$$S_{corrected} = 0.160 - 0.005 = 0.155 \Rightarrow C_{Ca} = 0.6092\,\mu g/mL$$

The concentration of Ca in the spiked aliquot (considering the sample treatment procedure and the mass of the aliquot) is $69.53\,\mu g/g$. The recovery is calculated as:

$$R(\%) = \frac{C_{Ca_{spiked\ aliquot}} - C_{Ca_{yoghurt}}}{C_{Ca_{spiked}}} \cdot 100$$

$$= \frac{(69.53 - 51.52)\,\mu g/g}{18.26\,\mu g/g} \cdot 100 = 99\,\%$$

It can be concluded that the recovery of the procedure is satisfactory.

8. A bottle with 50 g of an ore sample was received in a laboratory to determine copper by ICP-MS using scandium as internal standard. A 0.1712 g aliquot was weighted, digested with 10 mL of an acid mixture and the solution obtained diluted to 50.00 mL with ultrapure water. In order to eliminate the acid excess, a 10.00 mL aliquot of this solution was evaporated to dryness and the residue was dissolved and diluted to 25.00 mL with ultrapure water. The relative signal obtained for 5.00 mL of this solution was 7.35. A procedural blank provided a relative signal of 0.50. The calibration is shown in the following table.

C_{Cu} (μg/L)	0.00	10.00	20.00	30.00	40.00	50.00	60.00	70.00	80.00
Relative signal	0.20	2.24	4.52	6.63	9.01	10.94	13.71	15.49	17.91

To evaluate the recovery of the procedure, a 0.1683 g sample aliquot was spiked with 10 μL of a 250 mg/L Cu standard solution and prepared as described above, leading to a relative signal of 11.70.

Calculate the concentration of Cu in the ore sample and evaluate the recovery of the procedure.

SOLUTION:

This is another example of the internal standard calibration method. In this case, the relative signal is already provided so the calibration can be done directly. Good linearity was observed and outliers were not visualized (Figure 18). The calibration equation was:

$$S = (0.076 \pm 0.1059) + (0.222 \pm 0.0022) \cdot C_{Cu}$$

The intercept was checked first and it was found that $t_{experimental} = 0.71 < t_{tabulated(95\%,7)} = 2.365$, which indicates that the intercept is statistically zero. Therefore, the concentration of copper is calculated by direct interpolation of the corrected sample signal ($S_{sample} - S_{procedural\ blank}$) in the calibration. The corrected signal is:

$$S_{corrected} = 7.35 - 0.50 = 6.85 \Rightarrow C_{Cu} = 30.514\,\mu g/L$$

The sample treatment procedure (shown in Figure 19) is considered to calculate the concentration of copper in the ore sample.

$$C_{Cu} = 30.514\,\frac{\mu g}{L} \cdot \frac{1\,L}{10^3\,mL} \cdot \frac{25\,mL}{10\,mL} \cdot \frac{50\,mL}{0.1712\,g} = 22.28\,\mu g/g$$

To calculate the recovery of the procedure, the spiking amount has to be calculated:

$$C_{Cu_{spiked}} = \frac{10\,\mu L \cdot \dfrac{1\,mL}{10^3\,\mu L} \cdot 250\,\dfrac{\mu g}{mL}}{0.1683\,g} = 14.58\,\mu g/g$$

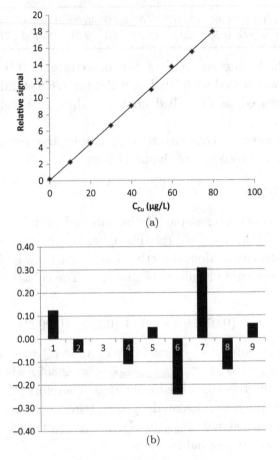

(a)

(b)

Figure 18.

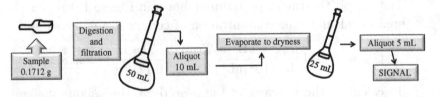

Figure 19.

Observe that the volume of the spike ($10\,\mu\text{L}$) was considered negligible compared to the mass of the sample aliquot analyzed.

The concentration of copper in this aliquot was evaluated:

$$S_{\text{corrected}} = 11.50 - 0.50 = 11.00 \Rightarrow C_{\text{Cu}} = 49.18\,\mu\text{g/L}$$

The concentration of copper in the spiked aliquot (considering the sample treatment procedure and the mass of the aliquot) is $36.53\,\mu\text{g/g}$.

Then, recovery is calculated as:

$$R(\%) = \frac{C_{\text{Cu}_{\text{spiked aliquot}}} - C_{\text{Cu}_{\text{ore sample}}}}{C_{\text{Cu}_{\text{spiked}}}} \cdot 100$$

$$= \frac{(36.53 - 22.28)\,\mu\text{g/g}}{14.85\,\mu\text{g/g}} \cdot 100 = 96\,\%$$

It can be concluded, therefore, that the recovery of the procedure is satisfactory.

9. The determination of nickel in marine water samples requires a preconcentration step as its concentration is usually very low. For this, a 100.00 mL sample aliquot was passed through a column with a cation exchange resin and the metal eluted with 10.00 mL of 0.5 % nitric acid. The standard addition method was used to quantify the concentration of nickel by ETAAS. Thus, 1.00 mL aliquots of the eluate were spiked with known volumes of a 0.5 mg/L Ni standard solution and diluted to 5.00 mL. Calculate the concentration of nickel in the marine water sample considering the data in the following table, expressed as $\mu\text{g/L}$.

Volume of Ni standard (mL)	0.00	0.10	0.20	0.30	0.40	0.50
Absorbance	0.070	0.170	0.330	0.450	0.550	0.680

SOLUTION:

In this exercise, the standard addition method is utilized to calculate the concentration of nickel in marine water. This approach is mandatory when matrix components interfere with the analyte determination. The standard addition calibration experimental procedure is depicted in Figure 20 (how to deal with the calibration was addressed in Chapter 2 — Figure 3).

The spiked concentration should be calculated from the standard volumes:

$$C_{Cu_{spiked}}(\mu g/L) = \frac{0.5\,\dfrac{mg}{L} \cdot \dfrac{10^3\,\mu g}{1\,mg} \cdot Volume\,(L)}{5\,mL\,\dfrac{1\,L}{10^3\,mL}}$$

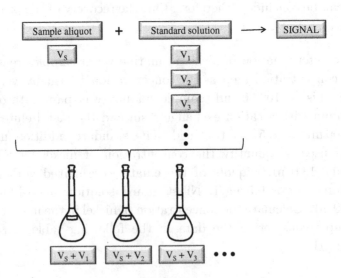

Figure 20.

The concentrations are shown in the following table:

C_{Ni} (μg/L)	0.00	10.00	20.00	30.00	40.00	50.00
Absorbance	0.070	0.170	0.330	0.450	0.550	0.680

The standardization line showed a good linear relationship, without outliers (Figure 21):

$$A = (0.0671 \pm 0.01131) + (0.0123 \pm 0.00037) \cdot C_{Ni}$$

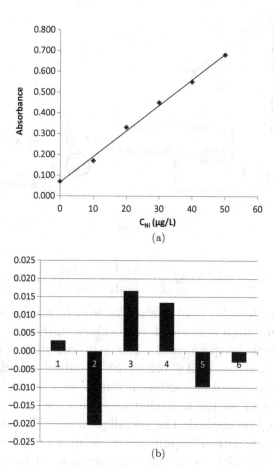

(a)

(b)

Figure 21.

As detailed in Chapter 2, interpolation in the least-squares fit equation yields the analyte concentration in the unknown test solution:

$$2 \cdot A = a + b \cdot C_{Ni}$$

$$C_{Ni} = \frac{(2 \cdot 0.070) - 0.0671}{0.0123} = 5.927\ \mu g/L$$

The concentration of Ni in the marine water sample is calculated after considering its treatment, which is depicted in Figure 22:

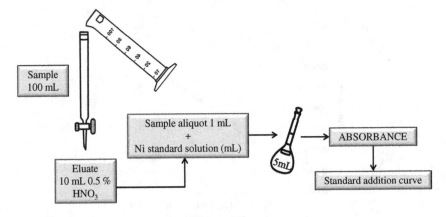

Figure 22.

$$C_{Ni_{sample}} = 5.927\ \frac{\mu g}{L} \cdot \frac{5\,mL}{1\,mL} \cdot \frac{10\,mL}{100\,mL} = 2.96\ \mu g/L$$

10. To determine the total content of iron in marine sediments, a digestion with a mixture of hydrochloric, nitric and hydrofluoric acids must be performed. The solutions thus obtained can be analyzed by FAAS using the standard addition method. A 0.2537 g aliquot of a sediment sample was weighted in a Teflon vessel, 10 mL of the acid mixture were added and a microwave oven program applied. The acid digests were filtered and diluted to 50.00 mL with ultrapure water. Solution aliquots of 5.00 mL were spiked with increasing concentrations of iron and diluted to 25.00 mL. The procedural blank provided a signal of 0.008.

Calculate the concentration of iron in the sample expressed as mass percent (%) using the data presented in the following table.

C_{Fe} (mg/L)	0.00	2.00	5.00	10.00	15.00	20.00	25.00
Absorbance	0.085	0.113	0.163	0.232	0.304	0.360	0.453

SOLUTION:

In this exercise, a sediment sample is subjected to a digestion procedure in a microwave oven to calculate the concentration of iron by means of the standard addition method. A procedural blank was prepared following exactly the same procedure, whose signal should be subtracted from those for the sample aliquots. The corrected signals are presented in the following table:

C_{Fe} (mg/L)	0.00	2.00	5.00	10.00	15.00	20.00	25.00
Absorbance	0.077	0.105	0.155	0.224	0.296	0.352	0.445

When the calibration line is studied (Figure 23a), there is a rather good straight-line relationship, although with a clear outlier (see Figure 23b), which has to be discarded. The new straight-line calibration is:

$$A = (0.0776 \pm 0.00149) + (0.0147 \pm 0.00012) \cdot C_{Fe}$$

Interpolation in the least-squares fit equation (see Chapter 2) yields the concentration we are looking for:

$$2 \cdot A = a + b \cdot C_{Fe} \quad C_{Fe} = \frac{(2 \cdot 0.077) - 0.0776}{0.0147} = 5.197 \, \text{mg/L}$$

Taking into account the sample treatment procedure illustrated in Figure 24, the concentration of iron in the sediment sample can be obtained.

$$C_{Fe_{sediment}} = 5.197 \, \frac{mg}{L} \cdot \frac{1 \, L}{10^3 \, mL} \cdot \frac{25 \, mL}{5 \, mL} \cdot \frac{50 \, mL}{0.2537 \, g} = 5.121 \, \text{mg/g}$$

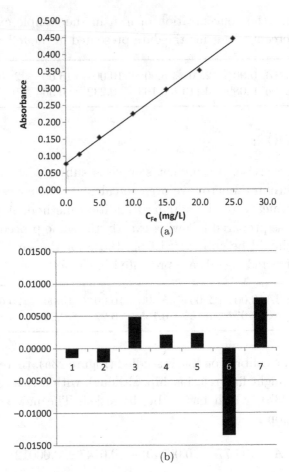

(a)

(b)

Figure 23.

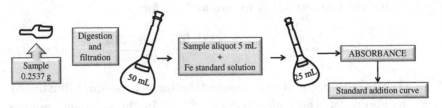

Figure 24.

The mass percent of iron in the sediment sample is calculated as follows:

Mass percent of Fe in the sediment sample

$$= 5.121 \, \frac{mg}{g} \cdot \frac{1\,g}{10^3\,mg} \cdot 100 = 0.51\,\%$$

11. Magnesium in vegetables can be quantitated by ICP-AES using scandium as internal standard. For this, a lyophilized spinach sample (50 g) was received in a laboratory. A 8.0789 g aliquot was weighted, treated with 10 mL of concentrated nitric acid and diluted to 50.00 mL. In order to reduce the acid content, a 25.00 mL aliquot of this solution was diluted to 100.00 mL. Then, four aliquots of 4.00 mL each were spiked with increasing concentrations of magnesium and diluted to 20.00 mL. The emission intensities were obtained for 5 mL of these solutions. Another two replicates of 8.7325 g and 8.6976 g each were treated in the same way, obtaining relative signals of 0.33 and 0.35, respectively; a procedural blank was prepared, yielding a 0.09 relative signal. Calculate the concentration of magnesium in the vegetable sample.

C_{Mg} (mg/mL)	0.00	2.00	4.00	6.00	8.00
Relative signal	0.30	1.40	2.40	3.40	4.50

SOLUTION:

This exercise presents an application of the standard addition method when several replicate aliquots of a sample are analyzed. First, the concentration of Mg in the aliquot used to prepare the standard addition calibration (aliquot 1) is calculated as usual. Then, the concentration of Mg in the other aliquots is obtained by the addition calibration method (Chapter 2). The original intercept, a_1, of the least-squares fit is replaced by new values (a_2 and a_3) calculated from the corresponding signal of aliquots 2 and 3.

Before proceeding, the signal of the procedural blank has to be subtracted from each solution of the standard addition curve. The corrected signals are presented in the following table:

C_{Mg} (mg/mL)	0.00	2.00	4.00	6.00	8.00
Relative signal	0.21	1.31	2.31	3.31	4.41

The calibration plot shows a good behavior without outliers (see Figure 25).

The least-squares fit is: Relative signal $= (0.230 \pm 0.0283) + (0.520 \pm 0.0058) \cdot C_{Mg}$; wherefrom the concentration of Mg in

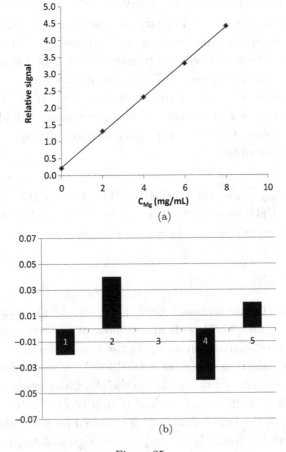

(a)

(b)

Figure 25.

aliquot 1 is calculated by interpolation:

$$2 \cdot S_1 = a_1 + b \cdot C_{Mg_1} \qquad C_{Mg_1} = \frac{(2 \cdot 0.21) - 0.23}{0.52} = 0.365 \, \text{mg/mL}$$

To calculate the concentration of Mg in aliquots 2 and 3, the addition calibration method is used, so *intercepts* a_2 and a_3, are calculated as follows:

Do not forget to subtract the signal of the procedural blank from the signals of the sample aliquots!

Aliquot 2: Corrected relative signal:

$$S_2 = 0.33 - 0.09 = 0.24$$

$$a_2 = a_1 - S_1 + S_2 = 0.23 - 0.21 + 0.24 = 0.26$$

Note that S_1 is the absorbance of the first aliquot of the original calibration, whereas a_1 is the intercept of the initial calibration line.

Aliquot 3: Corrected relative signal:

$$S_3 = 0.35 - 0.09 = 0.26$$

$$a_3 = a_1 - S_1 + S_3 = 0.23 - 0.21 + 0.26 = 0.28$$

Now the concentration of Mg in aliquots 2 and 3 is calculated by interpolation:

$$C_{Mg_2} = \frac{(2 \cdot S_2) - a_2}{b} = \frac{(2 \cdot 0.24) - 0.26}{0.52} = 0.423 \, \text{mg/mL}$$

$$C_{Mg_3} = \frac{(2 \cdot S_3) - a_3}{b} = \frac{(2 \cdot 0.26) - 0.28}{0.52} = 0.461 \, \text{mg/mL}$$

Finally, the concentration of Mg in the spinach sample is calculated considering the overall sample treatment shown in Figure 26:

$$C_{Mg_1} = 0.365 \, \frac{\text{mg}}{\text{mL}} \cdot \frac{20 \, \text{mL}}{4 \, \text{mL}} \cdot \frac{100 \, \text{mL}}{25 \, \text{mL}} \cdot \frac{50 \, \text{mL}}{8.0789 \, \text{g}} = 45.179 \, \text{mg/g}$$

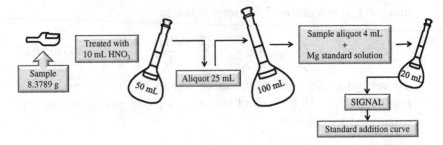

Figure 26.

$$C_{Mg_2} = 0.423 \, \frac{mg}{mL} \cdot \frac{20 \, mL}{4 \, mL} \cdot \frac{100 \, mL}{25 \, mL} \cdot \frac{50 \, mL}{8.7325 \, g} = 48.440 \, mg/g$$

$$C_{Mg_3} = 0.461 \, \frac{mg}{mL} \cdot \frac{20 \, mL}{4 \, mL} \cdot \frac{100 \, mL}{25 \, mL} \cdot \frac{50 \, mL}{8.6976 \, g} = 53.003 \, mg/g$$

The average concentration of Mg in the sample should be reported (including its confidence interval), as well as its RSD(%).

So, the concentration of Mg in the spinach sample is $(48.87 \pm 9.76) \, mg/g$; with an RSD $= 8.0\%$.

12. VGAAS is widely used to determine metals that form volatile species. In this example, selenium is determined in blood serum by VGAAS using the standard addition method to quantitate. For this, a 5.00 mL aliquot of a sample is digested with 2 mL of nitric acid in a microwave oven at 400 W during 5 min and subsequently diluted to 10.00 mL with ultrapure water. The absorbance of a procedural blank was 0.023. The calibration was prepared as shown in the following table. Another two aliquots of blood serum were prepared in the same way obtaining 0.093 and 0.085 absorbances, respectively.

$C_{Se}(\mu g/L)$	0.00	1.00	2.00	3.00	4.00
Absorbance	0.099	0.407	0.751	1.075	1.399

Calculate the concentration of selenium in the serum sample, in $\mu g/L$. Assess the trueness of the procedure accepting that the true value for Se in the sample is 0.40 $\mu g/L$.

SOLUTION:

This exercise presents another application of the standard addition method when several aliquots of a sample are analyzed. The resolution steps are the same as in the previous example and only brief details will be repeated.

First, subtract the procedural blank from the standards, so that:

C_{Se} ($\mu g/L$)	0.0	1.0	2.0	3.0	4.0
Absorbance	0.076	0.384	0.728	1.052	1.376

The calibration reveals a satisfactory behavior without outliers (see Figure 27). The least-squares fit is:

$$A = (0.0696 \pm 0.00667) + (0.3268 \pm 0.00272) \cdot C_{Se}$$

The concentration of Se in aliquot 1 is:

$$C_{Se_1} = \frac{(2 \cdot 0.076) - 0.0696}{0.3268} = 0.252 \, \mu g/L$$

The concentration of Se in aliquots 2 and 3 is calculated using the addition calibration method:

Aliquot 2:

Corrected absorbance:

$$A_2 = 0.093 - 0.023 = 0.070$$

$$a_2 = a_1 - A_1 + A_2 = 0.0696 - 0.076 + 0.070 = 0.064$$

Note that A_1 is the absorbance of the first aliquot of the original calibration, whereas a_1 is the intercept of the initial calibration line.

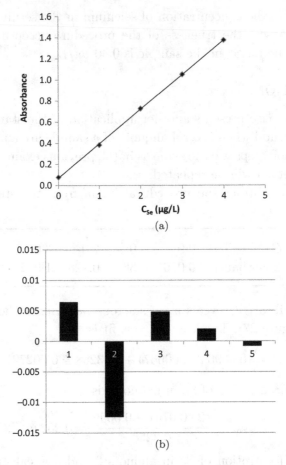

Figure 27.

Aliquot 3:

Corrected absorbance:

$$A_3 = 0.085 - 0.023 = 0.062$$

$$a_3 = a_1 - A_1 + A_3 = 0.0696 - 0.076 + 0.062 = 0.056$$

Now, the Se concentration is interpolated:

For aliquot 2:

$$C_{Se_2} = \frac{(2 \cdot S_2) - a_2}{b} = \frac{(2 \cdot 0.070) - 0.064}{0.3268} = 0.232 \, \mu g/L$$

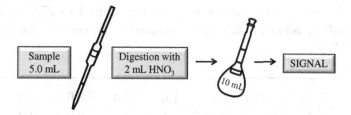

Figure 28.

For aliquot 3:

$$C_{Se_3} = \frac{(2 \cdot S_3) - a_3}{b} = \frac{(2 \cdot 0.062) - 0.056}{0.3268} = 0.208\,\mu g/L$$

The original concentrations are calculated considering the overall procedure depicted in Figure 28.

$$C_{Se_1} = 0.252\,\frac{\mu g}{L} \cdot \frac{10\,mL}{5\,mL} = 0.504\,\mu g/L$$

$$C_{Se_2} = 0.232\,\frac{\mu g}{L} \cdot \frac{10\,mL}{5\,mL} = 0.464\,\mu g/L$$

$$C_{Se_3} = 0.208\,\frac{\mu g}{L} \cdot \frac{10\,mL}{5\,mL} = 0.416\,\mu g/L$$

The average concentration of Se in the blood serum sample is $(0.48 \pm 0.07)\,\mu g/L$, with a good RSD = 5.9 %.

Trueness is evaluated by means of a Student's t-test intended to compare the average value to a theoretical one (Chapter 2). Here, $t_{experimental} = 2.6 < t_{tabulated(95\,\%,2)} = 4.30$ and, so, it can be concluded that the procedure is not biased.

13. Heavy metals in the atmosphere can be determined after collecting the particulate matter suspended in the air with a high volume dust sampler using quartz fiber filters, which retains the PM10 fraction (particulate matter with an aerodynamic diameter inferior to 10 μm). For the determination of antimony by ETAAS, a 3.14 cm^2 circular portion of the filter was weighted and it amounted to 1.2500 g. The mass of that filter portion before sampling was 1.0000 g. An extraction was performed with

0.5 % nitric acid and the extract was diluted to 25.00 mL. The standard addition line was obtained from one of the samples (PM10-1).

C_{Sb} (μg/L)	0.0	1.0	2.0	3.0	4.0
Absorbance	0.107	0.140	0.176	0.206	0.243

Another two samples, PM10-2 and PM10-3, were treated in the same manner, their data being:

	Filter circle		
	Area (cm^2)	Mass (g)	Absorbance
PM10-2	3.07	1.2382	0.095
PM10-3	3.21	1.2731	0.127

The procedural blank gave a 0.010 absorbance.

Calculate the concentration of antimony in the samples, expressed as μg/g and as ng/m^3, considering that the airflow through the sampler was constant at 68 m^3/h, the sampling time was 24 h and the total area of the filter was 445 cm^2.

SOLUTION:

This is another practice of the addition calibration method, analogous to previous examples. Hence, explanations will be resumed.

The concentration of Sb in the PM10-1 sample used to prepare the standard addition calibration is calculated by interpolation. The concentration of the PM10-2 and PM10-3 samples is calculated using the addition calibration method replacing the original intercept, a_1, by new values (a_2 and a_3) calculated from the corresponding signals of PM10-2 and PM10-3.

The corrected signals to perform the calibration are:

C_{Sb} (μg/L)	0.0	1.0	2.0	3.0	4.0
Absorbance	0.097	0.130	0.166	0.196	0.233

A good calibration, without outliers, was obtained (see Figure 29). The least-squares fit was:

$$A = (0.0968 \pm 0.00133) + (0.0338 \pm 0.00054) \cdot C_{Sb}$$

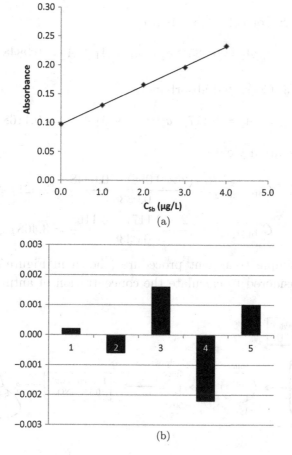

(a)

(b)

Figure 29.

wherefrom Sb in the PM10-1 sample is calculated by interpolation:

$$C_{Sb_{PM10-1}} = \frac{(2 \cdot 0.097) - 0.0968}{0.0338} = 2.876 \, \mu g/L$$

Samples PM10-2 and PM10-3 are quantitated using the addition calibration method:

Do not forget to subtract the signal of the procedural blank from PM10-2 and PM10-3! Recall that A_1 is the absorbance for the first aliquot of the original calibration.

PM10-2: Corrected absorbance:

$$A_2 = 0.085 \quad a_2 = a_1 - A_1 + A_2 = 0.0848$$

PM10-3: Corrected absorbance:

$$A_3 = 0.117 \quad a_3 = a_1 - A_1 + A_3 = 0.1168$$

Interpolation yields:

$$C_{Sb_{PM10-2}} = \frac{(2 \times 0.085) - 0.0848}{0.0338} = 2.521 \, \mu g/L$$

$$C_{Sb_{PM10-3}} = \frac{(2 \times 0.117) - 0.1168}{0.0338} = 3.468 \, \mu g/L$$

The sample treatment procedure (shown in Figure 30) has to be considered to calculate the concentration of antimony in the

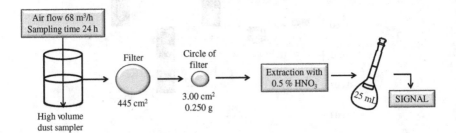

Figure 30.

particulate matter samples.

$$C_{Sb_{PM10-1}} = 2.876\,\frac{\mu g}{L} \cdot \frac{1\,L}{10^3\,mL} \cdot \frac{25\,mL}{0.2500\,g} = 0.29\,\mu g/g$$

$$C_{Sb_{PM10-2}} = 2.521\,\frac{\mu g}{L} \cdot \frac{1\,L}{10^3\,mL} \cdot \frac{25\,mL}{0.2382\,g} = 0.26\,\mu g/g$$

$$C_{Sb_{PM10-3}} = 3.467\,\frac{\mu g}{L} \cdot \frac{1\,mL}{10^3\,mL} \cdot \frac{25\,mL}{0.2731\,g} = 0.32\,\mu g/g$$

To calculate the concentration of antimony in ng/m^3, three parameters are to be considered:

— the airflow through the sampler: $68\,m^3/h$
— the sampling time: 24 h
— the total area of the filter: $445\,cm^2$

and the following calculations must be performed:

The mass of Sb in the circular portion of the filter can be obtained from the concentration of Sb in the test solution (calculated by interpolation in the standard addition calibration):

For PM10-1:

$$2.876\,\frac{\mu g}{L} \cdot \frac{1\,L}{10^3\,mL} \cdot 25\,mL = 0.07190\,\mu g \cdot \frac{10^3\,ng}{1\,\mu g} = 71.90\,ng$$

and then total amount of Sb retained in the overall filter is calculated considering its total area:

$$\frac{71.90\,ng\,Sb}{3.14\,cm^2} = \frac{X\,ng\,Sb}{445\,cm^2} \quad \text{Mass of Sb in PM10-1} = 10\,189\,ng$$

The total volume of air sampled can be calculated as:

$$68\,m^3/h \times 24\,h = 1\,632\,m^3$$

And, finally, the concentration of Sb, expressed as ng/m^3 is:

$$C_{Sb_{PM10-1}} = \frac{10\,189\,ng}{1\,632\,m^3} = 6.24\,ng/m^3$$

In the same way:

PM10-2:

$$2.521\,\frac{\mu g}{L} \cdot \frac{1\,L}{10^3\,mL} \cdot 25\,mL = 0.06302\,\mu g \cdot \frac{10^3\,ng}{1\,\mu g} = 63.02\,ng$$

$$\frac{63.02\,ng\,Sb}{3.07\,cm^2} = \frac{X\,ng\,Sb}{445\,cm^2}$$

$$C_{Sb_{PM10-2}} = \frac{9\,134\,ng}{1\,632\,m^3} = 5.60\,ng/m^3$$

PM10-3:

$$3.467\,\frac{\mu g}{L} \cdot \frac{1\,L}{10^3\,mL} \cdot 25\,mL = 0.08668\,\mu g \cdot \frac{10^3\,ng}{1\,\mu g} = 86.68\,ng$$

$$\frac{86.68\,ng\,Sb}{3.21\,cm^2} = \frac{X\,ng\,Sb}{445\,cm^2}$$

$$C_{Sb_{PM10-3}} = \frac{12\,016\,ng}{1\,632\,m^3} = 7.36\,ng/m^3$$

14. The determination of metals in petroleum products by atomic spectrometric techniques is not straightforward due to their high viscosity and high organic matter content. A methodology based on the preparation of emulsions was proposed to overcome the difficulties associated with these samples. For the determination of vanadium in fuel oil by ETAAS, the following procedure was applied: a 0.2500 g aliquot of fuel oil was dissolved with 10.00 mL of toluene; then a 0.25 mL aliquot of this solution was mixed with 0.10 mL of xylene, diluted to 1.00 mL with ultrapure water and manually agitated. A 4 μL aliquot of the emulsion was injected in the graphite furnace along with 16 μL of ultrapure water (total volume injected was 20 μL). A procedural blank prepared in the same way gave a 0.005 signal.

To investigate if the matrix interferes with the determination of vanadium, both calibration and standard addition solutions were prepared, see the following table.

C_V (μg/L)	0.00	10.00	20.00	30.00	40.00	50.00
Calibration	0.001	0.045	0.083	0.128	0.168	0.210
Standard addition	0.078	0.108	0.142	0.181	0.211	0.248

Determine whether the matrix affects the determination of vanadium and calculate its concentration on the unknown sample using the most adequate method of quantitation.

SOLUTION:

The effect of the matrix in the determination of the analyte can be visualized by comparing the slopes of the calibrations (see Chapter 2 for more details), by means of a Student's *t-test*.

First, the standard addition curve should be corrected for the procedural blank:

C_V (μg/L)	0.0	10.0	20.0	30.0	40.0	50.0
Addition	0.073	0.103	0.137	0.176	0.206	0.243

A representation of both calibrations reveals good linear relationships (see Figure 31) for both. The residuals plots showed that outliers are not present (these figures were not included here and the reader is encouraged to make such representations).

The least-squares fits were:

Calibration curve:

$$A = (0.0016 \pm 0.00109) + (0.0042 \pm 3.61 \cdot 10^{-5}) \cdot C_V$$

Standard addition curve:

$$A = (0.0708 \pm 0.00179) + (0.0034 \pm 5.91 \cdot 10^{-5}) \cdot C_V$$

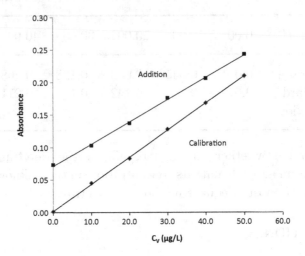

Figure 31.

The Fisher–Snedecor's F-test is now applied to compare the variances (standard errors) of the least squares fits:

> Recall that the $S_{y/x}$ statistic is obtained straightforwardly when using a spreadsheet to calculate the least-squares fit, although, otherwise, is very simple to calculate manually.

Calibration curve:

$(S_{y/x})_c = 0.00151$;   Addition curve: $(S_{y/x})_a = 0.00247$ and, so:

$$F = \frac{0.00247^2}{0.00151^2} = 2.682$$

The tabulated values for F are 6.36 and 15.98 at 95% and 99% confidence levels, respectively (4 dof for both calibrations). As $F_{\text{experimental}} < F_{\text{tabulated}}$, the variances are not statistically different. Hence, a pooled $S_{y/x}^2$ can be calculated (see Chapter 2 for details):

$$S_{x/y,\text{pool}}^2 = \frac{4 \cdot 0.00151^2 + 4 \cdot 0.00247^2}{6 + 6 - 4} = 4.19 \cdot 10^{-6}$$

And, now, the t statistic is calculated (the reader is strongly encouraged to program these equations in a spreadsheet to accelerate calculations):

$$t = \frac{0.0042 - 0.0034}{\sqrt{4.19 \cdot 10^{-6} \left(\dfrac{1}{1750} + \dfrac{1}{1750} \right)}} = 10.78$$

This experimental value is compared against the tabulated one (95 % confidence, and dof = 8); $t_{\text{tab (95 \%,8)}} = 2.306$.

In this case, since the experimental statistic is higher than the tabulated one, it must be concluded that the slopes of the calibration and the addition curves are statistically different. This indicates that the components of the matrix interfere with the determination of vanadium in the fuel oil samples. Therefore, the standard addition calibration must be used to calculate the concentration of this metal.

The concentration of V in the fuel oil aliquot is interpolated in the standard addition least-squares fit:

$$A = (0.0708 \pm 0.00179) + (0.0034 \pm 5.91 \cdot 10^{-5}) \cdot C_V$$

$$C_V = \frac{(2 \cdot 0.073) - 0.0708}{0.0034} = 22.12 \, \mu g/L$$

The concentration of V in the fuel oil sample is calculated considering the treatment procedure depicted in Figure 32.

$$C_{V_{\text{fuel oil}}} = 22.12 \, \frac{\mu g}{L} \cdot \frac{1\,L}{10^6 \, \mu L} \cdot \frac{20 \, \mu L}{4 \, \mu L} \cdot \frac{10^3 \, \mu L}{1\,mL} \cdot \frac{1\,mL}{0.25\,mL}$$

$$\cdot \frac{10\,mL}{0.2500\,g} = 17.70 \, \mu g/g$$

15. The determination of calcium in soil samples can be performed by FAAS. For this, a 1.0000 g aliquot of soil is dissolved with 10 mL of an acid mixture including hydrofluoric acid in a microwave oven. The solution thus obtained is filtered and diluted to 150.00 mL with ultrapure water. Aliquots of 10.00 mL of this solution were withdrawn to measure their absorbance. A calibration

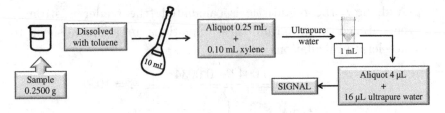

Figure 32.

and a standard addition curve were constructed. The results are shown in the following table. A procedural blank was prepared following the same steps and a 0.015 absorbance was obtained.

C_{Ca} (mg/L)	0.00	1.00	2.20	3.10	3.90	5.00	6.00
Calibration	0.010	0.100	0.204	0.283	0.347	0.450	0.530
Standard addition	0.105	0.185	0.295	0.367	0.418	0.525	0.605

Calculate the concentration of calcium in the soil sample using the correct quantitation method.

SOLUTION:

In this exercise, the effect of the matrix components in the determination of calcium in soil samples is studied. As in the previous example, the comparison of the slopes of the calibration and the standard addition method will inform about this effect.

First, correct the standard addition curve for the procedural blank.

C_{Ca} (mg/L)	0.00	1.00	2.20	3.10	3.90	5.00	6.00
Addition	0.090	0.170	0.280	0.352	0.403	0.510	0.590

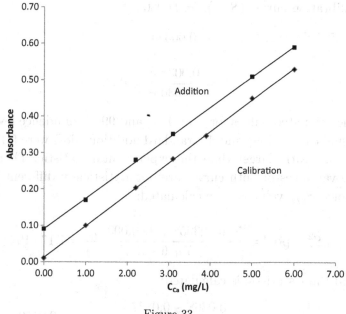

Figure 33.

The calibration plots show good linear relationships (Figure 33). The residuals plots reveal that outliers were not present in the usual calibration curve although the addition of 3.90 mg/L is an outlier for the standard addition line, therefore it was discarded for subsequent calculations (figures not included here).

The least-squares fits are:

Calibration curve:

$$A = (0.0120 \pm 0.00210) + (0.0868 \pm 0.00058) \cdot C_{\text{Ca}}$$

Addition curve:

$$A = (0.0907 \pm 0.00266) + (0.0837 \pm 0.00074) \cdot C_{\text{Ca}}$$

The F-test is now applied to compare the variances (standard errors) of the least squares fits.

Calibration curve: $(S_{y/x})_c = 0.00305$

Addition curve: $(S_{y/x})_a = 0.00384$

$$F = \frac{0.00384^2}{0.00305^2} = 0.631$$

The tabulated values for F at 95 % and 99 % confidence levels (5 (usual calibration) and 4 (standard addition) dof) were 5.19 and 11.39; clearly larger than the experimental statistic. Therefore, the variances of both curves are not statistically different and a pooled $S_{y/x}$ value can be calculated:

$$S^2_{x/y}, \text{pool} = \frac{5 \cdot 0.00305^2 + 4 \cdot 0.00384^2}{7 + 6 - 4} = 1.17 \cdot 10^{-5}$$

And the t statistic is calculated:

$$t = \frac{0.0868 - 0.0837}{\sqrt{1.17 \cdot 10^{-5} \left(\dfrac{1}{27.454} + \dfrac{1}{26.568} \right)}} = 0.0159$$

This is lower than the tabulated value, $t_{\text{tabulated}(95\,\%,9)} = 2.262$, and therefore the slopes of the calibration and the standard addition calibrations are not statistically different. We conclude that the components of the matrix do not interfere in the determination of calcium in these soil samples. The calibration curve can be used to calculate the concentration of Ca.

Recall that in order to take account of the procedural blank, it is also necessary to check whether the intercept of the calibration curve is statistically zero. The Student's t-test can be applied to address this issue.

$$t_{\text{experimental}} = \frac{|0 - a|}{S_a} = \frac{0.012}{0.00210} = 5.71$$

In this case, the experimental t value is higher than the tabulated one, $t_{\text{tabulated}(95\,\%,5)} = 2.571$, so the intercept is statistically different from zero due to some (unknown) experimental bias. In this situation, the signal of the calibration blank is subtracted from the signals of all standards, and the new calibration curve is used to calculate the Ca concentration by interpolation.

C_{Ca} (mg/L)	0.00	1.00	2.20	3.10	3.90	5.00	6.00
Standard solution	0.000	0.090	0.194	0.273	0.337	0.440	0.520

The new calibration:

$$A = (0.0020 \pm 0.00210) + (0.0868 \pm 0.00058) \cdot C_{\text{Ca}}$$

Recall that the absorbance provided by the aliquot of the unknown sample is the signal obtained for the first solution of the standard addition curve, corrected for the procedural blank.

$$C_{\text{Ca}} = \frac{0.090 - 0.002}{0.0868} = 1.014\,\text{mg/L}$$

The sample treatment procedure (Figure 34) leads to the concentration of calcium in the soil sample.

$$C_{\text{Casoil}} = 1.014\,\frac{\text{mg}}{\text{L}} \cdot \frac{1\,\text{L}}{10^3\,\text{mL}} \cdot \frac{150\,\text{mL}}{1.0000\,\text{g}}$$

$$= 0.152\,\frac{\text{mg}}{\text{g}} \cdot \frac{10^3\,\mu\text{g}}{1\,\text{mg}} = 152\,\mu\text{g/g}$$

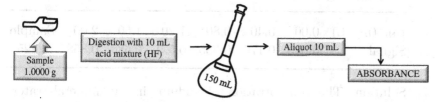

Figure 34.

EXERCISES PROPOSED TO THE STUDENT

16. The determination of manganese in drinking water can be performed by FAAS after a previous preconcentration step in a column with an appropriate ion exchange resin. For this, 200 mL of a water sample were passed through a column packed with a cation exchange resin. The manganese was eluted subsequently with 10 mL of 0.2 % sulfuric acid and the eluate was diluted to 25.00 mL with ultrapure water. A 10.00 mL aliquot of this solution provided an absorbance of 0.358. A series of manganese standard solutions were prepared and the absorbance values obtained are shown in the following table. Calculate the concentration of manganese in the drinking water sample.

C_{Mn}(mg/L)	0.00	0.10	0.20	0.30	0.40	0.50
Absorbance	0.00	0.15	0.28	0.40	0.55	0.70

Solution: The concentration of manganese in the drinking water sample is 32.25 mg/L.

17. Alkali metals in mineral water samples can be determined by flame atomic emission spectrometry (FAES) after an adequate dilution of the sample. Thus, 10.00 mL of sample were diluted to 100.00 mL with ultrapure water. Then, 5.00 mL of this solution were diluted to 50.00 mL also with ultrapure water before the addition of 40.00 mg/L of potassium, which is used as ionization suppressor. A series of sodium standard solutions were prepared containing 40.00 mg/L of potassium each. Calculate the concentration of sodium in the mineral water sample using the data in the next table.

C_{Na} (mg/L)	0.00	0.40	0.80	1.20	1.60	2.00	Sample
Signal		0.015	0.176	0.335	0.521	0.680	0.858	0.425

Solution: The concentration of sodium in the mineral water sample is 98.72 mg/L.

18. A coal sample (50 g) from a power plant was received in a laboratory for cobalt determination by ETAAS. A 2.5743 g aliquot of the sample was weighed and digested with aqua regia in a microwave oven. The solution obtained was filtered and diluted with ultrapure water to 250.00 mL. To perform the absorbance measurements, 10 μL of this solution were introduced in the atomizer along with 10 μL of ultrapure water (total volume injected 20 μL) providing an absorbance of 0.182. The signal obtained for the procedural blank was 0.013. The data for the calibration curve are in the following table. Calculate the concentration of cobalt in the coal sample.

C_{Co} (μg/L)	0.00	5.00	10.00	15.00	20.00	25.00	30.00
Absorbance	0.000	0.042	0.077	0.125	0.186	0.204	0.253

Solution: The concentration of cobalt in the coal sample is 3.97 μg/g.

19. An environmental quality control laboratory determines mercury in fish by cold vapor generation atomic absorption spectrometry (CVAAS). The following procedure is applied: three aliquots of 0.5215 g, 0.5034 g and 0.5187 g of fish tissue were weighed, digested with 5 mL of concentrated nitric acid and, after filtration, diluted to 100.00 mL with ultrapure water. Aliquots of 10.00 mL of these solutions were used for cold vapor generation whose absorbances were 0.241, 0.230 and 0.247, respectively. The procedural blank absorbance was 0.011. The standard solutions used to calibrate and the signals obtained are shown in the following table. Calculate the concentration of mercury in the fish tissue sample and check whether it can be quantitated reliably.

C_{Hg} (μg/L)	0.00	10.00	20.00	30.00	40.00	50.00	60.00
Absorbance	0.005	0.078	0.142	0.200	0.275	0.330	0.400

Solution: The concentration of Hg in the fish sample is $(6.71 \pm 0.40)\mu g/g$, RSD $= 2.4\%$. The LOQ of the procedure is $1.1\,\mu g/g$ (in classical terms) or $1.76\,\mu g/g$ (updated definition), so the concentration of Hg calculated in the sample can be quantitated reliably.

20. To determine iron in a granulated vitamin supplement, a commercial bottle of $50\,g$ was sent to a laboratory. Three aliquots of $0.1000\,g$ were weighed, dissolved in water and diluted to $20.00\,mL$. In order to reduce the matrix content, $2.00\,mL$ of this solution were diluted to $10.00\,mL$. The determination was performed by ETAAS introducing $20\,\mu L$ of the final solution in the atomizer. The masses of the aliquots and the absorbances obtained are in the following table.

	Mass (g)	Signal
Aliquot 1	0.1036	0.159
Aliquot 2	0.1254	0.167
Aliquot 3	0.1236	0.163

A procedural blank was prepared, which led to a 0.013 signal. The calibration was performed with iron standard solutions of 0.00, 5.00, 10.00, 15.00, 20.00 and $25.00\,\mu g/L$, obtaining signals of 0.002, 0.065, 0.120, 0.178, 0.230 and 0.295, respectively. Can the calculated concentration be quantitated reliably?

Evaluate the accuracy of the procedure whether the true concentration of iron in the supplement is $10.00\,\mu g/g$.

Solution: The concentration of iron in the vitamin supplement is $(10.85 \pm 2.26)\,\mu g/g$. RSD $= 8.4\%$. The LOQ of the procedure is $2.48\,\mu g/g$ (in classical terms) or $3.59\,\mu g/g$ (updated definition), so the concentration of Fe in the sample can be quantitated reliably. The accuracy of the procedure is satisfactory, since $t_{\text{experimental}}(1.62)$ is lower than $t_{\text{tabulated}(95\,\%,2)}(4.303)$.

21. A laboratory received an ore sample $(50\,g)$ for cadmium determination, which was treated with the following procedure: a

0.1712 g aliquot was weighed, digested with 4 mL of an acid mixture and diluted to 50.00 mL with ultrapure water. In order to eliminate the acid excess, a 10.00 mL aliquot of this solution was evaporated to dryness and the residue was dissolved with 25.00 mL of ultrapure water. The determination was performed by FAAS, by introducing 5.00 mL of the final solution in the nebulizer, which yielded a 0.179 absorbance signal. A procedural blank provided a signal of 0.017. The cadmium standard solutions and the signals obtained are shown in the following table.

C_{Cd} (mg/L)	0.00	2.00	3.80	5.80	8.00	9.60	10.00	11.20
Absorbance	0.003	0.053	0.104	0.160	0.220	0.260	0.300	0.310

Another aliquot of the sample (0.1683 g) was weighed, spiked with 100 μL of a cadmium standard solution of 2 500 mg/L and subjected to the same procedure, providing a 0.235 absorbance.

Calculate the concentration of cadmium in the ore sample, expressed as mg/g and mass percent, and evaluate the recovery of the procedure.

Solution: The standard solution of 10.00 mg/L was eliminated as anomalous from the calibration. The concentration of cadmium in the ore sample is 4.30 mg/g (0.43 %). The recovery of the procedure is 107 %.

22. The determination of tin in seawater can be performed by VGAAS. However, as its concentration is very low, a preconcentration step must be performed. For this, four 150.00 mL aliquots of the sample were passed through a column packed with an appropriate resin. The elution was performed with 15 mL of 1.0 % sulfuric acid and the eluates were diluted to 25.00 mL with ultrapure water. Aliquots of 5.00 mL of the final solutions were withdrawn for vapor generation and measurement. The signals obtained for the four aliquots analyzed were 21.4, 22.0, 21.7 and 22.3. The data for the calibration are in the following table. Calculate the concentration of tin in the seawater sample. Evaluate the accuracy of the procedure assuming

that the true concentration of tin in the seawater sample is $2.20 \, \mu g/L$.

C_{Sn} ($\mu g/L$)	0.00	5.00	10.00	15.00	20.00	25.00
Absorbance	0.10	8.30	16.90	25.20	33.50	42.00

Solution: The concentration of tin in the seawater sample is $(2.16 \pm 0.06) \, \mu g/L$. RSD = 1.8%. The accuracy of the procedure is satisfactory, since $t_{experimental}(2.0)$ is lower than $t_{tabulated(95\%,3)}$ (3.182).

23. To evaluate metal contamination in a river near an industrial area, sediment samples were collected. Aliquots of 0.2500 g were weighed and dissolved with 5 mL of an acid mixture (nitric and hydrofluoric acids). After filtering, the solutions were diluted to 50.00 mL with ultrapure water. A 10.00 mL aliquot of this was evaporated until 2 mL remain, diluting subsequently to 25.00 mL with ultrapure water. Procedural blanks were prepared in the same way. The determination of lead was performed by ICP-MS using bismuth as internal standard. The signals obtained are in the following tables. Calculate the concentration of lead in the river sediment samples.

C_{Pb} ($\mu g/L$)	10.0	20.0	30.0	40.0	50.0
Pb signal	2315	3800	5900	8680	11600
Bi signal	6150	5900	5740	6150	6730

	Sample 1	Sample 2	Sample 3	Procedural blank
Mass (g)	0.2505	0.2487	0.2579	—
Pb signal	5570	7721	6850	1235
Bi signal	6050	6120	5950	5924

Solution: The concentrations of lead in the river sediment samples are: $10.26 \,\mu g/g$, $15.28 \,\mu g/g$ and $13.19 \,\mu g/g$, for samples 1, 2 and 3, respectively.

24. Calcium can be determined in biological samples by ICP-AES using strontium as internal standard. Three 1.00 mL aliquots of a blood serum sample were treated with 5 mL of 0.5 % nitric acid and diluted to 10.00 mL with ultrapure water. Aliquots of 6.00 mL of these solutions were introduced in the nebulizer of the ICP-AES instrument. The relative signals (ratio between the analyte signal and the internal standard signal) obtained for the sample aliquots analyzed are 8.31, 7.50 and 7.96, respectively; whereas the calibration standards are compiled in the following table. Another 1.00 mL aliquot of the serum sample was spiked with $10 \,\mu L$ of a calcium standard solution containing 300 mg/L and treated as the other aliquots, obtaining a relative signal of 14.00. Calculate the concentration of calcium in the blood serum sample and evaluate the recovery of the analytical procedure.

C_{Ca} (mg/L)	0.00	0.10	0.20	0.30	0.40	0.50	0.60	0.70	0.80
Relative signal	0.20	2.24	4.52	6.63	9.01	10.94	13.71	15.49	17.91

Solution: The concentration of calcium in the blood serum sample is (3.53 ± 0.45)mg/L, RSD $= 5.1\%$. The recovery of the procedure is 91 %.

25. The determination of zinc in drinking waters can be determined by performing a liquid–liquid extraction and the subsequent measurement of the absorbance of the extracts by FAAS. The standard addition method was utilized to quantitate. For this, 100.00 mL aliquots of a drinking water sample were introduced in a funnel, spiked with increasing volumes of a standard

zinc solution of 10.00 mg/L, mixed with 10 mL of pyrrolidine dithiocarbamate (PDCA), shaken for 1 min and the organic phase separated. A new 10 mL portion of PDCA was added to the funnel and shaken again for 1 min. The two organic extracts were mixed and diluted to 25.00 mL also with PDCA. The signals obtained when aspirating these solutions in the nebulizer of the instrument are in the following table. Calculate the concentration of zinc in the drinking water sample.

Volume of Zn standard (mL)	0.00	4.00	8.00	12.00	16.00
Absorbance	0.061	0.112	0.157	0.206	0.255

Solution: The concentration of zinc in the drinking water sample is 0.50 mg/L.

26. The control of the residual waters discharged by an industry was performed through several analyses. One of them consists of the determination of the arsenic content by VGAAS using the standard addition method. Samples were collected in 1 L vessels. Aliquots of 15.00 mL of the samples were treated with 3 mL of concentrated nitric acid and diluted to 100.00 mL with ultrapure water. Five aliquots of 7.00 mL each were taken from this solution, spiked with 0.00, 1.00, 2.00, 3.00 and 4.00 mL, respectively, of a 10.00 mg/L arsenic standard solution and diluted to 25.00 mL with ultrapure water. Aliquots of the final solutions were withdrawn (10.00 mL) for vapor generation. The signals obtained were 0.080, 0.165, 0.242, 0.335 and 0.415. Two additional 15.00 mL aliquots of the residual water were subjected to the acid treatment and diluted to 100.00 mL. The 7.00 mL aliquots diluted to 25.00 mL provided absorbance values of 0.085 and 0.077, respectively. The signal obtained for the procedural blank was 0.006. Calculate the concentration of arsenic in the residual water sample.

Solution: The concentration of arsenic in the residual water sample is (8.53 ± 1.14) mg/L. RSD = 5.4%.

27. The concentration of chromium in a contaminated soil is determined by ETAAS. A 0.2135 g sample aliquot was digested with an acid mixture in a microwave oven. The solution obtained was filtered and diluted to 25.00 mL with ultrapure water. Aliquots of 2.00 mL of this solution were spiked with increasing amounts of a chromium standard solution and diluted to 10.00 mL; 5 μL of each solution were introduced in the atomizer together with 15 μL of ultrapure water (total volume injected = 20 μL). The absorbance values are shown in the following table. Two additional portions of sample (0.2097 g and 0.2042 g, respectively) were analyzed following the same procedure (except for the spiking step) providing signals of 0.100 and 0.106. The absorbance value obtained for the procedural blank was 0.009. Assess the accuracy of the procedure knowing that the true concentration of chromium in the contaminated soil is 20.00 μg/g.

C_{Cr} (μg/L)	0.0	5.0	10.0	15.0	20.0	25.0	30.0
Absorbance	0.095	0.145	0.200	0.263	0.320	0.372	0.434

Solution: The concentration of chromium in the contaminated soil sample is (19.72 ± 2.76) μg/g. RSD = 5.6 %. The accuracy of the procedure is satisfactory, since $t_{experimental}$(0.44) is lower than $t_{tabulated(95\%,2)}$(4.303).

28. In order to determine magnesium in almonds, several samples from different origins, were received in a laboratory. Aliquots of 0.2500 g of each sample were crushed and magnesium was extracted with 15 mL of an adequate organic solvent. The extracts were diluted to 50.00 mL with the same solvent. A standard addition calibration was constructed with 5.00 mL aliquots of the extract obtained from sample 1, which were spiked with increasing amounts of a magnesium standard solution and diluted to 25.00 mL. The solutions obtained from the other samples were diluted in the same proportion. The determination of magnesium was performed by FAAS. Calculate the concentration of magnesium in the almond samples using the following data.

	Mass (g)	Absorbance
Sample 1	0.2504	*
Sample 2	0.2478	0.110
Sample 3	0.2526	0.200

C_{Mg} (mg/L)	0.00	0.20	0.40	0.60	0.80	1.00
Absorbance	0.150	0.228	0.312	0.384	0.447	0.530

Solution: The concentrations of magnesium in the samples are: 0.39 mg/g, 0.28 mg/g and 0.52 mg/g for samples 1, 2 and 3, respectively.

*Recall that the absorbance of sample 1 is the signal obtained for the first solution of the standard addition curve.

29. The determination of vanadium in atmospheric particulate matter surrounding an industrial area was performed using a high volume sampler and glass fiber filters. An average flow of 43 m³/h was employed during 48 h. Filter portions of 9 cm² were weighed giving masses around 1.000 g. Vanadium was extracted with 3 mL of 1 % nitric acid. The extracts were diluted to 50.00 mL with ultrapure water. A standard addition calibration was obtained from one of the filter portions and diluting aliquots of 5.00 mL to 10.00 mL, after adding the final concentrations of V shown in the following table. All test solutions were diluted in the same way. Vanadium determination was performed by ICP-AES using cesium as internal standard. Calculate the concentration of vanadium in the particulate matter samples expressed in $\mu g/g$ and ng/m³.

Data: Total area of the filter = 350 cm²

C_V (μg/L)	0.00	10.00	20.00	30.00	40.00	50.00
Relative signal	0.90	1.40	1.85	2.46	2.88	3.45

	Mass (g)	Relative signal
Sample 1	1.025	*
Sample 2	1.087	1.03
Sample 3	0.987	1.15

Solution: The concentrations of vanadium in the atmospheric particulate matter samples are:

Sample 1: 1.75 μg/g; 33.38 ng/m^3
Sample 2: 1.89 μg/g; 38.67 ng/m^3
Sample 3: 2.32 μg/g; 43.12 ng/m^3

*Recall that the absorbance of sample 1 is the signal obtained for the first solution of the standard addition curve.

30. The determination of calcium in foodstuffs can be performed by FAAS after appropriate dissolution of the sample. A 1.000 g sample aliquot was digested with nitric acid in a microwave oven and its solution diluted to 50.00 mL with ultrapure water. A procedural blank provided a 0.015 absorbance. In order to evaluate the effect of the sample matrix in this determination, calibration and standard addition curves were constructed (see the following table). Calculate the concentration of calcium in the foodstuff sample using the correct quantitation method.

C_{Ca} (mg/L)	0.00	2.00	4.00	6.00	8.00	10.00
Calibration	0.006	0.052	0.095	0.136	0.174	0.221
Standard addition	0.130	0.165	0.195	0.231	0.264	0.300

Solution: The standard addition curve is used to calculate the concentration of calcium in the foodstuff sample, because the matrix components interfere in the Ca determination as showed statistically: $t_{experimental} = 11.90$, $t_{tabulated(95\%,8\ dof)} = 2.306$. The concentration of calcium in the foodstuff sample is 0.29 mg/g.

31. Iron can be determined directly by ETAAS in biological samples, including a dilution step only. Nevertheless, the effect of the components of the matrix in the absorption properties of iron must be evaluated. For this, calibration and addition solutions were prepared (see the following table). A 0.50 mL sample aliquot was diluted to 10.00 mL with ultrapure water. Aliquots of 1.00 mL of this solution were spiked with increased concentrations of an iron standard solution and diluted to 5.00 mL. Check whether the matrix of the biological sample interferes in the determination of iron and calculate the concentration of this metal using the correct quantitation method.

C_{Fe} (μg/L)	0.00	10.00	20.00	30.00	40.00	50.00
Calibration	0.004	0.068	0.134	0.188	0.248	0.305
Standard addition	0.083	0.148	0.207	0.258	0.315	0.375

Solution: The usual calibration curve is used to calculate the concentration of iron in the biological sample, since the matrix components do not interfere in the Fe determination: $t_{experimental} = 1.90$, $t_{tabulated(95\%,8\ degrees\ of\ freedom)} = 2.306$. The concentration of iron in the biological sample is 1.25 mg/L.

CHAPTER 6

CHROMATOGRAPHIC TECHNIQUES

María del Carmen Prieto-Blanco

OBJECTIVES AND SCOPE

This chapter aims at introducing quantitative analysis in chromatographic methods using both common calibration procedures (frequently termed 'external' calibrations in this field) and internal calibrations, as well as normalization. It stresses how to deal with calculations when the combination of chromatographic methods and multistage sample treatment methods occurs, as it is usually the case in many routine analyses.

The chapter examines briefly how to evaluate and optimize the separations by modifying chromatographic parameters like retention time, retention factor, resolution, etc. Some common working strategies are presented as well because of their relevance in qualitative chromatographic analysis. Most exercises are case studies derived from the need to quantify organic compounds in some relevant food, industrial and environmental fields.

Please note that it was assumed that you received already a minimum basic background on chromatography because many concepts cannot be explained in detail here.

1. INTRODUCTION

Chromatography is an analytical technique devoted to separate mixtures of compounds. Such mixtures are studied usually to know what compounds they contain (identification) and/or to what extent they are present (quantitation). The separation is carried out by two immiscible phases, one is immobilized over/into an inert support (stationary phase) and another flows through the former (mobile phase). The mixture of compounds, solubilized/mixed into the mobile phase, is moved across the stationary phase by the mobile phase. In order for the separation to be successful, the compounds should establish different interactions and to a different extent with the two phases. Thus, the compounds which become more retained within the system have more affinity for the stationary phase than the less retained compounds. The physical and chemical characteristics (solubility, volatility, hydrophobicity, etc.) of the compounds to be separated under the chromatographic conditions are responsible for the interactions with the two phases. A detailed examination of thermodynamic and kinetic aspects of the chromatographic separation is beyond the scope of this chapter. Readers are kindly forwarded to any general textbook on analytical chemistry [1–3].

Different mobile and stationary phases are possible. Initially, a literature search has to be done to select them as well as the most suitable experimental conditions; standard protocols and experience are also very important to success in this step. Nevertheless, they usually must be optimized further in order for the particular compounds present in the test mixture to move at different velocities among them and with respect to the mobile phase. Moreover, an additional phenomenon has to be taken into account: for each compound, the velocities with which the molecules move with respect to the mobile phase show a certain variability. As a consequence, a *chromatographic band* is formed. In general, the longer the time a compound is within the column, the larger the width of the chromatographic band will be. In addition, when a compound elutes from a column, its band is wider than at the very beginning of the elution process.

The main types of chromatography can be established according to the nature of the stationary and mobile phases. The two most relevant are *liquid chromatography* (LC) and *gas chromatography* (GC). Although there are huge differences in the instrumentation, the chromatographic process is similar for both. The mixture of compounds is introduced in the system (injected) and mixed with the mobile phase which transports it throughout the column (elution). The compounds are separated in the column according to the basic notions mentioned above. Then, each compound of the mixture elutes with the mobile phase at different time, passes through the detector and a signal is recorded for further treatment.

Since the late 1980s the conventional LC consists of passing the mobile phase through the stationary phase formed by the very small particles (around 5 μm diameter) at high pressure. Therefore, high-performance (or pressure) liquid chromatography (HPLC) is the usual term to denominate it. In chromatography (in particular HPLC), the molecular interactions between a compound and the chromatographic phases imply different types of forces (van der Waals, hydrophobic, ionic, etc.) all of them considered broadly under the term polarity. Thereby, a classification of the HPLC according to the relative polarity between the two phases (stationary phase and mobile phase) is very common. When the stationary phase is more polar than the mobile phase, it is called the *normal-phase mode* (historically, the first that was developed and used). Typical stationary phases are made from silica, or silica bonded with polar groups (diol, amine, etc.); while the mobile phase is constituted by solvents, such as hexane, toluene, etc. that are used in this mode. On the contrary, the *reversed-phase mode* (RP or RP-HPLC) works with stationary phase which is less polar than the mobile phase (mixtures of water and acetonitrile, methanol, tetrahydrofuran, etc.) being the HPLC type with more applications. In this case, the most common phase is the so-called C_{18} phase or octadecylsilane (ODS), in which an alkyl chain of 18 carbon atoms is bonded covalently to silica particles and arranged perpendicularly to its surface. Different lengths of the alkyl chains (C_4, C_8 either alone or combined with functional groups like such as nitrile or phenyl) allow modulating the polarity of the stationary phase.

2. CHROMATOGRAPHIC PARAMETERS

In general, chromatographic separations are monitored with a detector located at the end of the column. Thereby, a plot of the response of the detector (situated in the y axis) as a function of the so-called elution time (x axis) can be obtained, and this is called a *chromatogram*. The signal recorded while an analyte is not detected is called *baseline*. Whenever the detector responds physically or chemically to a substance, the signal changes (in general, it increases) and this is shown in the chromatogram in the form of a peak. The analyte does not yield always a very narrow peak because it is eluted as a band as mentioned in the previous section. Indeed, the shape of the peak represents the variability of the movement of the molecules of that substance around the average value for the majority of the molecules.

Two parameters are useful to characterize a chromatographic peak, namely the retention time and the peak width. The retention time (t_r) is the time elapsed from the instant of the sample injection to the detection of a substance by the detector, in particular to the maximum signal (which corresponds to the majority of the molecules of that substance). The width of the peak characterizes the chromatographic band, and it can be measured either at its baseline or at its half-height $(w_{1/2})$. Ideally, the peak is symmetric and, so, it could be defined as a Gaussian peak but an asymmetric shape is found usually, as well as a front (peak fronting) or a tail (peak tailing). The latter can be due to the interactions of molecules of the substance with residual groups of the stationary phase. Overloading of sample into the chromatographic column can cause peak fronting. An asymmetry factor can be defined as $A_S = b/a$, a and b being the distances from each side of the peak to a vertical line which passes through the maximum of the peak measuring a and b at 10% of the peak height (see Figure 1).

It is relevant to define the time necessary to elute an unretained component, i.e., a compound which does not interact with the stationary phase. This is called the *dead time* (t_0). It allows to determine the effective time during which a retained compound

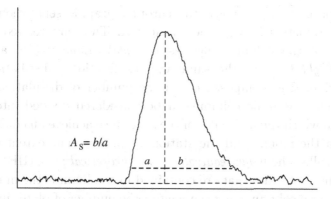

Figure 1. Asymmetry factor of a chromatographic peak.

interacts with the stationary phase, which is the *adjusted time* (t_r'). Mathematically, $t_r' = t_r - t_0$.

Thus, the ratio between the time a substance remains in the stationary phase and the time the same substance is in the mobile phase yields the retention factor (k), see Equation (1).

$$k = \frac{t_r'}{t_0} = \frac{t_r - t_0}{t_0} \tag{1}$$

Note that the retention factor is dimensionless.

High values of the retention factor imply a large retention of the component into the stationary phase. Values of k from 1 to 10 are recommended to get good separations.

Another parameter related to the retention of a substance is the separation factor or selectivity. It measures the ability of the chromatographic system to distinguish between two compounds. Mathematically, it can be defined as the ratio between the retention factors of two adjacent peaks (peaks 1 and 2), see Equation (2).

$$\alpha = \frac{k_2}{k_1} \tag{2}$$

k_2 being the retention time of the more retained compound. Thus, selectivity is always higher than 1, except when the retention times of the two compounds are identical.

In order to fully assess the chromatographic separation, some more parameters have to be considered. They are related to the efficacy of the column, such as the *plate number* (N) and the *plate height* (H). In the same way as fractional distillation can be considered as composed of a large number of distillation units, the chromatographic column can be considered divided into small zones, in which the molecules of the substance achieve an equilibrium between the mobile and the stationary phases. The extent of each zone is called the *height equivalent to a theoretical plate* (HETP) and the number of zones that are contained in a column is the number of plates. For a column of a given length, the number of plates increases when the plate height decreases. As a consequence, a high N implies more zones of equilibrium and this yields to lower band broadening, lower dispersion of molecules of substance and narrower peaks (see Equation (3)).

$$N = 16 \left(\frac{t_r}{w} \right)^2 = 5.545 \left(\frac{t_r}{w_{1/2}} \right)^2 \tag{3}$$

Finally, the *resolution* is a parameter which allows for a quantitative measure of the separation between two peaks, taking into account both the retention time and the width of the peak. In this way, resolution can be calculated by ratioing the difference between the retention time of two peaks and the peaks' average width (Equation (4)).

$$R_s = \frac{(t_{r2} - t_{r1})}{(w_1 + w_2)/2} \tag{4}$$

Another expression of the resolution is given by the Purnell equation. It relates many of the above mentioned parameters and it is especially useful to improve a chromatographic separation. It is calculated as the product of three terms (Equation (5)) including the retention factor (k), the selectivity (α) and the efficiency (N).

$$R_s = \left(\frac{(N)^{1/2}}{4} \right) \left(\frac{\alpha - 1}{\alpha} \right) \left(\frac{k}{k+1} \right) \tag{5}$$

3. OPTIMIZATION OF THE CHROMATOGRAPHIC SEPARATIONS

An adequate chromatographic separation is achieved only as a trade-off between the time of analysis and the resolution. This is so because some parameters which improve the resolution imply a very long analysis time, as for instance, an increase on the retention factor or on the column length.

The overall optimization of the separation is obtained by optimizing the experimental variables related to the mobile phase and/or the stationary phase. The effect of these variables on the resolution can be examined through chromatographic parameters such as selectivity, the retention factor and those related to efficiency. Using the resolution definition (Equation (5)), the effect of some of these parameters can be foreseen as long as the other parameters are kept constant. The retention factor of the compound more retained is included in one of the equation's terms. The higher the k values are, the larger the resolution will be. However, the influence is more significant for values ranging from 1 to 5 than for values greater than 5.

The experimental variables which have to be considered essentially in chromatography are: the solvent strength (for LC) and temperature (for GC) because they directly impact the retention factor. Low solvent strength and low temperature increase the resolution but yield longer analysis time. Unfortunately, when they have to be fixed to separate the first pair of peaks, the k for the latter peaks will increase and, consequently, the overall analysis time. A solution to this problem consists of modifying the mobile phase and/or the temperature with time. The resolution of the first compounds is not modified although the elution time of the last peaks gets reduced. This is called the *gradient elution* in LC or *temperature programme* in GC.

Selectivity is another term in Equation (5) with a huge influence on resolution. Small increases in selectivity result in significant resolution improvements. Depending on the type of chromatography, a change in the parameters of the stationary phase (GC) or of the mobile and stationary phases (LC) modifies the selectivity. In reversed-phase LC, the addition of specific compounds to the mobile phase (such as ion-pair reagents, buffers) or changing the pH of the mobile phase constitutes effective solutions. As a matter of example, the retention of the analytes containing ionizable groups is controlled by pH. Besides, this parameter is decisive to improve the selectivity of acidic and basic analytes, as well as that of neutral and ionic analytes. Other parameters related to the mobile phase like the organic solvent, temperature or solvent strength influence selectivity. In addition, the use of stationary phases containing different functional groups is another powerful tool to increase both selectivity and resolution.

Finally, in Equation (5), it is seen that resolution increases with the square root of N. The experimental variables which increase N are related to the characteristics of the column and the flow rate of the mobile phase. Using longer columns, higher N values are obtained although the analysis time is increased. A reduction in the flow rate increments the efficiency. In packed columns, a smaller diameter of the particles diminishes the plate height and higher plate numbers are obtained.

4. QUALITATIVE ANALYSIS: IDENTIFYING COMPOUNDS

In chromatography, the first step for interpreting the data is identifying the compound that gives rise to the chromatographic peak. For this purpose, it is necessary to obtain chromatograms of standard solutions under the same experimental conditions as for the samples. These conditions have to be established during the optimization of the separation. The most typical parameters to be optimized are related to both the mobile phase and the stationary phase but it is important to study the operational conditions of the injection system and the detector as well.

As introduced above, the retention time is the main identification criterion although it is not the only one as it will be seen below. Using one or several standard solutions, the retention time of each analyte can be obtained for the experimental conditions. Then, the retention time of each peak in the sample has to be compared to those obtained for the standard solutions. When doing this comparison, remember that the retention time (usually reported with three decimal digits) always shows small variations between the different injections, both within or between days. The differences can be larger when different instruments and/or different laboratories are considered. Several strategies to circumvent this difficulty are possible:

- To perform a minimum of three consecutive injections or sequences of standard/sample, important differences on the concentrations of the analyte in the standard and the sample should be avoided to minimize differences due to peak shape. The average retention times and the standard deviations of the same peak of the sample and the standard solutions are calculated. To positively identify which compound is responsible for a peak in the chromatogram of the unknown sample, check which average retention time is within one of the confidence intervals (average retention time and standard deviation) deduced for the peaks of the standards.
- If an internal standard (IS) is added to the sample, the ratio of the retention time of the analyte and that of the IS (relative retention time) must be similar to that obtained for a standard solution. In European Commission Decision 2002/657/EC, a tolerance of ±0.5 % for GC and ±2.5 % for LC is admitted.
- Addition of a known amount of a standard of an analyte to the sample which is likely to contain it. If the resulting chromatogram shows a larger peak, the analyte of the standard can be present in the sample. The enhanced peak area (or height) must be equivalent to the added amount of standard. On the contrary, a new peak, or a shoulder, or tailing in the existing peak indicates that analyte was not in the sample. In European Commission Decision 2002/657/EC, the differences before and after the addition of the standard were set as: 'the peak width at half-maximum height shall be within the 90–110 % range of the original width, and the retention times shall be identical within a margin of 5 %'. This

strategy is usually employed for the confirmation of a positive identification.

- Standards and samples are analyzed using different chromatographic conditions. The mobile or the stationary phases are changed so that the conditions allow to separate the analyte of interest. If under such different conditions, the retention time of the peak of interest of the sample corresponds to those obtained in the standard, the presence in the sample can be confirmed.
- Sometimes, the presence of a compound can be confirmed using chemical derivatization. If the derivatized product can also be separated in the sample, the disappearance of a peak (original compound) and the appearance of another derived peak (the derivatized compound) constitutes another secondary way to confirm the identification of the analyte in the unknown sample.

Note that in an identification process, positive conclusions (positive confirmation) are not 100 % definitive, but negative conclusions (no match) are definitive.

It is out of the scope of this chapter to discuss on the identification of analytes by advanced detectors such as diode array detector (DAD) or mass spectrometer. It suffices to mention here that the spectra of the sample peaks are compared with those of a collection of standards stored in a library in the computer controlling the chromatograph. This is studied in more detail in Chapter 7. Another relevant aspect that has to be addressed with the information provided by the detectors is the coelution of other compounds with the analyte. Whenever DAD is used, mathematic algorithms can be applied to the spectra obtained in the chromatographic peaks in order to evaluate the peak purity (i.e., if the peak is due to a single compound or more than one).

5. QUANTITATIVE ANALYSIS

Quantitative analysis should only be performed once the identification of a compound has been positive. The first step consists of the *integration* of the identified peaks. Although the peaks' integration is automatically provided by the chromatographic software, the peak area can be integrated manually. In order to measure the peak area, a baseline is drawn from the onset of the peak to its end. The peak height is measured from the apex to baseline. Different ways of integration (*integration events*) can be selected according to the different resolution of the peaks. If resolution does not reach the baseline, an integration with a tangent line or a vertical line drawn to the baseline is possible. It is important to fix the same integration event for standards and samples. However this is not always possible because when the sample is subject to a previous multistage treatment, a noisier baseline, shoulders or tailing, can appear and, so, the peaks of the sample are not exactly as those of the standards.

Both peak area and peak height can be used for quantitation although the former often yields more reproducible results. However, when integrating narrow and symmetric peaks, or with small peaks, peak height is also a good option. Small variations in the injection step influence both the area and the height and they are corrected using an IS.

5.1. External calibration, internal standard calibration and normalization

The following step for the quantitative analysis is to compare the integrated area or height of the peak of the sample with that obtained for the standards under the same chromatographic conditions. This requires the construction of a calibration or standardization using standard solutions containing different concentrations of the analyte. Thus, an empirical relationship between the peak area (or height) and the concentration is obtained. In most cases, a straight-line response is found in an adequate working range. Finally, the concentration of the analyte in the unknown peak is calculated by interpolation. A calibration curve per analyte is necessary and so, the quantitation process is very laborious if a high number of compounds (40–50) are separated.

Two main types of calibration are currently used for quantitative analysis in chromatography. They are referred to as 'external calibration' and 'internal calibration'. They were described in Chapter 2. The former term refers to a classical calibration procedure using standard solutions, whereas the latter involves the addition of a compound (an IS) to the solutions in order to calculate a ratio. The calibration line usually covers the overall lineal range of the detector (for the analyte of interest). Each signal is the average of the several replicates. Statistical analyses were detailed in Chapter 2.

In an IS calibration, a standard is added to both the samples and standard solutions. This type of calibration is to correct the variation in the signal caused by the sample treatment or the chromatographic process, and, is typically, very common in GC, when variations in the injection step are observed. A major drawback is to find a compound which meets the ideal requirements; namely, it should have a retention time close to that of the analyte, but it should be well resolved from any component of sample (including the analytes); it should not be present in the sample (although pre-existing 2% of added amount of IS can be admitted [4]); it should be available with high purity and should not react with any component of the sample. When GC or LC is coupled to mass spectrometry, the IS can be an isotopically labeled analyte (with a different mass than the analyte). Isotopically IS can be differentiated from the analyte in the detection step although sometimes it cannot be resolved perfectly.

To calibrate, a known and constant amount of an IS is added to the standard solutions. The ratio between the areas (or heights) of the analyte and the IS is plotted against the ratio of concentrations (analyte/IS). As the same amount of IS is added to the unknown samples, the amount of analyte present in the sample is calculated easily.

Normalization is not a common procedure but it can be used when the response of the detector is considered to be the same for a group of compounds. An example could be the quantitation of a series of homologues which present small differences in an alkyl chain. For each peak, quantitation is based on the percentage of area of the peak of interest with respect to the sum of the areas of all peaks. Standards are not required but all components of the group of interest present in the sample must be separated and detected under the same working conditions.

5.2. Multistage sample treatment in combination with chromatographic methods

The concentration of the analyte obtained by interpolation on the calibration is not of course the concentration we are looking for in the original sample since a multistage treatment is performed currently before the chromatographic analysis. The number and kind of stages depend on the chromatographic parameters (sensibility, selectivity, etc.); the nature of the matrix (solid, liquid or gas), and many other circumstances like the concomitants present in the sample, the concentration of analyte, etc. Extraction, purification, concentration, change of solvent and dilution are a few stages that may be performed.

Recall that the final stage of a sample treatment methodology has to provide a test aliquot with a suitable range of concentrations and a compatible solvent for it to be chromatographed.

Thus, the combination of sample treatment and chromatographic methods has quantitative and qualitative implications; among them (in relation to calibration) is the calculation of the performance parameters (mainly, limit of quantitation, limit of detection, etc.). In this sense, as mentioned in Chapter 2, it is of paramount importance to state clearly if the reported values correspond to instrumental or overall method limits.

To calculate the amount of the analyte in the original sample, several ways are possible, although a good option is to choose the shortest way (Chapter 1 presents some typical options and how to deal with the calculations). But the more important advice we can give you is to fully understand 'what occurs with the analyte in each stage' and keep in mind what an aliquot is.

REFERENCES

[1] Kellner, R.; Mermet, J.-M.; Otto, M.; Valcárcel, M.; Widmer H. M. (Eds.), (2004). *Analytical Chemistry: A Modern Approach to Analytical Science*, 2nd ed. Wiley-VCH, Weinheim.

[2] Skoog, D. A.; Holler, F. J.; Nieman. T. A. (2007). *Principles of Instrumental Analysis*, 6th ed. Cengage Learning, Santa Fe.

[3] Harris, D. H. (2007). *Quantitative Chemical Analysis*, 7th ed. W.H Freeman and Co. New York.

[4] Dolan, J. W. (2012). When should an internal standard be used? *LCGC*, 30(6), 474–480.

WORKED EXERCISES

1. Tanabe and Kawata [1] described a method to determine 1,4-dioxane in cosmetic products and household detergents using solid-phase extraction combined with gas chromatography/mass spectrometry (GC/MS). 1,4-dioxane is a toxic compound formed as a by-product in the industrial synthesis of some ethoxylated surfactants which can be used in the formulation of different cosmetic products. The treatment method allows sample weights in the 0.1–1.0 g range. In a typical analysis, a 0.9560 g sample of hair conditioner from trademark A was dissolved in 100.00 mL of a 10 % NaCl solution. This solution is passed at a 10 mL min^{-1} flow through two C_{18} and activated carbon cartridges and the analyte is eluted with 2 mL of acetone and 5 mL of dichloromethane (DCM). Then, the eluate was concentrated to 3 mL under a nitrogen stream. The chromatographic analysis was performed injecting 1 μL of the resulting solution. A 0.4 μg/mL concentration was obtained by interpolating the area of the corresponding peak in a calibration line (working range 0.01–1.0 μg/mL). The questions to be addressed are:

    a. Determine the 1,4-dioxane content in the hair conditioner, in mg/kg.

    b. A total of 10 samples of hair conditioner from another trademark B must be analyzed using the above mentioned method. In a previous monitoring campaign, values of 1,4-dioxane in hair conditioner for trademark B were found to be between 1.7 and 28 mg/kg. Justify numerically whether you would apply the procedure described for sample A or whether you would perform some modification.

SOLUTION:

Question a:

The sample treatment, as shown in Figure 2, consists of three steps: dilution, solid-phase extraction and concentration. It can be considered that the overall amount of analyte present in the 0.9560 g sample does not suffer from significant losses throughout the three stages (dilution, extraction and concentration). Besides, the concentration obtained by interpolation in the regression line corresponds to the concentration of the final solution (C_2) since an aliquot of this solution is taken. Given these considerations the calculations are simple:

$$0.4 \, \frac{\mu g}{mL} \cdot \frac{3 \, mL}{0.956 \, g} \cdot \frac{1 \, mg}{10^3 \, \mu g} \cdot \frac{10^3 \, g}{kg} = 1.2 \, mg/kg$$

Another option for the calculation is based on monitoring the concentration of analyte at each step of the sample treatment (depicted above). Thus, if we write the concentration at each

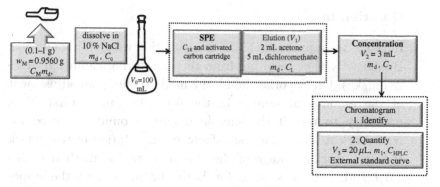

Figure 2.

step as a function of that in the preceding stage, we obtain:

$$C_0 = C_M \cdot \frac{W_M}{V_0}$$

$$C_1 = C_0 \cdot \frac{V_0}{V_1}$$

$$C_1 = C_2 \cdot \frac{V_2}{V_1}$$

being C_0 the concentration of the analyte in the dilution stage, C_1 the concentration of the analyte in the extraction stage, and C_2 the concentration of the analyte in the concentration stage. Therefore, equating the two expressions for C_1

$$C_2 \cdot V_2 = C_0 \cdot V_0$$

Substituting C_0 and solving for C_M

$$C_2 \cdot V_2 = \frac{C_M \cdot W_M}{V_0} \cdot V_0$$

$$C_M = \frac{C_2 \cdot V_2}{W_M}$$

The simplified expression is the same as that obtained for the first calculation option.

Question b:

Initially, we suppose that the concentration range for trademark B is exactly that resulting from the monitoring campaign (1.7–28 mg/kg). On the other hand, the method allows an adjustment of the amount of sample in the 0.1–1.0 g range. First, it is necessary to test if the sample weight (around 1 g) used for trademark A allows for a satisfactory quantitation of trademark B in the working range of the chromatographic method (0.01–1.0 µg/mL). This is done for both the lower and the upper limits of the estimated concentration range (1.7 mg/kg and

28.0 mg/kg).

$$1.7\,\frac{\mu g}{g}\cdot\frac{1.0\,g}{3\,mL}=0.57\,\mu g/mL$$

$$28\,\frac{\mu g}{g}\cdot\frac{1.0\,g}{3\,mL}=9.93\,\mu g/mL$$

For the lower limit, the concentration of analyte is within the working range but for the upper limit, it is not.

At least for the more concentrated samples (around $28\,\mu g/g$), you should modify the procedure. An option is to consider the lowest sample mass in both cases and check whether the expected concentrations are within the working range:

$$1.7\,\frac{\mu g}{g}\cdot\frac{0.1\,g}{3\,mL}=0.057\,\mu g/mL$$

$$28\,\frac{\mu g}{g}\cdot\frac{0.1\,g}{3\,mL}=0.93\,\mu g/mL$$

Now, it seems that the concentration range for C_2 is within the calibration working range. Therefore, a sample size of 0.1 g should be recommended for all samples of trademark B.

Note: The structure of the compounds referred to in this exercise is shown below:

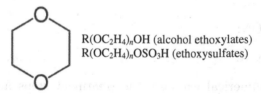

R(OC$_2$H$_4$)$_n$OH (alcohol ethoxylates)
R(OC$_2$H$_4$)$_n$OSO$_3$H (ethoxysulfates)

2. Miroestrol is a phytoestrogen found in *Pueraria candollei* var. *mirifica*, a plant from Leguminosae family used in traditional Thai medicine. According to a modified method reported by Yusakul *et al.* [2], miroestrol can be determined by HPLC-UV using an adequate sample treatment. Initially, 20 g of plants *Pueraria candollei* specimens from a representative sampling were dried at 50 °C and finely pulverized yielding a total weight of 5 g after drying. A 1.5203-g sample aliquot was washed with 5 mL

hexane and extracted four times with 5 mL chloroform–ethyl acetate (1:3, v/v). The extracts were combined and transferred to a 25.00 mL volumetric flask and diluted to volume. A 10.00 mL aliquot was evaporated and redissolved in 0.50 mL of ethanol. When 20 µL of the latter solution were analyzed by HPLC, 0.25 µg/mL miroestrol was reported. The student is requested to calculate:

a. Calculate the content of miroestrol in the sample in µg/g dry weight and µg/g wet weight.

b. Five samples were analyzed with the proposed HPLC method and with another enzyme-linked immunosorbent assay (ELISA) method for miroestrol determination. The values obtained (expressed in µg/g miroestrol, dry weight) are shown in the following table. Justify numerically whether the two methods are comparable.

	Concentration of miroestrol (µg/g, dry weight)				
ELISA method	194	90	75	30	0.1
HPLC method	198	96	70	29	0.09

SOLUTION:

Question a:

In this numerical exercise, the treatment steps are extraction, dilution and concentration (see Figure 3). Besides, to report the concentrations in the original sample, the division of the sample to gather an aliquot has to be considered. The calculation can be performed monitoring either the concentration or the amount of the analyte. In both ways, the expression to be obtained is the same:

$$C_M = C_{HPLC} \cdot V_3 \cdot V_1 \cdot \frac{1}{V_2} \cdot \frac{1}{W_s}$$

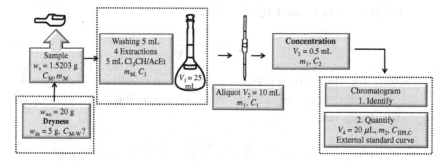

Figure 3.

Getting it is not difficult; as the mass of miroestrol (m_1) is determined from the reported concentration:

$$m_1 = C_{\text{HPLC}} \cdot V_3$$

Of course, the amount of miroestrol in the aliquot is lower than its total amount in the solution. The latter can be calculated immediately:

$$\frac{m_M}{V_1} = \frac{m_1}{V_2}$$

Substituting m_1 and solving for m_M:

$$m_M = C_{\text{HPLC}} \cdot V_3 \cdot V_1 \cdot \frac{1}{V_2}$$

This is the amount of miroestrol extracted from the weighted sample, w_s. Therefore, the concentration we were looking for, C_M is:

$$C_M = C_{\text{HPLC}} \cdot V_3 \cdot V_1 \cdot \frac{1}{V_2} \cdot \frac{1}{W_s}$$

If you prefer to make the calculations by taking into account the concentration of miroestrol in each step, the concentration obtained from the HPLC analysis is related to that present in the first dilution step C_1:

$$C_1 = C_{\text{HPLC}} \cdot V_3 \cdot \frac{1}{V_2}$$

Besides, C_1 is related to C_M

$$C_1 = C_M \cdot \frac{w_s}{V_1}$$

Equating the two expressions for C_1 and solving for C_M, we obtain the same equation. Now substituting the values, we already know:

$$C_{M(\text{dry wt.})} = 0.25\,\frac{\mu g}{mL} \cdot \frac{0.5\,mL}{10\,mL} \cdot \frac{25\,mL}{1.5203\,g} = 0.20\,\mu g/g$$

To evaluate the concentration of miroestrol in the original wet sample, we need to take into consideration the dryness step. We can calculate the percentage of the dry weight in the total weight:

$$\% \text{ dry weight} = \frac{5\,g \text{ dry wt.}}{20\,g \text{ wet wt.}} \cdot 100 = 25\,\%$$

In other words, there is 0.25 g dry weight per each gram of wet weight.

$$C_{M(\text{wet wt.})} = 0.20\,\frac{\mu g \text{ miroestrol}}{g \text{ dry wt.}} \cdot \frac{0.25\,g \text{ dry wt.}}{1\,g \text{ wet wt.}} = 0.05\,\mu g/g$$

Note that the result is logical because dryness is a concentration step. Therefore, the concentration of miroestrol in the dry weight is greater than in wet one.

Always try to be critical with the results you calculated.

Question b:

The two methods are comparable if the concentrations obtained for several levels of analyte correlate well and the slope of the regression line is statistically unity. The miroestrol values obtained with ELISA were represented against those with HPLC. The least squares equation is $y_{\text{ELISA}} = (0.97 \pm 0.09)x_{\text{HPLC}} + (2.11 \pm 9.32)$ with correlation coefficient of 0.9975. The confidence interval of the slope includes unity and, so, both methods yield comparable results.

Note: The structure of the compound referred to in this exercise is shown below:

3. Guan *et al.* [3] applied the modified Luke's method combined with GC with a nitrogen–phosphorus detector for the analysis of organophosphate and organochlorine pesticides in fruits and vegetables. Before the chromatographic analysis, the samples were treated as follows: 200 mL of acetone were added to 100.0000 g of chopped fruit, blended for 2 min and filtered. 50.00 mL of this solution were extracted with a mixture of solvents (50 mL of methylene chloride, 50 mL of petroleum ether and 10 mL of a saturated NaCl solution). Two other extractions were made with 50 mL of methylene chloride. The organic phases were modified and passed through 300 g of sodium sulfate. The final extract was brought to 240 mL with acetone. This solution was divided into two equal aliquots, one to analyze organophosphates and carbofuran plus malathion on each. The latter aliquot was evaporated to 2 mL and brought to 5.00 mL using acetone. Finally, the analysis of an unknown sample yielded a chromatogram in which the peak areas for carbofuran and malathion were 6 300 and 10 800, respectively. Calculate their concentrations (in ng/g) in the fruit. The calibration straight lines and the classical instrumental quantitation limits (IQLs) (see Chapter 2) are given in the following table:

Compound	Calibration	IQL (ng/mL)
Carbofuran	Peak area $= 167.9 C_{GC} + 30.1$	50
Malathion	Peak area $= 320.6 C_{GC} + 1982.9$	20

Where C_{GC} represents the concentration of analyte obtained by GC, in ng/mL.

SOLUTION:

For quantitation purposes, we must consider whether the experimental concentration is greater or lower than the limit of quantitation. To verify this issue, two straightforward options are possible: one consists of comparing the concentration of the test aliquot interpolated in the calibration line with the instrumental quantitation limit, whereas the other calculates the method quantitation limit (MQL) and compares it with the concentration of the analyte in the original sample. If the experimental value is lower than the quantitation limit, any reported value should be considered with special caution (see Chapter 2).

First, we have to interpolate the peak area of the test aliquots in the calibration line:

$$C_{GC(\text{Carbofuran})} = \frac{6\,300 - 30.1}{167.9}$$

$$= 37.3 \, \frac{\text{ng}}{\text{mL}} < IQL_{\text{Carbofuran}}$$

$$C_{GC(\text{Malathion})} = \frac{10\,800 - 1\,982.9}{320.6}$$

$$= 39.87 \, \frac{\text{ng}}{\text{mL}} > IQL_{\text{Malathion}}$$

It results that only malathion is quantifiable with an acceptable precision. The value for the carbofuran might be used as an indicative value.

To calculate the pesticide concentration in the fruit sample, the treatment stages have to be considered (see Figure 4). You can proceed as in exercise 2, although two extraction processes are performed here and two aliquots are drawn. Therefore, another factor (V_{T1}/V_{AQ1}) needs to be included in the expression:

$$C_M = C_{GC} \cdot V_5 \cdot \frac{V_3}{V_4} \cdot \frac{V_1}{V_2} \cdot \frac{1}{W_s}$$

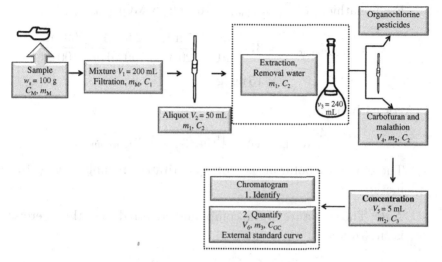

Figure 4.

Note that although in the second stage three extractions are done, finally only one solution is obtained.

$$C_{M(\text{Malathion})} = 39.87 \, \frac{\text{ng}}{\text{mL}} \cdot \frac{5 \, \text{mL}}{120 \, \text{mL}} \cdot \frac{240 \, \text{mL}}{50 \, \text{mL}} \cdot \frac{200 \, \text{mL}}{100 \, \text{g}}$$

$$= 16 \, \text{ng/g}$$

If the second way is preferred, it is necessary to calculate the MQL from the IQL and the sample treatment. Thus, MQL becomes:

$$\text{MQL} = \text{IDL} \cdot V_5 \cdot \frac{V_3}{V_4} \cdot \frac{V_1}{V_2} \cdot \frac{1}{W_s}$$

Considering that the lowest reliable concentration should coincide with the IQL, $C_{\text{GC}} = \text{IQL}$

$$\text{MQL}_{\text{Carbofuran}} = 50 \, \frac{\text{ng}}{\text{mL}} \cdot \frac{5 \, \text{mL}}{120 \, \text{mL}} \cdot \frac{240 \, \text{mL}}{50 \, \text{mL}} \cdot \frac{200 \, \text{mL}}{100 \, \text{g}}$$

$$= 20 \, \text{ng/g}$$

$$\text{MQL}_{\text{Malathion}} = 20 \, \frac{\text{ng}}{\text{mL}} \cdot \frac{5 \, \text{mL}}{120 \, \text{mL}} \cdot \frac{240 \, \text{mL}}{50 \, \text{mL}} \cdot \frac{200 \, \text{mL}}{100 \, \text{g}}$$

$$= 8 \, \text{ng/g}$$

For malathion, $C_{M(\text{Malathion})} = 16\,\text{ng/g} > \text{MQL}_{\text{Malathion}}$

$$C_{M(\text{Carbofuran})} = 37.3\,\frac{\text{ng}}{\text{mL}} \cdot \frac{5\,\text{mL}}{120\,\text{mL}} \cdot \frac{240\,\text{mL}}{50\,\text{mL}} \cdot \frac{200\,\text{mL}}{100\,\text{g}}$$

$$= 15\,\text{ng/g}$$

For carbofuran,

$$C_{M(\text{Carbofuran})} = 15\,\text{ng/g} < \text{MQL}_{\text{Carbofuran}}$$

Therefore, carbofuran is not quantitated reliably using this method.

Note: The structure of the compounds referred to in this exercise is shown below:

Carbofuran   Malathion

4. A sediment sample from a marine ecosystem nearby an urban area was searched for phthalate acid esters by liquid chromatography coupled to mass spectrometry (LC-MS) [4]. A sample treatment method (A) consisted of the following steps: 2.0000 g of sediment were extracted three times with 50 mL of a (1:1, v/v) DCM and hexane mixture in an ultrasonic water bath during 10 min. The combined extracts were concentrated to 5 mL under a nitrogen stream. The concentrate was passed through a column containing deactivated alumina. First, the interferent compounds were eluted and removed with 30 mL of hexane. Then, the analytes were eluted with a (1:1, v/v) DCM and hexane mixture. Finally, this eluate was evaporated to dryness and the residue reconstituted in 0.50 mL of methanol. A 3 µL aliquot was injected into the liquid chromatograph. The following table presents the peak areas obtained to get the calibration line. The linear range of the chromatographic method was previously found to be in the 2–240 ng/mL range.

C_{DnBP} (ng/mL)	Peak area (arbitrary units)
2	320
25	5 369
95	19 172
240	49 087

The exercise asks you to calculate:

a. The di-*n*-butyl phthalate (D*n*BP) content in sediment in ng/g if the D*n*BP peak area of the analyzed test aliquot was 2 482.
b. Would it be reliable enough to quantify DnBP in this sample using the sample treatment method (B) detailed next? Justify your answer numerically.

   Sample treatment method B: 10 g of a sediment sample are extracted under the same conditions as method A. The combined extracts are evaporated to dryness and reconstituted in 2.00 mL of methanol. A 3 μL aliquot is injected into the liquid chromatograph.
c. Calculate the selectivity parameter for D*n*BP and diethyl phthalate (DEP) if the dead time was estimated as 0.9 min and the retention times for DEP and D*n*BP were 4.1 and 8.9 min, respectively.

SOLUTION:

Question a:

First, the calibration straight line is calculated by the least-squares fit criterion as usual (see Figure 5).

The calibration equation is:

$$\text{Peak Area}_{DnBP} = (204.2 \pm 6.1)C_{DnBP} + (3.6 \pm 780)$$

Good linearity is obtained (e.g., studying the residual plot, as detailed in Chapter 2) with a satisfactory correlation coefficient of 0.9999.

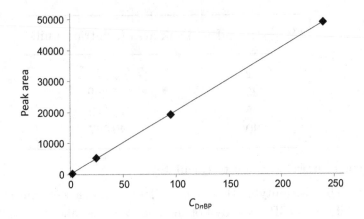

Figure 5.

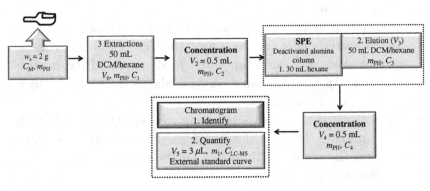

Figure 6.

Interpolating for the test aliquot:

$$C_{\text{LC-MS}(DnBP)} = \frac{2\,482 - 3.6}{204.2} = 12.1\,\text{ng/mL}$$

To report this concentration as ng/g of $DnBP$ in the sediment, the overall sample treatment is considered (see Figure 6).

The simplified calculation obtained in exercise 1 can be applied in the present one also, because:

— The concentration $C_{\text{LC-MS}}$ is equal to C_4 (see Figure 6) since $C_{\text{LC-MS}}$ solution is an aliquot of C_4 solution.

— The amount of D*n*BP (m_{PH}) remains constant (no losses are reported) throughout the treatment steps including solid–liquid extraction, cleanup and concentration.

Substituting in the simplified expression:

$$C_M = \frac{C_{\text{LC-MS(D}n\text{BP)}} \cdot V_4}{W_s}$$

$$C_M = 12.1 \frac{\text{ng}}{\text{mL}} \cdot \frac{0.5\,\text{mL}}{2\,\text{g}} = 3\,\text{ng/g}$$

Question b:

The optional treatment B consists of extraction and concentration steps. If we extract 10.0000 g of sediment with an average content of 3 ng/g, we obtain 30 ng of D*n*BP. They are dissolved in 2 mL, which leads to a 15 ng/mL concentration (slightly higher than treatment 'A').

Question c:

In the introductory section of this chapter, some chromatographic parameters were reviewed and the relation between retention factor and selectivity was presented:

$$k_{\text{D}n\text{BP}} = \frac{8.9 - 0.9}{0.9} = 8.89$$

$$k_{\text{DEP}} = \frac{4.1 - 0.9}{0.9} = 3.55$$

$$\alpha = \frac{k_{\text{D}n\text{BP}}}{k_{\text{DEP}}} = 2.5$$

Note: The structure of the compounds referred to in this exercise is shown below:

5. A sample of milk is analyzed to determine a pesticide, sodium monofluoroacetate (MFA) [5]. The previous treatment method involves removing proteins and lipids and concentrating the sample in order to get a suitable concentration for the test aliquot to be injected into the chromatographic liquid-chromatography tandem mass spectrometry (LC-MS/MS) device. Thus, 1.0000 g of sample is weighted in a propylene tube, fortified with 50 μL of 0.2 μg/mL $^{13}C_2$-MFA IS and dissolved in 20.00 mL of water. To remove proteins, 8 mL of acetonitrile are added to the fortified sample and the mixture is centrifuged for 5 min. Lipids are then eliminated by washing the supernatant with 10 mL of hexane and centrifuged again for 5 min. The hexane phase is removed and the other phase is acidified, a mixture of salts is added and centrifuged once more. The supernatant is evaporated to 500 μL. A 20 μL aliquot is injected into the chromatographic system. Four MFA standards containing the same concentration of IS as the sample yield the peak area ratios (analyte/IS) shown in the following table:

$C_{MFA}/C^{13}C_2$-MFA(ng/mL)	$A_{MFA}/A^{13}C_2$-MFA
0.5	0.41
1.0	0.85
1.5	1.24
2.0	1.65

Questions to be solved are:

a. The concentration of the IS ($^{13}C_2$-MFA) in the fortified milk sample in μg/kg.

b. The concentration of ($^{13}C_2$-MFA) in the solution injected into the chromatographic system.

c. The content of MFA in the sample of milk, in μg/kg, considering that its peak area ratio (analyte/IS) is 0.92.

SOLUTION:

Question a:

The amount of IS added to the sample was:

$$W_{IS} = 0.2 \frac{\mu g}{mL} \cdot \frac{1\,mL}{10^3\,\mu L} \cdot 50\,\mu L = 0.01\,\mu g$$

Therefore, its concentration was:

$$C_{IS} = \frac{0.01\,\mu g}{1\,g} \cdot \frac{10^3\,g}{kg} = 10\,\mu g/kg$$

Note that the amount of IS is considered negligible to calculate the total mass (term in the denominator).

Question b:

The method has to be optimized so that MFA and IS are not lost when removing proteins, lipids and concentration (see Figure 7). Following, the $0.01\,\mu g$ of IS added in the first stage should also be in the $500\,\mu L$ final solution, thus:

$$C_{2IS} = \frac{0.01\,\mu g}{500\,\mu L} \cdot \frac{10^3\,ng}{1\,\mu g} \cdot \frac{10^3\,\mu L}{mL} = 20\,ng/mL$$

Question c:

The calibration straight-line equation (depicted in Figure 8) ($r^2 = 0.9994$) was:

$$\frac{A_{MFA}}{A_{^{13}C_2\text{-MFA}}} = (0.822 \pm 0.06)\frac{C_{LC\text{-MS(MFA)}}}{C_{LC\text{-MS}(^{13}C_2\text{-MFA})}} + (0.01 \pm 0.08)$$

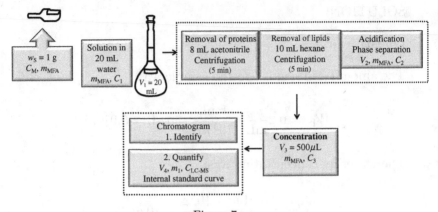

Figure 7.

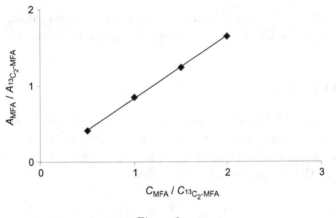

Figure 8.

Solving for $C_{\text{LC-MS(MFA)}}$

$$C_{\text{LC-MS(MFA)}} = \frac{(A_{\text{MFA}}/A_{^{13}\text{C}_2\text{-MFA}}) - 0.01}{0.822}$$
$$\cdot\, C_{\text{LC-MS}(^{13}\text{C}_2\text{-MFA})}$$

Moreover, $C_{\text{LC-MS(MFA)}} = m_{\text{MFA}}/V$ and $C_{\text{LC-MS}(^{13}C_2\text{-MFA})} = m_{^{13}\text{C}_2\text{-MFA}}/V$. Since V is the same volume in both expressions, it is straightforward to derive the final calculations:

$$m_{\text{MFA}} = \frac{(A_{\text{MFA}}/A_{\text{13}C_2\text{-MFA}}) - 0.01}{0.822} \cdot m_{\text{13}C_2\text{-MFA}}$$

As the method was optimized not to lose MFA or IS, m_{MFA} is the amount of MFA before sample treatment. Expressed as a concentration, as it was required:

$$C_{\text{MFA}} = \frac{(A_{\text{MFA}}/A_{\text{13}C_2\text{-MFA}}) - 0.01}{0.822} \cdot \frac{m_{\text{13}C_2\text{-MFA}}}{W_s}$$

$$C_M = \frac{0.92 - 0.01}{0.822} \cdot \frac{0.01\,\mu g}{1\,g} \cdot \frac{1\,g}{10^3\,kg} = 11.1\,\mu g/kg$$

Note: The structure of the compounds referred to in this exercise is shown below:

6. The synthesis of a quaternary compound, triethylbenzylammonium chloride (TEBA) is monitored in order to control the by-products [benzyl alcohol (BZA) and benzaldehyde (BZ)] and one of the reagents [benzyl chloride, (BZC)] in an industrial process [6]. An RP-HPLC separation for the four compounds (TEBA, BZA, BZ, BZC) is optimized using a cyano-propyl silica column of dimensions 125 mm long × 4.6 mm diameter, 5 μm particle size. Three mobile phases were tested: mixtures of eluent 'A' and acetonitrile; eluent 'B' and acetonitrile; and eluent 'C' and acetontrile; at 1 mL/min flow rate. The retention factors (*k*) are shown in the following table. No differences for the *k* values were found for BZA, BZ, BZC using different aqueous eluents (A, B, C). Questions to be addressed are:

   a. Which parameters of the mobile phase influenced the retention of the four compounds?

b. Using the k values displayed in the table, what conditions would you select to separate the compounds?

c. Report the elution order of the compounds under the conditions selected in the previous question.

Additional information: eluent 'A': 100 mM sodium acetate adjusted with acetic acid at pH = 5, eluent 'B': 100 mM sodium acetate adjusted with acetic acid at pH = 7, eluent 'C': 35 mM sodium acetate adjusted with acetic acid at pH = 7.

		Acetonitrile %			
k	45	50	55	60	Eluent
k_{TEBA}	0.88	0.87	0.88	0.9	A
k_{TEBA}	1.41	1.44	1.39	1.38	B
k_{TEBA}	2.65	2.61	2.66	2.60	C
k_{BZA}	0.54	0.49	0.47	0.45	–
k_{BZ}	0.81	0.70	0.66	0.58	–
k_{BZC}	1.47	1.15	0.97	0.80	–

SOLUTION:

Question a:

Some parameters associated to the mobile phase have no influence on the retention factor of every compound. The retention factor of the by-products and the reagent did not change at different pH and concentrations of sodium acetate. This is because they have no ionizable groups in their molecule. However, TEBA increased its retention when the pH was increased (from 5 to 7) and $NaCH_3COO$ concentration was decreased (100 mM to 35 mM).

With respect to the acetonitrile (ACN) %, it did not influence the retention factor of TEBA, although the retention factors of BZA, BZ and BZC diminished when ACN % increased. The following table summarizes the effects of the three parameters on the studied compounds.

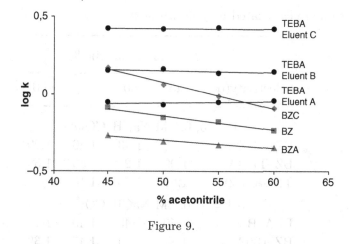

Figure 9.

	TEBA	BZA	BZ	BZC
[NaCH$_3$COO]	+	−	−	−
pH	+	−	−	−
ACN %	−	+	+	+

Figure 9 indicates the trends derived from the results above. Note that k was replaced by log k in order to observe linear trends. For BZA, BZ and BZC, there is a negative relationship. Moreover, there are conditions under which the k values for two compounds are similar (e.g., benzyl chloride and TEBA at pH = 5 and 55 % ACN).

The separation factor or selectivity has to be studied now. As selectivity is calculated for each pair of adjacent peaks, the elution order is deduced first from the retention factors observed at each setup. As it is known, the retention factor increases with the retention time. Studying the tables and the graph in the previous paragraphs, there are two different trends depending on the pH. Thus, the elution order at pH = 5 is BZA, BZ, TEBA and BZC. But at pH = 7, the TEBA retention factor increases and the elution order is BZA, BZ, BZC and TEBA. Taking into account the elution order to get the adjacent peaks, the selectivity factors are calculated and shown next:

	Acetonitrile %			
Selectivity(α)	45	50	55	60
pH = 5, 0.1 M NaCH$_3$COO				
BZA–BZ	1.50	1.43	1.40	1.29
BZ–TEBA	1.09	1.21	1.33	1.55
TEBA–BZC	1.67	1.35	1.10	—
pH = 7, 0.1 M NaCH$_3$COO				
BZA–BZ	1.50	1.43	1.40	1.29
BZ–BZC	1.81	1.64	1.47	1.39
BZC–TEBA	1.04	1.25	1.40	1.71
pH = 7, 0.035 M NaCH$_3$COO				
BZC–TEBA	1.80	2.27	2.74	3.25

TEBA is more retained than BZC at pH $= 5, 0.1$ M and 60 % acetonitrile, although at pH $= 7, 0.1$ M and 45 % acetonitrile, TEBA is less retained than BZC.

Question b:

To select the best conditions, the selectivity calculated for each setup for three pairs of compounds is evaluated. Those conditions with selectivity values from 1 to 1.1 were rejected because they indicate the worst separations. Thus, at pH $= 5$, the best conditions are those in which the mobile phase contains 50 % acetonitrile. In general, the selectivity factors were higher at pH $= 7$. A significant improvement on selectivity was achieved using a lower buffer concentration. In this case, the selectivity (BZA–BZ and BZ–BZC) is favored for low % acetonitrile, but the BZC–TEBA pair shows best selectivity with high % acetonitrile. As a consequence, the best conditions should contain 45 % acetonitrile, which yields the best trade-off for the three pairs of peaks.

However, it is remarkable to bear in mind that the selectivity and the retention factor provide only partial information on the separation. As TEBA is the major compound (most concentrated) in the samples, the others are at trace levels, the peak widths would also be considered.

Question c:

For 45 % acetonitrile and pH = 7, the elution order (from less to more retained) is

$$BZA < BZ < BZC < TEBA$$

Note: The structure of the compounds referred to in this exercise is shown below:

7. Calculate the resolution between the pairs of adjacent peaks discussed in exercise 6 under 45 % acetonitrile and pH = 7 conditions using the data of the following table. Discuss the influence of selectivity and peak width on the quality of the separation. Under those conditions, the retention time for a non-retained compound is 1.23 min.

	BZA	BZ	BZC	TEBA
w (min)	0.083	0.077	0.1	2.1

SOLUTION:

The optimization of many chromatographic separations is performed fixing the concentration levels for all compounds of the mixture in a similar level. However, in the separation studied in exercises 6 and 7, the concentrations were very different because the objective was the monitoring of the by-products and a reagent after obtaining TEBA (mg/mL versus μg/mL). In this situation, only BZA, BZ and BZC are quantitated whereas TEBA is 'in excess'. This fact explains the difference between the widths observed for TEBA (2.1 min) and for the other compounds (around 0.1 min, see table above).

The retention time of each compound is calculated considering the dead time and the retention factor provided in exercise 6. Solving for t_r, we obtain:

$$t_r = kt_0 + t_0 = t_0(1 + k)$$

The following t_r values are obtained:

	BZA	BZ	BZC	TEBA
t_r (min)	1.85	2.17	2.96	4.38

And, accordingly, the following resolution values are calculated.

	BZA–BZ	BZ–BZC	BZC–TEBA
Resolution	4.05	8.95	1.29

Differences are observed when the resolution values are compared to the selectivity ones in exercise 6 (which is also a measure of the quality of the separation). The BZ–BZC pair has better selectivity than the BZA–BZ one and also best resolution (more than twice). The BZ–BZC pair and the BZC–TEBA pair has similar selectivities (around 1.80), although the resolution of the BZC–TEBA pair is seven times lower, due to the width of the TEBA peak.

8. Dimethyl phthalate (DMP) and di-2-ethylhexyl phthalate (DEHP) are separated using an RP-HPLC-UV method. Some chromatographic parameters, and the calibration equations are

shown in the following table:

Compound	Retention time (min)	$W_{1/2}$	Calibration
DMP	3.60 ± 0.17	0.13	Peak area $= 869C_{DMP} + 50.1$
DEHP	7.22 ± 0.2	0.11	Peak area $= 227C_{DEHP} + 50.6$

A sample of water is analyzed with this method after its treatment (extraction, cleanup and concentration). Their characteristics are tabulated next.

		Retention time (min)	$W_{1/2}$ (min)	Peak
Sample	Peak 1	3.58 ± 0.12	0.13	21 750
	Peak 2	7.42 ± 0.08	0.12	5 700
Fortified	Peak 1	3.65	0.14	43 075
sample	Peak 2	7.25	0.11	6 700
	Peak 3	7.50	0.15	5 980

An aliquot of the sample, fortified previously with $25\,\mu g/L$ DMP and $30\,\mu g/L$ DEHP was analyzed as well. The results are collected in the table above.

a. Can you identify which compounds the peaks correspond to? Justify your answer.
b. Calculate the concentration of the compound(s) you identified in the test aliquot.

SOLUTION:

Question a:

The only criterion to identify the peaks in this exercise is the retention time as the detection is not performed using a detector which provides additional information (mass spectrometer or diode array detector). If we compare retention times of DMP and DEHP in the standards with the two sample peaks of the test aliquot, the first can be DMP since its average retention time and its standard deviation lie well within those calculated for the standard (see table below). This is validated considering the fortified test aliquot because a single and larger peak appears at 3.6 min. Moreover, the differences between the retention time and width at its half-maximum height $(W_{1/2})$ before and after the standard addition were checked (Commission Decision 2002/657/EC) [7]. The retention time had a difference of only 0.07 min, which represents 1.95 % (within the 5 % margin recommended). Regarding the $W_{1/2}$, it was 107 % in the fortified test aliquot, within the 90–110 % range. Hence, the first peak can be assigned to DMP.

The second peak has a retention time close to that for DEHP on the standard, but it presents a retention time quite deviated from the upper limit of the standard. Two peaks found in the fortified test aliquot at 7.25 min and 7.5 min did not permit to identify unequivocally the second peak of the test aliquot to DEHP.

Question b:

The concentration of the positively identified compound, DMP is calculated using its corresponding calibration straight line:

$$\text{Peak area} = 869 C_{\text{DMP}} + 50.1$$

$$C_{\text{DMP}} = \frac{21\,750 - 50.1}{869} = 25.0\,\mu\text{g/L}$$

Its concentration on the fortified test aliquot, to which 25 μg/L DMP has been added, was:

$$C_{\text{DMP}} = \frac{43\,275 - 50.1}{869} = 49.7\,\mu\text{g/L}$$

And, therefore, the overall recovery has been:

$$R = \frac{49.7 - 25.0}{25} \cdot 100 = 99.1\,\%$$

9. The separation of several polycyclic aromatic hydrocarbons (PAHs) is optimized using a C_{18} column and isocratic elution. The retention times of chrysene and benzo[a]pyrene under different mobile phase conditions are tabulated next. A good straight-line relationship between the natural logarithm of the retention time, Ln t_r, and the percentage of acetonitrile was observed. The retention time for another PAH (benzo[a]anthracene) was measured at 6.9 min using a 85 % acetonitrile and 15 % water mobile phase. What is the elution order for a mixture of chrysene, benzo[a]anthracene and benzo[a]pyrene under these conditions? Justify your answer numerically.

Acetonitrile %	$t_{r_{\text{chrysene}}}$ (min)	$t_{r_{\text{benzo}[a]\text{pyrene}}}$ (min)
100	3.38	6.06
95	4.40	8.43
90	5.54	11.58
80	10.1	22.93

SOLUTION:

Part of the solution is given in the enunciate as, first, the natural logarithms of the retention times for the two compounds whose values are tabulated have to be calculated. The least-squares fits are:

$$\text{Ln } t_{r_{\text{chrysene}}} = (-0.0546 \pm 0.0008) \cdot (\% \text{ acetonitrile})$$
$$+ (6.6571 \pm 0.8482)$$

$$\text{Ln } t_{r_{\text{Benzo}[a]\text{pyrene}}} = (-0.0665 \pm 0.0025) \cdot (\% \text{ acetonitrile})$$
$$+ (8.4417 \pm 0.2339)$$

If the relationships hold, it is possible to evaluate the retention times of the compounds under different conditions, at least in the range at which the relation is verified. Thus, the retention times for the two compounds at 85 % acetonitrile are:

$$\text{Ln}\, t_{r_{\text{chrysene}}} = (-0.0546) \cdot 85 + 6.6571 = 2.02$$

$$\text{Ln}\, t_{r_{\text{benzo}[a]\text{pyrene}}} = (-0.0665) \cdot 85 + 8.4417 = 2.79$$

And, so:

$$t_{r_{\text{chrysene}}} = 7.54\, \text{min and}\; t_{r_{\text{benzo}[a]\text{Pyrene}}} = 16.36\, \text{min}$$

$$t_{r_{\text{benzo}[a]\text{antrhacene}}} < t_{r_{\text{chrysene}}} < t_{r_{\text{benzo}[a]\text{pyrene}}}$$

The requested elution order is: benzo[a]anthracene, chrysene and benzo[a]pyrene.

EXERCISES PROPOSED TO THE STUDENT

10. To monitor fenamiphos (an organophosphorus insecticide) in vegetables sold at a local market, an analytical method based on (GC-MS) was selected [8]. The working procedure was: 10.0000 g of a grinded lettuce sample were extracted with 30 mL of acetonitrile in an ultrasound bath for 10 min. Magnesium sulfate (3 g) was added to remove water and a 15 mL aliquot was concentrated to 2 mL using a rotary evaporator. For cleanup, these 2 mL were loaded in a solid-phase extraction cartridge and fenamiphos was eluted with 20 mL of acetonitrile/toluene (3:1). The eluate was concentrated and the residue reconstituted in 0.5 mL of acetone. A 2 μL aliquot was injected into the gas chromatograph. Considering that the concentration obtained by interpolation in the calibration function was 0.5 μg/mL, calculate the content of fenamiphos in the original sampled lettuce in mg/kg.

    Solution: The calculated concentration of fenamiphos in lettuce was 0.05 mg/kg.

11. Two PAHs, naphthalene (Naph) and fluorene (Flu), were analyzed in river water using solid-phase extraction and GC-MS [9]. The procedure was as follows: 250.00 mL of river water were loaded in an extraction cartridge at 5 mL/min flow rate. Then, vacuum was applied to the cartridge in order to remove residual water. The PAHs' elution was performed with 15 mL of hexane

at 1 mL/min flow rate. The eluate was evaporated to dryness and reconstituted in 0.5 mL of hexane. The calibration lines for the two compounds were: $y_{Naph} = 226.6\,x_{Naph} + 15.8$, and $y_{Flu} = 157.9\,x_{Flu} - 8.6$ where all concentrations were in ng/L. The next questions have to be solved:

a. If the peak areas for Naph and Flu were 36 045 and 12 623, respectively, calculate the contents of Naph and Flu in the river water, in ng/L.

b. Under the chromatographic conditions of the method, the retention factors for Naph and Flu were 3.5 and 6.8, respectively. If the composition of the mobile phase changed, the number of theoretical plates remained constant for the two peaks, but the new retention factors for Naph and Flu were 4.5 and 9.4. Decide under what conditions the peaks of Naph and Flu are resolved best.

Solution: (a) The calculated concentration of Naph and Flu in river water were 0.32 and 0.16 ng/L, respectively.

(b) It is possible to relate the resolution obtained for the experimental setups using the Purnell's equation (see theoretical part); in this case: $R_1 = 0.96R_2$. Therefore, the second conditions (R_2) provide best resolution between the compounds.

12. A brominated flame retardant, 1,2-bis(2,4,6-tribromophenoxy) ethane (BTBPE), was analyzed in indoor dust by GC-MS [10]. The sample treatment applied was: 50.0 mg of a sample were fortified with a volume (V_1) of a 0.2 μg/mL solution of 3,3′,4,4′-tetrabromodiphenyl ether (BDE) (as IS). Then they were extracted with 2 mL of hexane:acetone (3:1) in an ultrasound bath for 5 min, and centrifuged for 10 min. The supernatant was evaporated to dryness with a nitrogen stream and reconstituted in 1.00 mL of hexane. A cleanup was performed with a silica cartridge and the analyte eluted with 10.00 mL of DCM. This latter eluate was submitted to a second cleanup step, in a Florisil cartridge; 5.0 mL of hexane and 5.0 mL of DCM were used. After eluting BTBPE, the second eluate was evaporated to dryness and

solubilized in $50\,\mu\text{L}$ of isooctane for the GC analysis. Questions to be addressed are:

a. Calculate the volume of solution of IS (V_1) required to obtain a final concentration of $50\,\text{ng/mL}$ in the solution to be injected into GC-MS.

b. Determine the content of BTBPE in indoor dust (in ng/g) if the measured peak area ratio (analyte/IS) was 1.2.

$C_{\text{BTBPE}}/C_{\text{IS}}$ (ng/mL)	$A_{\text{BTBPE}}/A_{\text{IS}}$
0.4	0.1
3.6	0.6
6.8	1.1
12.2	1.9
16.0	2.5

Solution: (A) The volume of the IS to be used is $12.5\,\mu\text{L}$. (B) The content of BTBPE was $377\,\text{ng/g}$.

13. *N*-nitrosodiethanolamine (NDELA) is a toxic substance which may occur in cosmetic products through reaction between diethanolamine and some preservatives [11]. To determine it in two shampoo samples ($M1$, whose due date had expired and $M2$ within its expiration date), 0.9571 g $M1$ and 0.9825 g $M2$ were weighted. The samples were dissolved by stirring them in 19 mL of water for 15 min. Then 2 mL of each solution were passed through a C_{18} cartridge at 10 mL/min flow rate to remove interferences. NDELA was eluted with 4 mL of an organic solvent. Then, the eluate was evaporated and solubilized in 0.2 mL and a $20\,\mu\text{L}$ aliquot was injected into the liquid chormatograph. You are requested to calculate:

a. The content of NDELA in both shampoos (expressed as $\mu\text{g/g}$ NDELA) considering that $M1$ and $M2$ yield peak areas of 248738 and 804, respectively. The calibration equation was: Peak Area $= 49786\,C_{\text{NDELA}} - 192$, where C_{NDELA} was in $\mu\text{g/mL}$. The linear range was found to be $0.03-10\,\mu\text{g/mL}$.

b. The retention factor and selectivity for NDELA and an interfering compound considering that their retention times were 5.3 min and 6.1 min, respectively (dead time was 1.04 min).

Solution: (a) The content of NDELA in $M2$ cannot be quantified because its interpolated value is not within the linear range ($0.02\,\mu g/mL$). The content of NDELA in $M1$ was $9.92\,\mu g/mL$. (b) $k_{NDELA} = 4.1$, $k_{INTERFERENT} = 4.9$, Selectivity = 1.19.

14. A laboratory is determining benzo[a]pyrene in fresh waters in a concentration range between 1 and $100\,ng/L$ using a chromatographic method whose linear range is 0.1–200 ng/mL. The current sample treatment is to be replaced for any of the two methods described below:

Method A: $200.00\,mL$ of water are extracted with $30.00\,mL$ of DCM. The organic fraction is passed through a sodium sulfate column to remove residual water and, then, evaporated to dryness. The residue is solubilized in $1.00\,mL$ of acetonitrile for the chromatographic analysis.

Method B: $25.00\,mL$ of water sample are loaded into a solid-phase cartridge. The analyte is eluted with $10\,mL$ of a DCM:methanol mixture. The eluate is evaporated to dryness. The residue is solubilized in $0.5\,mL$ of acetonitrile for chromatographic analysis.

a. What would the more suitable treatment method be?
b. Due to operative constraints, the linear range was restricted to the 0.1–$15\,ng/mL$. What would the more suitable treatment method be?

Solution: (a) Method A is adequate because the samples whose concentrations of analyte range from $1\,ng/L$ to $100\,ng/L$ will become concentrated within the linear range ($0.2\,ng/mL$–$20\,ng/mL$).
(b) Any of the two methods will concentrate the samples within the stated linear range.

EXERCISE REFERENCES

[1] Tanabe A.; Kawata, K. (2008). Determination of 1,4-dioxane in household detergents and cleaners, *Journal of AOAC International*, 91(2): 439–444.

[2] Yusakul, G.; Udomsin, O.; Juengwatanatrakul, T.; Tanaka, H.; Chaichantipyuth, C.; Putalun, W. (2013). High performance enzyme-linked immunosorbent assay for determination of miroestrol, a potent phytoestrogen from *Pueraria candollei*, *Analytica Chimica Acta*, 785: 104–110.

[3] Guan, H.; Brewer, W. E.; Garris, S. T.; Craft, C.; Morgan. S. L. (2010). Multiresidue analysis of pesticides in fruits and vegetables using disposable pipette extraction (DPX) and micro-Luke method. *Journal of Agricultural and Food Chemistry*, 58: 5973–5981.

[4] Lin, Z. P.; Ikonomou, M. G.; Jing, H.; Mackintosh, C.; Gobas, F. P. C. (2003). Determination of phthalate ester congeners and mixtures by LC/ESI-MS in sediments and biota of an urbanized marine inlet, *Environment Science Technology*, 37: 2100–2108.

[5] Bessaire, T.; Tarres, A.; Goyon, A.; Mottier, P.; Dubois, M.; Tan, W. P.; Delatour, T. (2015). Quantitative determination of sodium monofluoroacetate "1080" in infant formulas and dairy products by isotope dilution LC-MS/MS, *Food Additives & Contaminants: Part A*, 32(11): 1885–1892.

[6] Prieto-Blanco, M. C.; López-Mahía, P.; Prada-Rodríguez, D. (2006). Analysis of residual products in triethylbenzylammonium chloride. Study of retention mechanism, *Journal of Chromatographic Science*, 44(4): 187–192.

[7] Commission Decision of 12 August 2002 implementing Council Directive 96/23/EC concerning the performance of analytical methods and the interpretation of results (2002/657/EC). *Official Journal of the European Union*, L 221/8.

[8] González-Rodríguez, R. M.; Rial-Otero, R.; Cancho-Grande, B.; Simal-Gándara, J. (2008). Determination of 23 pesticide residues in leafy vegetables using gas chromatography–ion trap mass spectrometry and analyte protectants, *Journal of Chromatography A*, 1196–1197: 100–109.

[9] Ma, J.; Xiao, R.; Li, J.; Yu, J.; Zhang, Y.; Chen, L. (2010). Determination of 16 polycyclic aromatic hydrocarbons in environmental water samples by solid-phase extraction using multi-walled carbon nanotubes as adsorbent coupled with gas chromatography–mass spectrometry, *Journal of Chromatography A* 1217, 5462–5469.

[10] Ali, N.; Harrad, S.; Muenhor, D.; Neels, H.; Covaci, A. (2011). Analytical characteristics and determination of major novel brominated flame retardants (NBFRs) in indoor dust, *Analytical and Bioanalytical Chemistry*, 400(9): 3073–3083.

[11] Ghassempour, A.; Abbaci, M.; Talebpour, Z.; Spengler, B.; Römpp, A. (2008). Monitoring of N-nitrosodiethanolamine in cosmetic products by ion-pair complex liquid chromatography and identification with negative ion electrospray ionization mass spectrometry, *Journal of Chromatography A*, 1185: 43–48.

CHAPTER 7

NUCLEAR MAGNETIC RESONANCE AND MASS SPECTROMETRIES

Miguel Angel Maestro-Saavedra

OBJECTIVES AND SCOPE

The main objective of this chapter is to present students a concise, basic and practical overview of two powerful techniques for the structural characterization of organic and inorganic chemical structures; namely, nuclear magnetic resonance (NMR) spectrometry and mass spectrometry (MS). Explanations range from essential background to descriptions of key concepts to some practical applications. A selected collection of exercises show how these key concepts are applied and how a structural elucidation from the NMR and MS spectra can be obtained.

To explain how to interpret a mass spectrum is anything but simple because there are no fixed rules which can be followed. Therefore, in this chapter, it is attempted to present the very basics of the MS technique and give some general guidances, along with some examples.

PART A: NUCLEAR MAGNETIC RESONANCE SPECTROMETRY

1. INTRODUCTION TO NUCLEAR MAGNETIC RESONANCE

Nuclear magnetic resonance (NMR) spectrometry is a fundamental tool for the structural determination of organic and inorganic molecules. It employs low-energy radiation, in the radio frequencies (RF) region, in order to gather information on the structure of alkyl groups and other hydrogen-containing elements. Then, the presence of functional groups in the molecule is deduced.

1.1. Basic principles

Many atomic nuclei behave as if they were spinning around themselves (*nuclear spin*). When a charged particle (nucleus) moves or rotates, it creates a magnetic field. The orientation of H is random in the space until the nuclei is inserted in an external magnetic field H_0, which causes H to get aligned with H_0. This can occur in two forms: aligning the magnetic moment H in the direction of the H_0 field (energetically favorable) or otherwise, counterclockwise to H_0 (it requires the input of energy as it is energetically unfavorable). The two possibilities are referred to as the nuclear spin states α and β, respectively.

The existence of these different energy states is a required condition for NMR spectrometry. The irradiation of a sample with a certain energy frequency that provides the exact energy difference between the states (α and β) leads to the resonant absorption of energy that allows a nucleus with an α spin, to reverse its spin to a β mode. Upon excitation, the nucleus relaxes and returns to the initial state. When there is resonance, continuous excitation and relaxation occurs.

The energetic difference (ΔE) between the α and β states depends directly on the intensity of the external magnetic field H_0 and they are related by Equation (1):

$$\Delta E = \gamma(h/2\pi)B_0 \tag{1}$$

where h is the Planck's constant, B_0 is the strength of the external magnetic field and γ is the gyromagnetic constant core observed. The higher the gyromagnetic constant is, the greater the energetic difference between the α and β states will be. The frequency at which absorption occurs is proportional to H_0. For instance, the frequency for ^1H is 200 MHz at 4.7 T (Tesla), 300 MHz at 7.05 T and 500 MHz at 11.75 T. The energy required for a transition from α to β can be calculated typically as Equation (2):

$$\Delta E_{\beta-\alpha} = h\nu \qquad (2)$$

In the case of a proton (^1H) in a field of 7.05 T (300 MHz) $\Delta E_{\beta-\alpha}$ is 2.9 kcal/mol, so that the exchange between states is very fast. Therefore, within such a magnetic field only slightly more than half the nuclei will be in an α state, whereas the remaining ones will be in a β state.

Hydrogen is not the only nucleus that can be observed by NMR. In general, nuclei with an odd number of protons (as hydrogen isotopes, ^{14}N, ^{19}F, ^{31}P) or an odd number of neutrons (such as ^{13}C) are magnetically active and thus observable by NMR. However, when both the number of protons and number of neutrons are even, as for ^{12}C and ^{16}O, the nucleus is not magnetically active and, accordingly, cannot be observed by NMR.

Active nuclei which are magnetically different resonate at different frequencies when introduced within a magnetic field (Figure 1). For example, if we record the NMR spectrum of a sample of chlorofluorodeuteromethane, the following resonance frequencies are observed at 7.05 T:

^1H	300 MHz
^{19}F	282.31 MHz
^{13}C	75.43 HZ
^2H	46.05 MHz
^{35}Cl	29.40 MHz
^{37}Cl	24.47 MHz

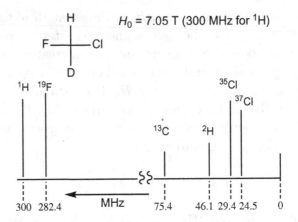

Figure 1.   NMR: different nuclei resonance frequencies.

Nowadays, the widespread application of NMR in organic and inorganic chemistry is due to several breakthroughs that took place in analytical technology. They allowed obtaining high-resolution spectra which can differentiate between nuclei of the same element. Advances in power and stability of the magnetic field, along with improvements in sensitivity and stability of the detection systems, as well as the availability of powerful computer systems capable of processing large amounts of information constitute the basis of current state-of-the-art high-resolution NMR spectrometry (HR-NMR). Noteworthily, when a spectrum of HR-NMR is recorded, there is no single signal for each nuclei, as the signal becomes split.

As an example, the ^{1}H-NMR of chloromethyl methyl ether recorded on a 300 MHz apparatus between 300 000 000 Hz and 300 006 000 Hz yields two different signals. On the other hand, the ^{13}C-NMR spectrum for this same compound when registered between 75 430 000 Hz and 75 580 860 Hz also presents two different signals (Figure 2). The presence of two different signals is a clear sign of the existence of different types of carbon and hydrogen in a molecule. Hence, HR-NMR is a powerful tool for the structural elucidation of hydrocarbon molecules because it remarkably distinguishes the different structural environments of the hydrogens and the carbons.

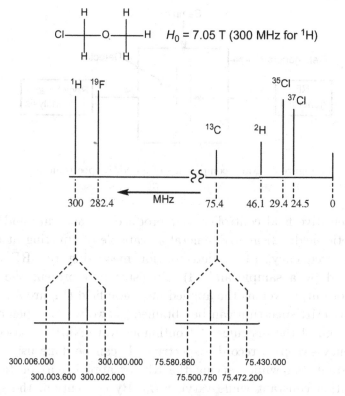

Figure 2. Examples of HR-NMR frequencies.

1.2. Basic concepts of the NMR spectrometers [1]

NMR spectrometry is employed routinely since 1960. The first ^{1}H-NMR systems had a 1.41 T electromagnet that provided an ^{1}H resonance frequency of 60 MHz. At present, more powerful field spectrometers allow access to ^{1}H frequencies up to 900 MHz, with a corresponding considerable improvement in resolution and detection. The most common systems nowadays correspond to 300 MHz devices. These high magnetic fields (200–900 MHz) are based in superconducting magnets, which are kept at low temperatures using liquid helium (4 K or $-269\,°C$).

An NMR spectrometer consists conceptually of four elements: (1) a field generator (electromagnet or superconducting coil) with a

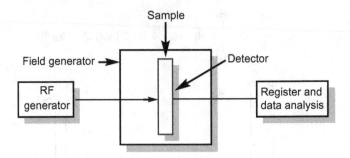

Figure 3. Main components of an NMR spectrometer.

very sensitive field controller that produces an accurate and stable magnetic field; (2) an RF generator, capable of emitting at a very precise frequency; (3) a detector for measuring the RF energy absorbed by a sample and (4) a register to represent the signal detector output versus the applied magnetic field (Figure 3).

An NMR spectrum can be obtained in two ways, depending on the design of the instrument. Continuous-wave NMR is based on a frequency sweep to record a spectrum. It can be done using either a constant H_0, while varying the RF, or applying a constant RF irradiation constant while varying H_0. By convention, the spectra are calibrated in Hz, representing the frequency as a function of a constant magnetic field H_0. Nowadays, most instruments employ a methodology based on Fourier transform (FT) to register the spectra. This is the same transform as that employed in infrared spectrometry (FT-IR).

Thus, instead of performing a sweep in the spectral region of interest, the sample is irradiated with one or more RF pulses, which excite all nuclei at a time. Then the data signals derived from the nuclei relaxation are acquired simultaneously and processed in a computer. These spectra are recorded faster, with a better signal-to-noise ratio since it is possible to accumulate several sweeps that are averaged in the computer. The FT-NMR spectra are acquired in the time domain by measuring the temporary relaxation of the signals, the FT algorithms transform the time domain to the traditional frequency domain.

The application of the FT methodology has been pivotal in the tremendous growth experienced by NMR as an analytical technique.

1.3. Hands-on: Acquisition of an NMR spectrum

Only few milligrams of a sample (0.1–10 mg) are required to acquire its ^{1}H-NMR spectrum, the stronger the magnetic field is, the smaller is the amount of sample that is required. The sample is dissolved in 0.2–1.0 mL (depending on the size of the container, a glass tube) of a solvent, which should preferably not absorb in the working range of the NMR study. Typical solvents currently used are carbon tetrachloride (CCl_4), deuterated solvents such as deuterochloroform (deuterotrichloromethane, Cl_3CD), deuterated water (deuterium oxide, D_2O), deuterated dimethylsulfoxide ($D_3CSO_2CD_3$), deuterated benzene (hexadeuterobenzene, C_6D_6), deuterated acetone (hexadeuteropropanone, D_3CCOCD_3), or deuterated tetrahydrofuran (octadeuterooxacyclopentane, C_4D_8O). Deuterated solvents are employed because they allow using the deuterium resonance frequency as a reference for the magnetic field homogeneity, or LOCK system. Once the sample is rotating within the magnetic field, local homogeneity of the magnetic field is adjusted using microcoils, which generate magnetic microfields which avoid local inhomogeneities of the magnetic field applied to the sample; the system is adjusted, and then the LOCK is activated. This allows for obtaining a better signal-to-noise ratio in the spectrum. Next, the sample is subjected to a series of RF pulses, and an RF receiver coil registers the energy released during the relaxation of the nuclei of the system. The emission and detector RF coils must be perpendicular to avoid interferences between transmission and reception.

The signals are processed by a computer and signal intensities are represented as a function of the resonance frequencies.

1.4. The chemical shift

The position of a signal in an NMR spectrum, which is called **chemical shift**, depends on the electron density around the nucleus, which in turn is controlled by the molecular structure in the vicinity of the nucleus (structural environment). NMR chemical shifts of the nuclei of a molecule are essential data to accurately determine its structure. When the most abundant nuclei are considered, ^1H, their NMR chemical shifts constitute essential data for determining its structure.

As an example, the high-resolution ^1H-NMR spectrum of chloro(methoxy)methane reveals that there are two types of hydrogen in the molecule as they show differentiated resonances. They are due to the electronic environments of the ^1H. Further, the ^1H nuclei of an organic molecule are not free because they are covalently bonded. The covalently bonded ^1H are surrounded by an electron cloud whose density varies with bond polarity, atom hybridization and the presence of donor groups or electron acceptors.

The chemical shift describes the position of an NMR signal and it is measured in ppm units.

When a nucleus surrounded by electrons is subjected to a magnetic field of intensity H_0, electrons move generating a small local magnetic field H_{local} that opposes H_0. Thus, there is a decrease in the intensity of the total magnetic field H in the vicinity of the hydrogen atom. It is said that this hydrogen nucleus is *shielded* against H_0 by the electron cloud. The degree of shielding depends on the electron density around the core. When there is a high electron density, shielding increases, otherwise deshielding appears. The effect of shielding makes necessary a more intense external field to produce a given resonance. By convention, the NMR spectra are plotted so that the field intensity increases from left to right, and we talk about 'displacement to high fields' when the signal is shifted to the right.

Chemically different protons have a unique pattern with characteristic resonances. Analogously, chemically equivalent protons have the same chemical shift and, so, chemically equivalent nuclei are not easy to differentiate. Molecular symmetry helps establish chemical equivalence. The existence of a plane or an axis of symmetry means that every hydrogen of a methyl group can occupy the position of the other two, without structural change. Therefore, for a methyl group that rotates freely, it is expected that all the hydrogens are equivalent and possess the same chemical shift. The interconversion between conformers can result in equivalence during the NMR time scale. Rotation can be decreased and interconversion between conformers can be slowed down by cooling the sample, but the rotation activation energy (E_a) is small (2.8 kcal/mol), so that the required temperature of 93 K ($-180\,^{\circ}$C) is not feasible because most solvents would freeze.

As an example, for cyclohexane, rapid conformational equilibration (E_a 10.8 kcal/mol) produces rapid interconversion between H_{eq} and H_{ax} in the NMR timescale (H_{ax} and H_{eq} represent the two sets of equivalent H in a cyclohexane chair conformation, H perpendicular to molecular plane, H_{ax}; H parallel to molecular plane, H_{eq}). Its ^{1}H-NMR spectrum shows one signal at 1.36 ppm at room temperature (Figure 4), but at 183 K ($-90\,^{\circ}$C) two resonance signals at 1.12 ppm (H_{ax}) and 1.60 ppm (H_{eq}) are observed.

How is NMR spectrometric information delivered? NMR signals are constituted by the frequencies of all resonances that appear when obtaining a spectrum. But this has the major problem that those frequencies will change when the spectrum is recorded with a different magnetic field, so the spectra are not comparable.

To avoid this problem, an independent field scale was established, **the chemical shift**, δ. It requires referring the spectra to an internal standard, the tetramethylsilane [(CH_3)$_4$Si, TMS]. The 12 equivalent

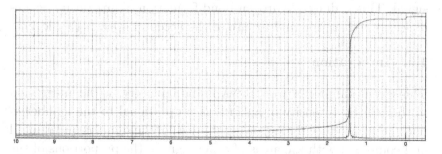

Figure 4. ^{1}H-NMR of cyclohexane (298 K).

protons of this compound are relatively shielded with respect to most ^{1}H present in the molecules, and, so, they resonate at higher fields than the usual spectral range. Then, the spectra are normalized with the frequency of the spectrometer. The δ chemical shift is thus obtained by dividing the difference between the signal frequency and the TMS frequency (in Hz) by the frequency of the spectrometer. The δ scale is indicated in units of parts per million (ppm), for which ^{1}H TMS has $\delta = 0.00$ ppm.

1.4.1. Characteristic chemical shifts of some selected functional groups [2, 3]

As the chemical shift of a proton is determined by its environment, it is possible to provide a table of approximate chemical shifts for many types of compounds according to their functional groups. NMR allows for the identification of different types of hydrogen in a molecule, [1–3] so it constitutes a powerful analytical technique for organic and inorganic chemistry.

In general, the ^{1}H signals of alkanes appear at relatively high fields (δ 0.8–1.7 ppm). Hydrogen in the proximity of a 'density attractor' group or electronegative atom (O, halogens, etc.) resonates downfield because these substituents deshield the neighboring nuclei. So, the more electronegative the atom is, the higher the deshielding effect will be. This effect is cumulative; it increases by increasing the number of electronegative atoms and it decreases rapidly with distance. In the case of exchangeable hydrogens (–OH, –SH, –NH–), they can absorb a wide range of frequencies, giving usually a broad signal. This is due to hydrogen bond formation, temperature dependence, concentration and nature of the acceptor species (Table 1).

1.5. Integration

This is a useful property of ^{1}H-NMR spectrometry, as the integration of the relative intensity of a signal is proportional to the number of nuclei that produce that signal. If the area of the signal is quantitated and compared with the areas of other signals, the proportions of the different types of hydrogen can be quantitatively estimated. Thus,

Table 1. ^{1}H-NMR: characteristic proton chemical shifts (δ) [4].

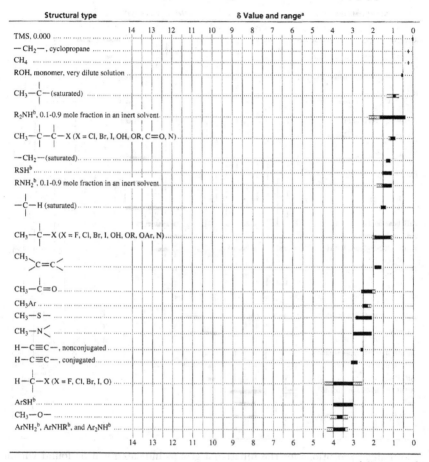

| Structural type | δ Value and range[a] |

	14	13	12	11	10	9	8	7	6	5	4	3	2	1	0

TMS, 0.000

—CH$_2$—, cyclopropane

CH$_4$

ROH, monomer, very dilute solution

CH$_3$—C—(saturated)

R$_2$NH[b], 0.1-0.9 mole fraction in an inert solvent

CH$_3$—C—C—X (X = Cl, Br, I, OH, OR, C=O, N)

—CH$_2$—(saturated)

RSH[b]

RNH$_2$[b], 0.1-0.9 mole fraction in an inert solvent

—C—H (saturated)

CH$_3$—C—X (X = F, Cl, Br, I, OH, OR, OAr, N)

CH$_3$\C=C/

CH$_3$—C=O

CH$_3$Ar

CH$_3$—S—

CH$_3$—N$\langle$

H—C≡C—, nonconjugated

H—C≡C—, conjugated

H—C—X (X = F, Cl, Br, I, O)

ArSH[b]

CH$_3$—O—

ArNH$_2$[b], ArNHR[b], and Ar$_2$NH[b]

(*Continued*)

Table 1. (*Continued*)

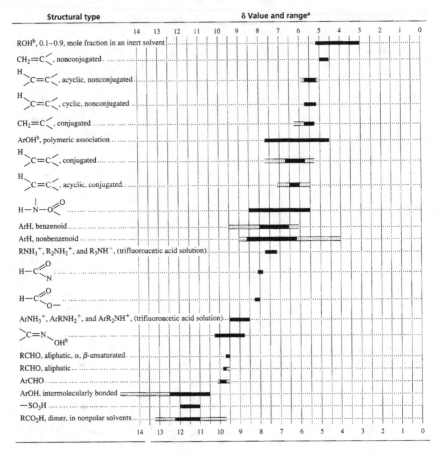

ᵃNormally, absorptions for the functional groups indicated will be found within the range shown in black shading. Occasionally, a functional group will absorb outside this range. Approximate limits are indicated by extended outlines.

ᵇAbsorption positions of these groups are concentration-dependent and are shifted to lower δ values in more dilute solutions.

the structure of a hydrocarbon molecule can be determined from the chemical shift and the integration of the signals.

As an example, the monochlorination products of 1-chloropropane ($CH_3CH_2CH_2Cl$) have the same molecular formula ($C_3H_6Cl_2$) and very similar physical properties.

1,1-Dichloropropane has three types of non-equivalent hydrogens and its ^{1}H-NMR shows three signals with a 3:2:1 integration; one hydrogen absorbs at low field (δ 5.93 ppm) for the cumulative 2 Cl deshielding, the other hydrogens resonate at high fields (δ 1.01 and 2.34 ppm).

1,2-Dichloropropane has three types of non-equivalent hydrogens and its ^{1}H-NMR presents three signals with a 3:2:1 integration; but they have quite different chemical shifts, two groups with integration 2 and 1 are connected to a chlorine atom and their signals appear at low (δ 3.68 ppm for 2H signal and δ 4.17 ppm for 1H) and high fields (δ 1.70 ppm). By integrating the signal that can be attributed to the third type of hydrogen, 3H, a methyl group appears.

1,3-Dichloropropane shows only two signals (δ 3.71 and 2.25 ppm) in a 2:1 ratio, clearly distinct from the other isomers.

1.6. Spin–spin coupling

The NMR spectra described so far are composed of signals in the form of sharp and isolated peaks, called singlets. Hence, the hydrogens responsible for these spectra are undistinguishable from other H separated by only an O or a C. This means that the H are isolated. When several H are present on adjacent carbons, the spectrum becomes more complicated as a result of spin–spin coupling (Figure 5). The signals can be split into doublets (two signals of intensity 1:1), triplets (three signals, 1:2:1), quadruplets (four signals, 1:3:3:1), quintuplets (five signals, 1:4:6:4:1), or sextuplets (six signals, 1:5:10:10:5:1). The cause of this effect is the behavior of the nuclei within an external magnetic field, as they act like little magnets that align in favor (α) or against (β) the field. The energetic difference between the two states is minimal at room temperature, with almost identical nuclei populations, resulting in the existence of two magnetic classes, half of the nuclei have a neighbor in state α whereas the other nuclei have a neighbor β.

Following, the signal for a proton splits into two signals. The signal aligned with H_0 is subjected to a higher total magnetic field and, therefore an external magnetic field with less intensity is required to get resonance (compared to the absence of neighbor). Thus, the signal is seen at lower fields. This observation is due to half the H_a nuclei, the other half has its neighbor in state b, aligned against the magnetic field, the magnetic field intensity around H_a decreases, so that the external magnetic field has to increase for achieving the resonance, so the signal appears at a higher intensity field.

Analogously, hydrogen H_b has two types of neighbor hydrogens, $H_{a(\alpha)}$ and $H_{a(\beta)}$. Consequently, the H_b signal appears as a doublet; it is said that H_a is coupled to H_b. The intensity of the mutual coupling is the same. The distance between the two resonance signals of hydrogen, measured in Hertz, is called spin–spin coupling constant (J) (Figure 5). The coupling constant is independent of the intensity of the external magnetic field, since it is only due to the contributions of neighboring nuclei.

Spin–spin coupling is currently observed only between hydrogens that are immediate neighbors. This could be because they are attached to the same carbon (geminal coupling) or to two adjacent carbons (vicinal or 1,2-coupling). Hydrogen nuclei separated by more than two carbon atoms are normally loosely coupled (1,3-coupling). Local field contributions due to more than one hydrogen are additive. In many cases, the $N + 1$ rule provides the multiplicity of the spin–spin coupling. The relative proportion of each peak in the

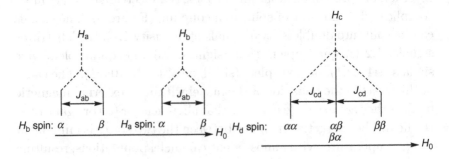

Figure 5. Coupling constants (J).

Table 2. ^{1}H-NMR: Spin–spin coupling constants (J); a = axial; e = equatorial [4].

Type	J, Hz	Type	J, Hz
>C< (H, H geminal)	12–15	>C=C< (H, C–H)	4–10
>CH—CH< with free rotation	2–9 ~7	>C=C< (C–H ; H)	0.5–2.5
—C—(—C—)$_n$—C— (H ... H)	~0	>C=C< (H ; C—H)	~0
CH$_3$—CH$_2$—X	6.5–7.5	>C=CH—CH=C<	9–13
CH$_3$, CH$_3$ >CH—X	5.5–7.0	>CH—C≡C—H	2–3
H—C—C—H, X Y (cyclohexane)	a,a 5–10 a,e 2–4 e,e 2–4	>CH—C<(H)(=O)	1–3
>C=C<(H, H)	0.5–3	>C=C<(H, C(=O)H)	6–8
H, H >C=C<	7–12	benzene	6–9 1–3 0–1
H >C=C< H	13–18		

multiplet can be seen in the mnemotechnic rule given by the Pascal triangle. The coupling between non-equivalent H is mutual (Table 2).

Figure 6 exemplifies the ^{1}H-NMR of 1-bromobutane where four sets of equivalent H are present, so that four different chemical shifts with coupling constants can be seen.

1.6.1. Spin–spin coupling: problems of interpretation

The resonance signals described above are slightly idealized because it was considered that (i) the peaks are symmetrical (Gaussian), (ii) the difference between the ^{1}H resonance frequencies is much higher than the observed coupling constant ($\Delta\delta \gg J$) and (iii) the

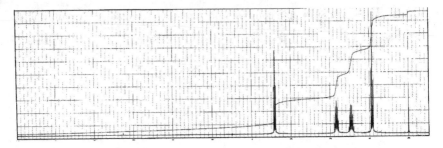

Figure 6. ^{1}H-NMR of 1-bromobutane.

intensities comply with the $N+1$ rule. Whenever this is true, it is said that the spectrum is of 'first grade'. When either the relative intensities of multiplets do not match the expected ideal relationship, or they are not symmetrical and tend to lean toward a side, or when the difference between the chemical shifts of the coupled signals approaches the constant coupling ($\Delta\delta \sim J$), problems arise in getting first-order spectra. Today, this problem is solved by the use of NMR simulation that uses software to simulate coupling constants, or by using more intense fields in the experiments ($\uparrow H_0$), so that a higher resolution is obtained and, so, $\Delta\delta \gg J$. These strategies tend to provide first-order spectra.

Coupling between different non-equivalent nuclei modifies the number of signals ($N + 1$ rule). Each signal is split by each of the non-equivalent nuclei, producing more complex signal patterns (multiplets).

When hydroxyls and amines are present in the molecule, the existence of fast proton exchange decouples the hydrogens, decreasing their signals or, even, making them disappear. The exchange can be slowed down when traces of water or acids are removed from the sample or if its solution is cooled.

2. CARBON-13 NMR SPECTROMETRY (^{13}C-NMR)

^{13}C-NMR uses a low natural abundance isotope, ^{13}C. The most abundant isotope of carbon, ^{12}C, is not detectable in NMR because it has an even number of protons and neutrons. ^{13}C has a natural abundance of 1.1 % and a similar magnetic behavior as ^{1}H. Acquisition of ^{13}C-NMR spectra is harder than the ^{1}H spectra due to the low natural abundance and the lower sensitivity of ^{13}C to magnetic fields (a specific property of each nucleus, the gyromagnetic constant (γ), is 600 times lower than for ^{1}H). Under similar experimental conditions, the ^{13}C signals are less intense than the ^{1}H ones. This difficulty has been solved by the routine use of FT-NMR, which can accumulate sweeps, and the use of more powerful superconducting magnets.

The **chemical shift**, δ, is defined as done for ^{1}H; i.e., relative to an internal standard, tetramethylsilane [$(CH_3)_4Si$, TMS] and the spectrum is referred to the residual signal of the deuterated solvent used. The range of chemical shifts for ^{13}C-NMR is wider than that for ^{1}H RMN. The ^{13}C resonance frequency, like that for ^{1}H, for a given magnetic field is related to the magnetic field strength (7.04 T: ^{1}H 300 MHz, ^{13}C 75.43 MHz; 11.74 T: ^{1}H 500 MHz, ^{13}C 125.72 MHz). ^{13}C spectra do not show ^{13}C–^{13}C couplings due to the low natural abundance of ^{13}C (the probability for an abundance of 1.1 % is 1/10000). However, ^{13}C couples to ^{1}H, although the ^{1}H–^{13}C coupling constant decreases rapidly with distance: $^{1,2}J$ (HC) $\sim$ 125–200 Hz, $^{1,3}J$ (HCC) $\sim$ 0.7–6 Hz. The existence of ^{1}H–^{13}C couplings decreases the intensity signals of the ^{13}C-NMR spectrum, which together with the low ^{13}C natural abundance makes the acquisition of ^{13}C-NMR difficult.

The sensitivity of the technique improves dramatically when the ^{1}H–^{13}C couplings are canceled out by irradiating all resonance frequencies of ^{1}H with intense RF while the ^{13}C signal is acquired (Figure 7a); this is called ^{1}H *broadband decoupling*, BB. The result is a ^{13}C spectrum without couplings, and each magnetically different carbon appears as a singlet. This simplifies the spectrum and is very useful for measuring complex molecules. Molecular symmetry allows for a further reduction of the signals displayed for chemically different carbons. If the spectrum is acquired under ^{1}H irradiation, the signal intensity is not directly proportional to the number of nuclei and the ^{13}C chemical shifts depend on the structural environment (Table 3). As for ^{1}H-NMR, electron-withdrawing groups deshield the signals. The study of the δ values and numbers of different C atoms provide important data for the elucidation of a molecular structure.

Table 3.　^{13}C-NMR chemical shifts in organic compounds (*) [4].

Structure	Name
$>C=O$	Ketone
$>C=O$	Aldehyde
$>C=O$	Acid
$>C=O$	Ester, amide
$>C=S$	Thioketone
$>C=N$	Azomethine
$-C\equiv N$	Nitrile
$>C=N$	Heteroaromatic
$>C=C<$	Alkene
$>C=C<$	Aromatic
$>C=C<$	Heteroaromatic
$-C\equiv C-$	Alkyne
$\geq C-C\leq$	(C Quaternary)
$\geq C-O$	
$\geq C-N<$	
$\geq C-S$	
$\geq C$—Halogen	
$>CH-C\leq$	(C Tertiary)
$>CH-O$	
$>CH-N<$	
$>CH-S$	
$>CH$—Halogen	
$-CH_2-C<$	(C Secondary)
$-CH_2-O$	
$-CH_2-N<$	
$>CH_2-S$	
$-CH_2$—Halogen	
$H_3C-C\leq$	(C Primary)
H_3C-O	
$H_3C-N<$	
H_3C-S	
H_3C—Halogen	
Resonances of common solvents	

$(CH_3)_2CO$　CS$_2$　　CF$_3$COOH　　C$_6$H$_6$　　CCl$_4$　CHCl$_3$　CH$_3$OH　$(CH_3)_2CO$
　　　　　　　　　　　　　　　CF$_3$COOH　　　1,4Dioxane　DMSO

ppm (TMS)　220 210 200 190 180 170 160 150 140 130 120 110 100 90 80 70 60 50 40 30 20 10 0

Cyclo-propane

*Relative to internal tetramethylsilane.

3. OTHER ADVANCED TECHNIQUES

FT-NMR is an extremely versatile technique because depending on the data acquisition method and the signal processing mode, it is possible to obtain important information about the molecule. So far, the advancements in the technique were based on the development of complex sequences of RF pulses, or the simultaneous and independent RF irradiation of different frequencies, leading to two-dimensional spectra (2D NMR), three-dimensional (3D NMR), etc. More recently, coupling was studied and, following, the connection between the neighboring nuclei can be established. The coupling between neighboring ^1H (homonuclear coupling) or between ^1H and chemically bound ^{13}C (heteronuclear coupling) allows us to study the molecular connectivity using the magnetic effect on other nuclei.

A pulse sequence method used widely in laboratory routine work to get ^{13}C-NMR spectra is the distorsionless enhancement by polarization transfer (DEPT) sequence. The importance of this method is that it allows you to decide which kind of carbon (CH_3, CH_2, CH or C) a signal belongs to (Figure 7b). The DEPT experiment is analyzed by comparing its spectrum with a normal BB-^{13}C spectrum. DEPT spectrum absorptions appear as positive signals when CH and CH_3 groups are present, while CH_2 groups yield negative absorption signals and quaternary atoms of C have no signal due to the lack of bonded protons (Figure 7).

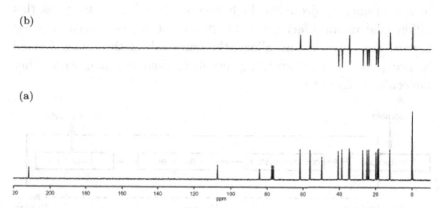

Figure 7. Example of the appearance of a ^{13}C-NMR spectrum when different recording techniques are used: (a) broadband decoupling (BB) and (b) DEPT

PART B: MASS SPECTROMETRY

4. INTRODUCTION TO MASS SPECTROMETRY

Mass spectrometry (MS) is a technique widely used in qualitative analysis for the determination of organic and inorganic structures, either as stand-alone or in combination with other spectrometric techniques, and also in quantitative analysis when coupled to a chromatograph.

MS is not a typical method where physical processes are non-destructive and the sample is recovered after analysis. On the contrary, in MS, chemical processes take place, and the sample is destroyed, still the amount of sample required is very small (μg).

Operationally, MS is about obtaining gaseous ions from organic or inorganic molecules in the gas phase or heavy ions from suitably dispersed very polar inorganic or organic molecules in liquid phase. After obtaining the ions, they are separated according to their mass and charge and, finally, recorded on a suitable detector.

4.1. Basic instrumentation for MS

In view of the processes that need to be produced in a mass spectrometer, its main parts are: (1) a sample introduction system, (2) a sample ionization system, (3) an analyzer system and (4) a detection system and ion recorder. All of them have to be enclosed in a container subjected to high vacuum (10^{-3}–10^{-6} torr), as this differential vacuum facilitates the movement of gas molecules (ions) toward the detector and allows the 'survival' of those ions formed in the path to the detector, preventing their collision with other molecules (Figure 8).

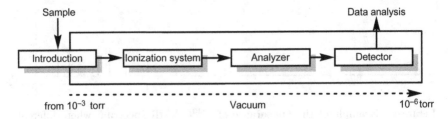

Figure 8.   Mass spectrometer diagram.

4.1.1. Introduction systems

A limiting factor in MS is the requirement that molecules should be analyzed in their gaseous phase. Thus, the primitive systems were based on the volatility of some small polar molecules by heating them at reduced pressures. The introduction of highly polar molecules or macromolecules as commonplace by using techniques based on the dispersion into small droplets of a solution containing the compound of interest in a volatile mobile phase came 40 years later.

4.1.2. Ionization systems

There are several types of ionization systems and they can be sorted according to their energy. In decreasing order of energy, we can list as follows:

(a) Electron impact ionization (EI): the gas phase molecules are ionized passing through an electron beam with high energy;
(b) Chemical ionization (CI): ionization of the molecules occurs by collision with a reactive gas ionized by an electron beam;
(c) Fast atom bombardment ionization (FAB): ionization of the molecules is produced by bombarding accelerated ions on a suspension of the compound of interest in a suitable matrix;
(d) Matrix-assisted laser desorption ionization (MALDI): ionization of the molecules is produced by irradiating a suspension of the molecule of interest with a laser in a suitable matrix;
(e) Electrospray ionization (ESI): ion production is achieved by generating a spray from an electrode subjected to a high voltage.

4.1.3. Ion analyzer systems

Of the several available options, the most important ones are, likely:

(a) Quadrupole analyzer: it is formed by four parallel metal bars with an exact circular or hyperbolic section, accurately arranged in a circular manner, at the center of which the ion beam incides from the source. The bars are subjected to a constant potential and an overlapping alternate potential RF is applied, which make mass sweeps possible very quickly. The quadrupole acts as a mass filter to select a mass that transmits (travels) to the detector. Relevant advantages are the possibility of very fast sweeps (0.01 s), high sensitivity for not requiring slits to focus beams, the large range of masses that can be measured with respect to the range of potentials applied, and it enables mass interpolations. Typical limitations are its restricted resolution (<1 000) and its limited capability to separate ions (it is said that its resolution is under 2 000).

Resolution for a single peak, corresponding to singly charged ions, at mass m can be calculated as the $m/\Delta m$ ratio, where Δm is the width of the peak at a height which is a 5 % fraction of the maximum peak height (this corresponds to IUPAC recommendations set in 2013).

(b) Magnetic analyzer: when the ions exit the ionization chamber, accelerated by the electric field to which they were subjected, they enter into a magnetic field perpendicular to their movement. They are forced to describe a circular movement whose radius depends on the magnetic field, the charge and the mass of the ions. This is the **mass/charge ratio**, m/z, a fundamental parameter in MS. The sweep of the magnetic field causes a sweep on the masses reaching the detector. It reaches resolutions greater than 10 000, enabling the determination of m/z values with up to five decimal figures (called exact mass). It can separate m/z ratios up to 4 000. However, it cannot perform rapid or frequent scans because the magnetic fields do not follow precisely the variations in the current of the electromagnet (hysteresis of magnetic core).

(c) Ion trap analyzer: it is used as chromatographic detector in gas chromatography with mass spectrometry detection (GC-MS) systems. It is a modification of the quadrupole analyzer, based on the use of an electromagnetic containment zone generated by two RF signals. It consists of three electrodes with an hyperbolic surface. The central is annular while the upper and lower electrodes form the closure of the ring ends. The cavity formed by the three electrodes is the space where the ionization and fragmentation mass analysis occurs. Its advantages are: a superior sensitivity to typical quadrupole analyzers, and very high scanning speed, so that its performance as chromatographic detector is very good. Main drawbacks are that it has a limited resolution (<1 000), a small m/z discrimination range for ions (1 000 or lower), and it generates mass spectra which looks different from other analyzers because bimolecular reactions are common between ions.

(d) Time of flight analyzer: ions from the source are accelerated by a pulse of electric potential and travel a specific length within the analyzer, the time of flight is directly proportional to the m/z ratio. The main problem of flight analyzers is that the ions need very little time to reach the detector, around microseconds, which demands very fast detection systems. Their advantages are a remarkably short analysis time and no limitations on the mass of the ions that can be separated. They are widely used in the analysis of substances of very high molecular weight.

4.1.4. Detectors

Detectors in MS need to be very sensitive because ion currents leaving the analyzer are very weak (10^{-8}–10^{-14} A), which represents a problem when the records have to be very fast and very accurate. There are three main types of detectors:

(a) Faraday box: it consists of a box within which an inclined plate is disposed to avoid reflection of the ions; when ions collide with the plate, they take out electrons to neutralize their charge, generating a current whose intensity is related exactly to the number of ions that reached the plate. Sophisticated electronics require accurate determination of the generated currents.

(b) Multipliers: they work similarly as typical photomultiplier tubes used in radiation detectors. This detector uses the kinetic energy of ions impinging on a plate whose surface is coated with rare earth oxides; when ions collide with the plate, a stream of electrons is released toward a second plate, and so forth. They usually employ 10–16 plates. Amplifications of 10^6 or greater are possible. It has the drawback that the initial and final streams do not exhibit good proportionality, which is required for use in very precise quantitative determinations.

(c) Photographic plates: they are of limited use, used only in equipments with extraordinarily high sensitivity and resolution. The plates are developed by standard techniques of photography and their reading is performed by densitometers.

From a practical viewpoint, to obtain a mass spectrum, we first need to introduce the sample (solid, liquid, gas) into the vacuum system. There, vacuum produces the gaseous molecules to pass to the ionization system. Ions are collimated before entering the analyzer system where the mass scanning occurs. The recorded mass spectrum is a 2D representation of the intensity of certain mass/charge (m/z) ratios.

5. INTERPRETING A MASS SPECTRUM

Mass spectra provide abundant information on the analyzed compounds. In addition, it is possible to identify a substance by comparing its experimental spectrum to a suite of mass spectra databases. Finally, a mass spectrum itself can be interpreted by studying the m/z records (although this is far more complicated). This sometimes is the only way to ensure the structural identification of some molecules. Also, it has to be noted that the information provided by the mass spectrum comes from chemical reactions experienced by the sample molecules in their excited state. Consequently, interpreting mass spectra requires knowledge of the reactions that can occur in the spectrometer and the ions that these reactions can generate. Experience is also required.

5.1. Types of characteristic ions

5.1.1. The molecular ion

Molecular ions are ions derived from a molecular fragmentation that presents the same mass as the neutral, parent molecule. They appear in the spectrum as a molecular peak (M^+) and their formation is the primary stage in the fragmentation process. The determination of the molecular mass by using this ion is the most accurate technique known so far.

The determination of the molecular weight of a substance from its mass spectrum is independent of impurities, as far as the impurity can be identified. The molecular peak provides information on the molecular weight of the molecule and serves as a starting point for the interpretation of the mass spectrum. This is why a large certainty when identifying the molecular peak is a must, which should meet the following requisites:

(a) The molecular peak is the highest peak that can be generated from a compound, regardless of isotopic peaks;

(b) The molecular ion is the ion appearing at the lowest potential;

(c) The molecular ion presents an even mass if there is either no nitrogen in the molecule or it possesses an even number of atoms of nitrogen. On the contrary, the molecular mass peak is odd if the molecule contains an odd number of atoms of nitrogen;

(d) The molecular ion must contain all the elements that can be identified in fragments;

(e) The molecular peak intensity is proportional to the amount of sample introduced into the ion source;

(f) The mass difference between the molecular peak and peak signals corresponding to other fragment ions should be chemically explainable for those signals, similar to the molecular peak, as it is presented below:

M-1: loss of hydrogen

M-2: loss of dihydrogen (H_2)

M-15: loss of a methyl group

M-16: loss of oxygen, an amine group or methane

M-17: loss of a hydroxyl group

M-18: loss of water

M-19: loss of fluoride

M-20: loss of hydrogen fluoride (HF)

M-26: loss of ethyne or a nitrile group

M-27: loss of a terminal vinyl group or hydrogen cyanide (CNH)

M-28: loss of carbon monoxide (CO)

M-29: loss of an ethyl group or a formyl group

M-31: loss of a methoxy group

M-45: loss of an ethoxy group or a nitro group

M-47: loss of a methyl sulfide

M-57: loss of a *tert*-butyl group or an ethyl ketone.

5.1.2. Isotopic ions

Almost all chemical elements in nature are formed by different isotopes. This is very important since all the molecules present mixtures of various isotopic compositions, whose abundance is related to the isotopes of each element.

Due to the existence of several isotopic species, the mass spectrometer will not detect a single molecular peak but several, each corresponding to the different isotopic compositions of the molecular ion. The relative abundances of isotopes of the most common elements found in organic compounds are summarized in Table 4.

From the isotopic peaks, the elemental composition of a compound can be approximated. The relative intensity of the first isotopic peak, usually due to a species with a ^{13}C, divided by 1.112 (the relative abundance of ^{13}C in nature) will provide the maximum number of atoms of carbon of the compound.

In case a molecule contains only C, H, N and O, the second isotopic peak would have a relative intensity between 0.5 % and 2.5 % that of the molecular peak. In the event that there is sulfur in the molecule, the relative intensity of the second isotopic peak will be about 5 %. If more sulfur is present, the intensity of the second isotopic peak would increase by 4.4 % per atom of sulfur present.

The presence of chlorine or bromine in the molecule produces a very intense and characteristic second isotopic peak, depending on the different isotopic abundance. Thus, the appearance of the spectrum allows identifying atoms of chlorine and/or bromine in the molecule depending on the intensity profile that is visualized (Figure 9). Figure 10 presents two examples where chlorine and bromine are substituents in the propane molecule.

Table 4. Mass and abundance of natural isotopes of the most common elements found in organic compounds.

Isotope	Mass	Abundance
^{1}H	1.007 8	100
^{2}H	2.014 1	0.015
^{12}C	12.000 0	100
^{13}C	13.003 4	1.112
^{14}N	14.003 1	100
^{15}N	15.000 1	0.367
^{16}O	15.994 9	100
^{17}O	16.999 1	0.038
^{18}O	17.999 1	0.200
^{19}F	18.998 4	100
^{23}Na	22.989 8	100
^{24}Mg	23.985 0	100
^{25}Mg	24.985 8	12.660
^{26}Mg	25.982 6	13.938
^{27}Al	26.981 5	100
^{28}Si	27.976 9	100
^{29}Si	28.976 5	5.063
^{30}Si	29.973 8	3.361
^{31}P	30.973 8	100
^{32}S	31.972 1	100
^{33}S	32.971 5	0.789
^{34}S	33.967 9	4.431
^{36}S	35.967 1	0.022
^{35}Cl	34.968 9	100
^{37}Cl	36.965 9	31.978
^{79}Br	78.918 4	100
^{81}Br	80.916 3	97.278
^{127}I	126.904 4	100

5.1.3. Fragments

Upon ionization, molecular ions can be decomposed in several ways. Each cleavage reaction depends on the energy required to reach the ionization and its activation energy. Theoretical models to foresee the internal energy of various ions intended to explain these

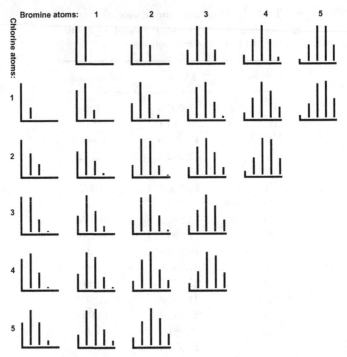

Figure 9.   Isotopic distributions of characteristic peaks of compounds containing chlorine and/or bromine.

reactions failed and that this motivates that the interpretation of the fragmentation of the molecular ion is explained currently by empirical rules.

Fragmentation reactions are not normal condensed phase reactions as they occur in very different conditions (gas phase, excited molecules). It can be said grossly that the decompositions of ions in the mass spectrometer hold some similarities with pyrolysis reactions or photochemical decompositions.

Some empirical rules were developed for the interpretation of decomposition reactions:

(a) Aliphatic hydrocarbons: carbon–carbon bonds in the molecule are preferably broken down at or close to the largest branching centers (Figure 11);

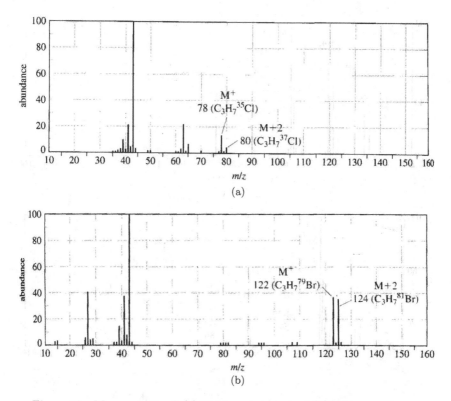

Figure 10. Mass spectra of: (a) 2-chloropropane and (b) 1-bromopropane.

(b) Olefins and linked π bond molecules (π-conjugated): double bond systems favor the breaking down of bonds in the allylic or benzylic position (Figure 12);

(c) Heteroatoms favor the rupture of the bond with the carbon to which they are attached (Figure 13);

(d) Double bonds and heteroatoms favor transpositions of hydrogen through cyclic transition states.

5.2. Determination of the molecular formula of a molecule by high-resolution MS

Technological advances allowed establishing the molecular formula of compounds using minute sample amounts (<1 mg), which is highly

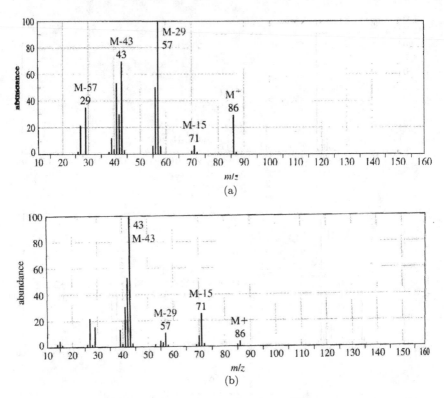

Figure 11. Mass spectra of: (a) *n*-hexane and (b) 2-methylpentane.

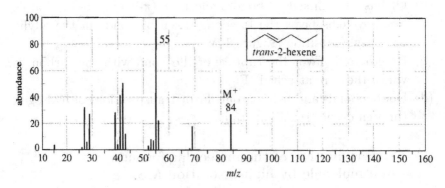

Figure 12. Mass spectra of *trans*-2-hexene.

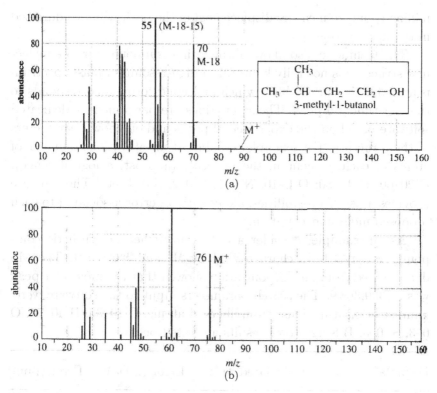

Figure 13. Mass spectra of: (a) 3-methyl-1-butanol and (b) ethylmethylsulfide.

advantageous for biological products. High-resolution MS does not require the unknown compound to be pure and it only requires the formation of a molecular ion (or a quasi molecular fragment) with sufficient intensity for determining the mass accurately. A high-resolution mass spectrometer, with resolution higher than 10000, is needed to establish the mass of an ion with four decimals, which is achieved by calibrating the scanning spectrometer with a reference substance possessing an ion in the vicinity of the ion of interest.

An example will be of help to understand the procedure: A synthetic compound has a $C_{16}H_{24}O_2$ molecular formula, theoretical mass $= 248.1771$, and molecular peak at 248.1772, the difference between the masses is 0.0001 and the error in the determination represents a deviation of $0.5\,ppm$. *Note*: the meaning of parts per million (ppm) to measure displacement of the peaks will be addressed in the next section.

The technique also allows formulating compounds whose molecular structure is not fully known. If a high-resolution mass spectrum is available, it can establish which molecular formula corresponds to the experimental peak. The simulations are performed by dedicated software based on the experimental peaks and on known parameters of the structure. For example: number of unsaturations; number of atoms of each element in the molecule (at least, a logical range): C 20–60; H 40–80; O 0–10, N 0–2, P 0–2; S 0–4, etc. The software provides a list of possibilities, along with an error associated to each proposed molecular equation.

As an example, consider a case study where a synthetic compound was not fully characterized by ^1H and ^{13}C NMR. The MS data derived from an ESI ionization showed that the molecular peak was at 399.368 9. The search parameters input to the software were: number of insaturations: 8; numbers of atoms: C 10–30, H 40–70, O 0–3, Si 0–3, B 8–12. Some results are listed below:

Formula	Calculated m/z	Error (mDa)	Error (ppm)
$C_{16}H_{41}B_{10}Si_2$	399.368 2	−2.385	−5.972
$C_{18}H_{40}B_9OSi$	399.369 1	−3.240 9	−8.115 1
$C_{11}H_{42}B_{11}O_3Si_2$	399.370 1	−4.259 2	−10.665 1
$C_{12}H_{45}B_{10}OSi_3$	399.371 4	−5.528 2	−13.842 5
$C_{14}H_{44}B_9O_2Si_2$	399.372 2	−6.384 1	−15.985 6
$C_{16}H_{43}B_8O_3Si$	399.373 1	−7.24	−18.128 8
$C_{11}H_{47}B_8O_3Si_3$	399.358 2	7.603	19.037 7
$C_{15}H_{47}B_8Si_3$	399.373 5	−7.653	−19.163
$C_{10}H_{48}B_9O_3Si_3$	399.375 4	−9.527 3	−23.856 1
$C_{15}H_{43}B_8O_2Si_2$	399.355 1	10.746 2	26.908 2
$C_{13}H_{44}B_9OSi_3$	399.354 2	11.602 1	29.051 3
$C_{12}H_{41}B_{10}O_3Si_2$	399.353	12.871	32.228 7

(*Continued*)

(*Continued*)

Formula	Calculated m/z	Error (mDa)	Error (ppm)
$C_{10}H_{42}B_{11}O_2Si_3$	399.352 1	13.726 9	34.371 9
$C_{17}H_{40}B_9Si_2$	399.351 1	14.745 3	36.921 8
$C_{15}H_{42}B_{11}Si_2$	399.385 4	−19.515 3	−48.865 9

As a consequence, the structure which had not been elucidated by NMR spectrometry was deduced by MS because the first choice of the list above (that with the lowest error) corresponds to the $[MH]^-$ molecular ion.

The use of modern software for structure elucidation is of paramount importance nowadays. To aid in the elucidation of unknown structures, 25 years ago, comprehensive lists of combinations of C, H, N and O molecular formulas, including their molecular weights were available. For example, a list covering between 12 and 250 amu (atomic mass units) occupied 20 pages with three columns each, and 55 molecular possibilities per column. Nowadays, we can take full advantage of web-based applications. The student can consult; for instance, **ChemCalc** (http://www.chemcalc.org/) [5].

Note: NMR and EPR Tables are from BRUKER Almanac, 2004.

REFERENCES

[1] Silverstein, R. M.; Bassler, G. C.; Morrill, T. C. (1991). *Spectrometric Identification of Organic Compounds*, 5th ed. John Wiley & Sons, New York, USA.

[2] Vollhardt, H. P. C.; Schore, N. E. (2000). *Organic Chemistry*, 3rd ed. W. H. Freeman & Co., New York, USA.

[3] Wade, L. G. (2006). *Organic Chemistry*, 6th ed. Pearson Prentice Hall, New Jersey, USA.

[4] Pretsch, E.; Simon, W.; Seibl, J.; Clerc T. (1989), *Tables of Spectral Data for Structure Determination of Organic Compounds*, 2nd English ed. Springer-Verlag, Berlin.

[5] Patiny, L.; Borel, A. (2013). ChemCalc: a building block for tomorrow's chemical infrastructure. *Journal of Chemical Information and Modeling*, 53(5): 1223–1228.

[6] Pouchert, C. J.; Behnke, J. (1993). *The Aldrich Library of* ^{13}C *and* ^{1}H
    FT NMR Spectra, 1st ed. Aldrich Chemical Company, Inc., USA.

Note: To get the NMR explanations, Tables 1, 2 and 3 have been
extracted from Wade's textbook; whereas Figures 4 and 6 (as
well as the NMR spectra displayed in the exercises) are from the
comprehensive Pouchert and Behnke's collection of NMR spectra.
The original MS Figures 10, 11, 12 and 13 are from Vollhardt and
Schore's book.

WORKED EXERCISES

In the following exercises, the following nomenclature will be used [6]:
s, singlet; d, doublet; t, triplet; c, quadruplet; q, quintuplet; h,
hexaplet; hp, heptaplet; oc, octaplet.

In the examples, the spectra and their interpretation are
organized in a box divided into two parts: the upper box
contains the ^{1}H-NMR, whereas the lower box contains the
^{13}C-NMR. The identification of the signals is presented below
and, finally, a graphical display of the molecule, along with an
ad-hoc interpretation of the ^{1}H-NMR and ^{13}C-NMR spectra
is given.

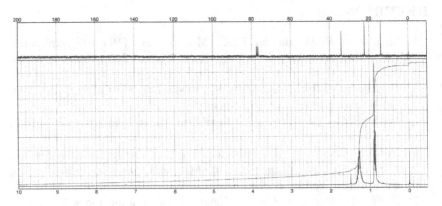

Figure 14.

1. Study the NMR spectrum of pentane (C_5H_{12}) (Figure 14).
   ^{1}H-NMR (300 MHz): δ 0.88 (6H, t, J = 6 Hz, –Me), 1.28 (6H, m) ppm.
   ^{13}C-NMR (75.43 MHz): δ 14, 23, 35 ppm.

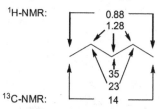

2. Study the NMR spectrum of 2-methylbutane (C_5H_{12}) (Figure 15).

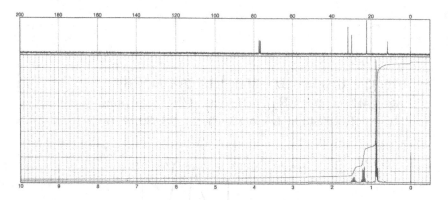

Figure 15.

^{1}H-NMR (300 MHz): δ 0.85 (3H, d, $J = 7$ Hz, –Me),
0.87 [6H, d, $J = 7$ Hz, –CH(Me)$_2$], 1.20 (2H, q, $J = 7$ Hz, CH$_2$),
1.55 [1H, oc, –CH(Me)$_2$] ppm.
^{13}C-NMR (75.43 MHz): δ 12, 22, 30, 32 ppm.

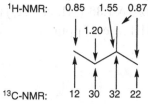

3. Study the NMR spectrum of 2-methylhexane (C_7H_{16}) (Figure 16).
   ^{1}H-NMR (300 MHz): δ 0.85 [6H, d, $J = 6$ Hz, –CH(Me)$_2$],

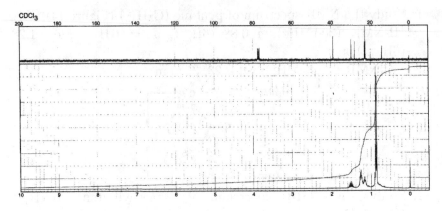

Figure 16.

0.86 [3H, d, $J = 6\,\text{Hz}$, –Me), 1.18 (2H, m, CH_2),
1.26 (4H, m, $2 \times CH_2$), [1H, oct, $-CH(Me)_2$] ppm.
^{13}C-NMR (75.43 MHz): δ 14, 23, 24, 28, 30, 39 ppm.

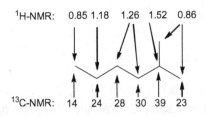

4. Study the NMR spectrum of cyclohexane (C_6H_{12}) (Figure 17).
   ^{1}H-NMR (300 MHz): δ 1.36 (1H, s, CH_2) ppm.
   ^{13}C-NMR (75.43 MHz): δ 27 ppm.

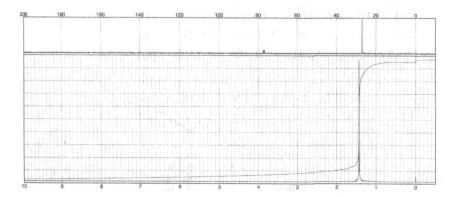

Figure 17.

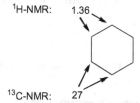

5. Study the NMR spectrum of *cis*-1,2-dimethylcyclohexane (C_8H_{16}) (Figure 18).

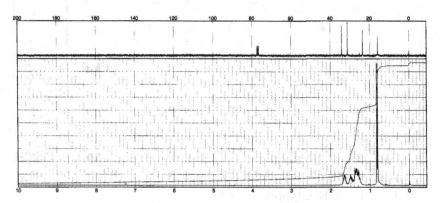

Figure 18.

^{1}H-NMR (300 MHz): δ 0.82 (3H, d, $J = 6\,\text{Hz}$, –Me),
1.35 (3H, m), 1.40 (1H, m), 1.65 (1H, m) ppm.
^{13}C-NMR (75.43 MHz): δ 16, 24, 31, 34 ppm.

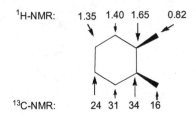

6. Study the NMR spectrum of 1-chlorobutane (C_4H_9Cl) (Figure 19).

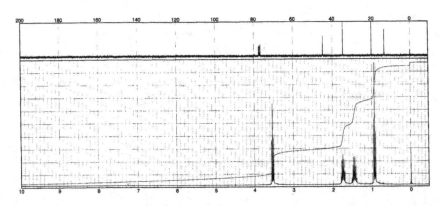

Figure 19.

^{1}H-NMR (300 MHz): δ 0.93 (3H, t, $J = 6\,\text{Hz}$, –Me), 1.45 (2H, hx, $J = 6\,\text{Hz}$, CH_2), 1.75 (2H, q, $J = 6\,\text{Hz}$, CH_2), 3.52 (2H, t, $J = 6\,\text{Hz}$, CH_2Cl) ppm.
^{13}C-NMR (75.43 MHz): δ 13, 20, 34, 55 ppm.

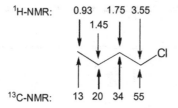

7. Study the NMR spectrum of 1-bromobutane (C_4H_9Br) (Figure 20).

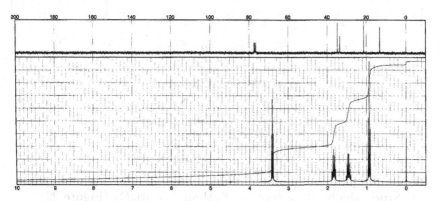

Figure 20.

^{1}H-NMR (300 MHz): δ 0.92 (3H, t, $J = 6$ Hz, $-$Me), 1.45 (2H, h, $J = 6$ Hz, CH_2), 1.85 (2H, q, $J = 6$ Hz, CH_2), 3.41 (2H, t, $J = 6$ Hz, CH_2Br) ppm.
^{13}C-NMR (75.43 MHz): δ 13, 22, 31, 53 ppm.

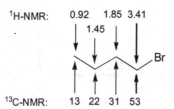

8. Study the NMR spectrum of 2-methyl-2-chloropropane (C_4H_9Cl) (Figure 21).
^{1}H-NMR (300 MHz): δ 1.95 (1H, s) ppm.
^{13}C-NMR (75.43 MHz): δ 30, 41 ppm.

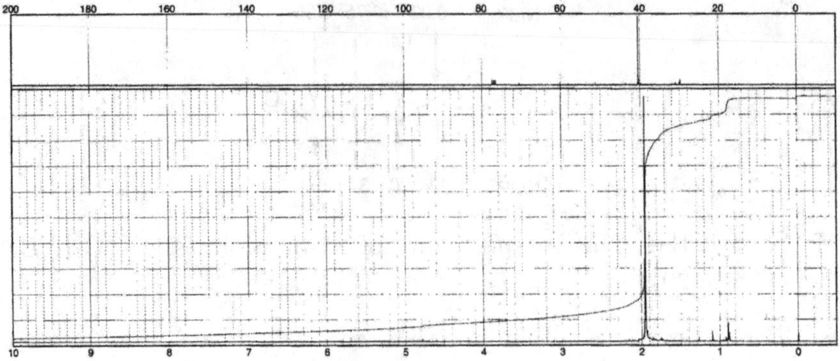

Figure 21.

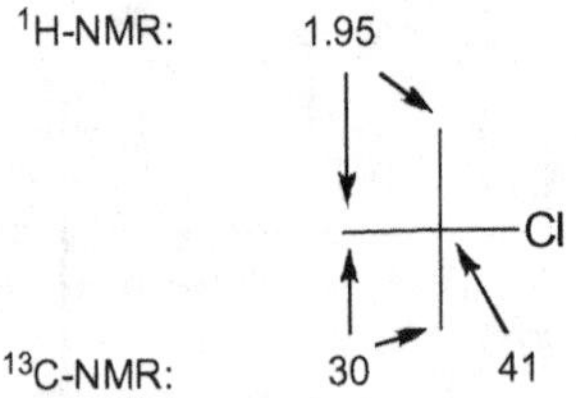

9. Study the NMR spectrum of ethanol (C_2H_6O) (Figure 22).

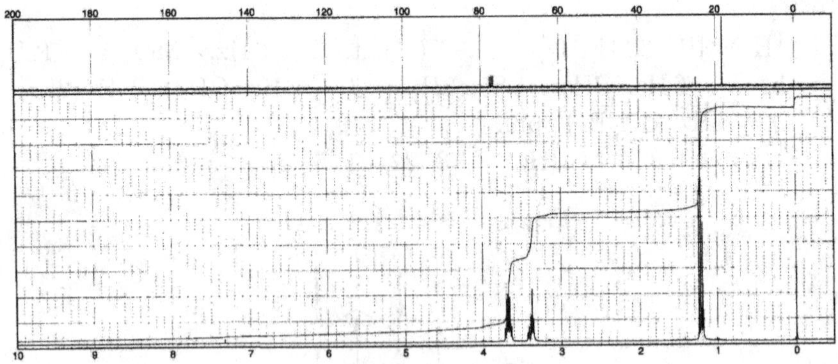

Figure 22.

^{1}H-NMR (300 MHz): δ 1.22 (3H, t, $J = 6$ Hz, CH$_3$),
3.37 (1H, t, $J = 6$ Hz, –OH), 3.68 (2H, q, $J = 6$ Hz, CH$_2$) ppm.
^{13}C-NMR (75.43 MHz): δ 18, 58 ppm.

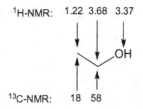

10. Study the NMR spectrum of 1-propanol (C$_3$H$_8$O) (Figure 23).

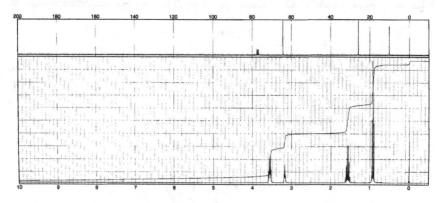

Figure 23.

^{1}H-NMR (300 MHz): δ 0.92 (3H, t, $J = 6$ Hz, CH$_3$),
1.66 (2H, hex, $J = 6$ Hz, CH$_2$), 3.18 (1H, t, $J = 6$ Hz, –OH),
3.56 (2H, c, $J = 6$ Hz, CH$_2$) ppm.
^{13}C-NMR (75.43 MHz): δ 11, 26, 64 ppm.

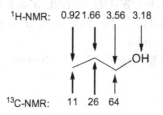

11. Study the NMR spectrum of 2-butanol ($C_4H_{10}O$) (Figure 24).

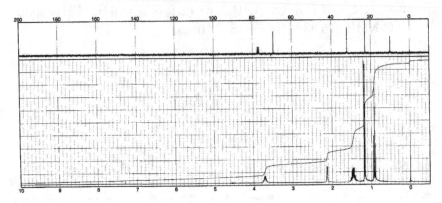

Figure 24.

^{1}H-NMR (300 MHz): δ 0.92 (3H, t, $J = 6\,Hz$, CH_3), 1.18 (3H, d, $J = 6\,Hz$, CH_3), 1.57 (2H, q, $J = 6\,Hz$, CH_2), 2.13 (1H, br s,–OH), 3.70 (1H, m, CH_2) ppm.

^{13}C-NMR (75.43 MHz): δ 11, 23, 32, 70 ppm.

^{1}H-NMR: 0.92 1.57 3.70 1.18 2.13

^{13}C-NMR: 11 32 70   23

12. Study the NMR spectrum of 2,4-dimethyl-3-pentanol ($C_7H_{16}O$) (Figure 25).

^{1}H-NMR (300 MHz): δ 0.92 [6H, d, $J = 6\,Hz$, –CH(Me)$_2$], 1.43 (1H, br s,–OH), 1.75 (2H, non, $J = 6\,Hz$, CH_3), 3.00 (1H, c, $J = 6\,Hz$, CH) ppm.

^{13}C-NMR (75.43 MHz): δ 17, 30, 82 ppm.

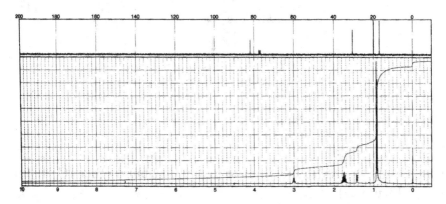

Figure 25.

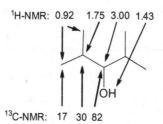

13. Study the NMR spectrum of ethoxyethane, diethyl ether ($C_4H_{10}O$) (Figure 26).

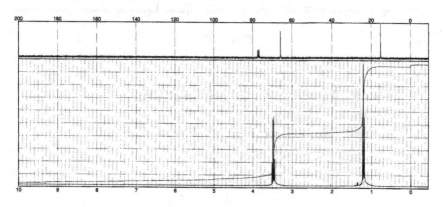

Figure 26.

^{1}H-NMR (300 MHz): δ 1.20 (3H, t, $J = 6\,Hz$, $-CH_3$), 3.46 (2H, c, $J = 6\,Hz$, CH_2) ppm.
^{13}C-NMR (75.43 MHz): δ 16, 66 ppm.

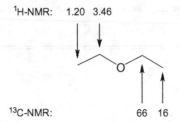

14. Study the NMR spectrum of methyl-*tert*-butyl ether ($C_5H_{12}O$) (Figure 27).

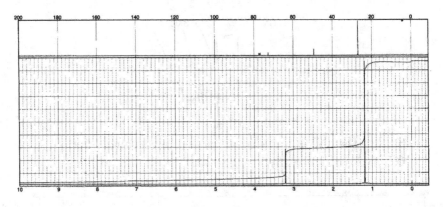

Figure 27.

^{1}H-NMR (300 MHz): δ 1.20 [9H, s, $-C(CH_3)_3$], 3.21 (1H, s, CH_3) ppm.
^{13}C-NMR (75.43 MHz): δ 27, 49, 73 ppm.

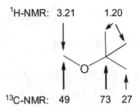

15. Study the NMR spectrum of propanone, acetone (C_3H_6O) (Figure 28).

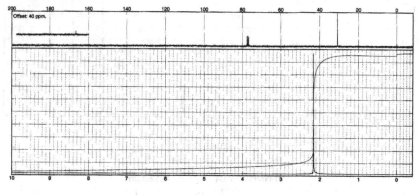

Figure 28.

^{1}H-NMR (300 MHz): δ 2.18 (3H, s, CH_3) ppm.
^{13}C-NMR (75.43 MHz): δ 31, 207 ppm.

^{1}H-NMR: 2.18

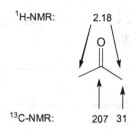

^{13}C-NMR: 207 31

16. Study the NMR spectrum of ethyl ethanoate $(C_4H_8O_2)$ (Figure 29).
^{1}H-NMR (300 MHz): δ 1.25 (3H, t, $J = 6$ Hz, CH_3),
2.04 (3H, s, CH_3), 4.12 (2H, c, $J = 6$ Hz, CH_3) ppm.
^{13}C-NMR (75.43 MHz): δ 15, 31, 60, 171 ppm.

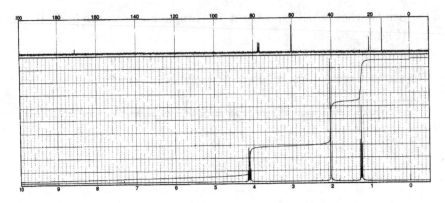

Figure 29.

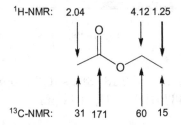

17. Study the NMR spectrum of pentanoic acid, valeric acid ($C_5H_{10}O_2$) (Figure 30).

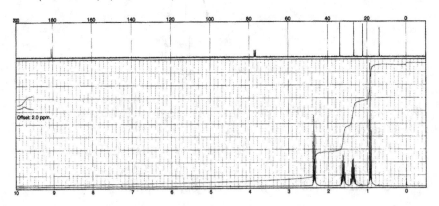

Figure 30.

^{1}H-NMR (300 MHz): δ 0.93 (3H, t, J = 6 Hz, CH$_3$),
1.38 (2H, h, CH$_2$), 1.62 (2H, q, J = 6 Hz, CH$_2$),
2.34 (2H, c, J = 6 Hz, CH$_2$), 11.75 (1H, br s, OH) ppm.
^{13}C-NMR (75.43 MHz): δ 14, 22, 27, 34, 181 ppm.

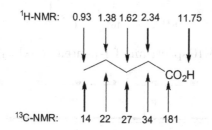

18. Study the NMR spectrum of propyl formate (C$_4$H$_8$O$_2$) (Figure 31).

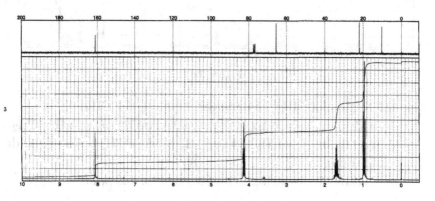

Figure 31.

^{1}H-NMR (300 MHz): δ 0.97 (3H, t, J = 6 Hz, CH$_3$),
1.70 (2H, h, J = 6 Hz, CH$_2$), 4.13 (2H, c, J = 6 Hz, CH$_2$),
8.07 (1H, s, HCO) ppm.
^{13}C-NMR (75.43 MHz): δ 11, 22, 65, 161 ppm.

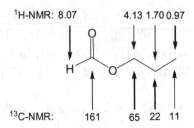

19. Study the NMR spectrum of 1-pentene (C_5H_{10}) (Figure 32).

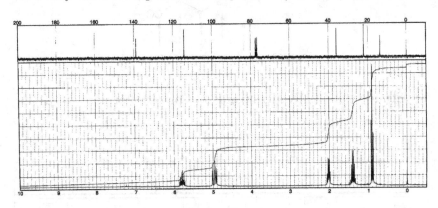

Figure 32.

^{1}H-NMR (300 MHz): δ 0.90 (3H, t, $J = 6$ Hz, CH_3), 1.40 (2H, h, $J = 6$ Hz, CH_2), 2.02 (2H, dt, $J = 6$ Hz, CH_2), 4.92 (1H, br d, $=CH_2$), 4.98 (1H, br d, $=CH_2$), 5.80 (1H, dt, $=CH$) ppm.
^{13}C-NMR (75.43 MHz): δ 14, 22, 36, 115, 139 ppm.

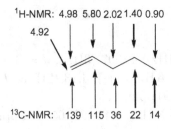

20. Study the NMR spectrum of *cis*-2-pentene (C_5H_{10}) (Figure 33).

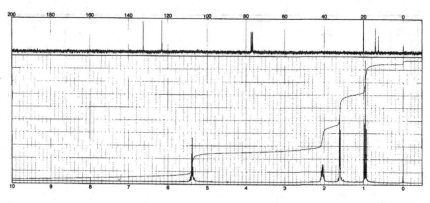

Figure 33.

^{1}H-NMR (300 MHz): δ 0.96 (3H, t, $J = 6$ Hz, CH_3),
1.46 (3H, d, $J = 6$ Hz, CH_3), 2.03 (2H, m, CH_2), 5.38 (3H, m,
$=CH$) ppm.
^{13}C-NMR (75.43 MHz): δ 13, 14, 20, 123, 133 ppm.

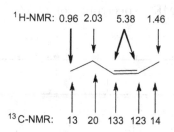

21. Study the NMR spectrum of *trans*-2-pentene (C_5H_{10})
(Figure 34).
^{1}H-NMR (300 MHz): δ 0.95 (3H, t, $J = 6$ Hz, CH_3),
1.63 (3H, m, CH_3), 1.98 (2H, m, CH_2), 5.44 (3H, m, $=CH$) ppm.
^{13}C-NMR (75.43 MHz): δ 14, 18, 26, 124, 133 ppm.

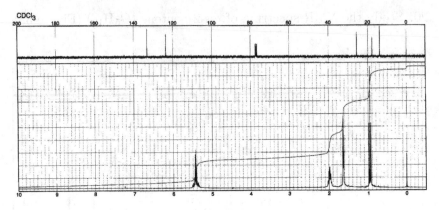

Figure 34.

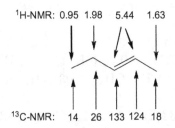

¹H-NMR: 0.95 1.98 5.44 1.63

¹³C-NMR: 14 26 133 124 18

22. Study the NMR spectrum of 2-ethyl-1-butene (C_6H_{12}) (Figure 35).

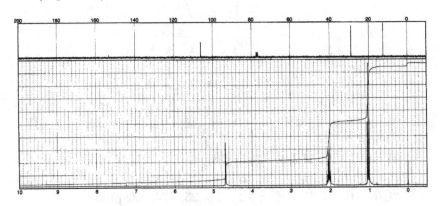

Figure 35.

^{1}H-NMR (300 MHz): δ 1.24 (3H, t, $J = 6$ Hz, CH$_3$),
2.04 (2H, c, $J = 6$ Hz, CH$_2$), 4.69 (1H, m, =CH$_2$) ppm.
^{13}C-NMR (75.43 MHz): δ 13, 29, 106, 154 ppm.

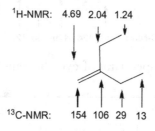

23. Study the NMR spectrum of *cis*-2-*trans*-4-hexadiene (C$_6$H$_{10}$) (Figure 36).

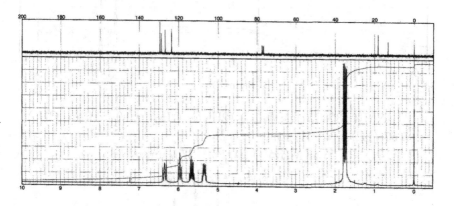

Figure 36.

^{1}H-NMR (300 MHz): δ 1.72 (3H, t, $J = 6$ Hz, CH$_3$),
1.77 (3H, t, $J = 6$ Hz, CH$_3$), 5.35 (1H, c, =CH),
5.65 (1H, dc, =CH), 5.95 (1H, t, =CH), 6.35 (1H, dd, =CH) ppm.
^{13}C-NMR (75.43 MHz): δ 13, 18, 124, 127, 129, 130 ppm.

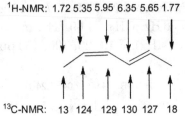

24. Study the NMR spectrum of cyclohexene (C_6H_{10}) (Figure 37).

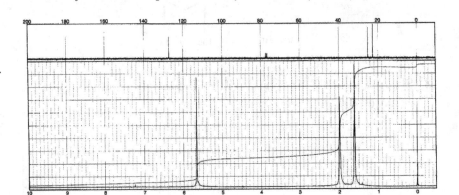

Figure 37.

^{1}H-NMR (300 MHz): δ 1.61 (2H, m, CH_2), 1.99 (2H, m, CH_2), 5.67 (1H, s, =CH) ppm.
^{13}C-NMR (75.43 MHz): δ 23, 25, 127 ppm.

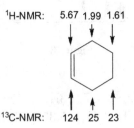

25. Study the NMR spectrum of 3-methyl-1-cyclohexene (C_7H_{12}) (Figure 38).

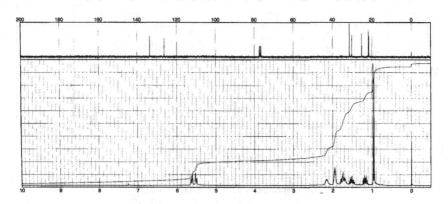

Figure 38.

^{1}H-NMR (300 MHz): δ 0.96 (3H, d, $J = 6$ Hz, CH_3), 1.18 (1H, m), 1.51 (1H, m), 1.75 (2H, m), 1.95 (2H, m), 2.15 (1H, m), 5.51 (1H, dd, =C), 5.62 (1H, ddd, =C) ppm.
^{13}C-NMR (75.43 MHz): δ 21, 22, 25, 30, 31, 126, 133 ppm.

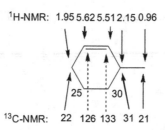

^{1}H-NMR: 1.95 5.62 5.51 2.15 0.96

^{13}C-NMR: 22 126 133 31 21

26. Study the NMR spectrum of 1-pentyne (C_5H_8) (Figure 39).
^{1}H-NMR (300 MHz): δ 1.00 (3H, t, $J = 6$ Hz, CH_3), 1.55 (2H, h, $J = 2$ Hz, CH_2), 1.95 (1H, t, $J = 2$ Hz, $\equiv$CH), 2.14 (2H, dd, $J = 2$ and 6 Hz, CH_2) ppm.
^{13}C-NMR (75.43 MHz): δ 13, 21, 23, 68, 85 ppm.

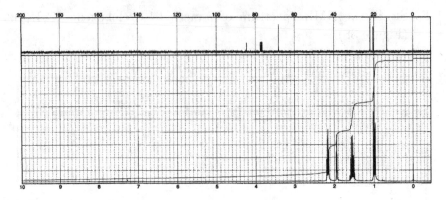

Figure 39.

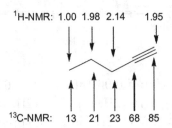

¹H-NMR: 1.00 1.98 2.14   1.95

¹³C-NMR: 13   21   23   68 85

27. Study the NMR spectrum of 2-hexyne (C_6H_{10}) (Figure 40).

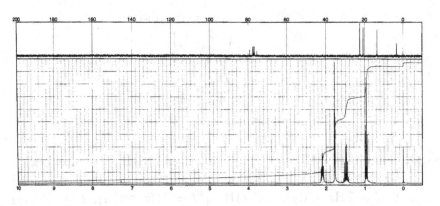

Figure 40.

^{1}H-NMR (300 MHz): δ 0.96 (3H, t, J = 6 Hz, CH$_3$), 1.50 (2H, h, J = 6 Hz, CH$_2$), 1.78 (3H, t, J = 6 Hz, CH$_3$), 2.10 (2H, m, CH$_2$) ppm.

^{13}C-NMR (75.43 MHz): δ 3, 13, 22, 24, 75, 79 ppm.

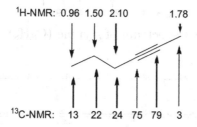

28. Study the NMR spectrum of *iso*-propylbenzene, cumene (C$_9$H$_{12}$) (Figure 41).

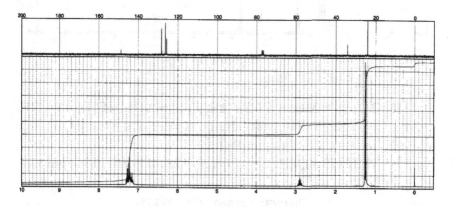

Figure 41.

^{1}H-NMR (300 MHz): δ 1.25 (3H, d, J = 6 Hz, CH$_3$), 2.89 (1H, oc, J = 6 Hz, CH$_2$), 7.21 (5H, m, Ph) ppm.

^{13}C-NMR (75.43 MHz): δ 24, 34, 126, 127, 129, 149 ppm.

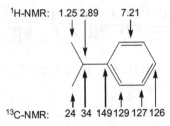

29. Study the NMR spectrum of styrene (C_8H_8) (Figure 42).

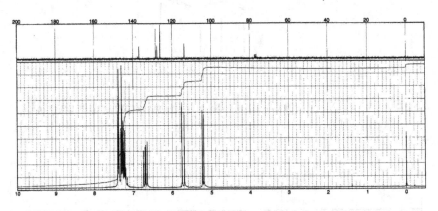

Figure 42.

^{1}H-NMR (300 MHz): δ 5.22 (1H, d, $J = 8$ Hz, =CH), 5.74 (1H, d, $J = 12$ Hz, =CH), 6.60 (1H, dd, $J = 12$ and 8 Hz, =CH), 7.25 (3H, m, Ph), 7.38 (2H, m, Ph) ppm.
^{13}C-NMR (75.43 MHz): δ 114, 126, 128, 129, 137, 138 ppm.

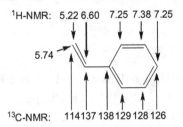

30. Study the NMR spectrum of allylbenzene (C_9H_{10}) (Figure 43).

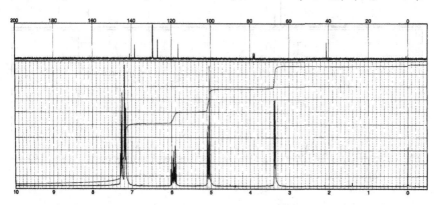

Figure 43.

^{1}H-NMR (300 MHz): δ 3.38 (1H, d, J = 6 Hz, CH_2), 5.05 (2H, m, $=CH_2$), 5.90 (1H, dt, J = 12 and 8 Hz, $=CH$), 7.18 (3H, m, Ph), 7.35 (2H, m, Ph) ppm.
^{13}C-NMR (75.43 MHz): δ 42, 116, 127, 130, 138, 141 ppm.

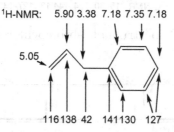

31. Study the NMR spectrum of 1-bromo-2-ethylbenzene (C_8H_9Br) (Figure 44).
^{1}H-NMR (300 MHz): δ 1.21 (3H, t, J = 6 Hz, CH_3),
2.74 (2H, d, J = 6 Hz, CH_2), 7.03 (1H, m, Ph),
7.20 (2H, m, Ph), 7.50 (1H, d, Ph) ppm.
^{13}C-NMR (75.43 MHz): δ 15, 30, 124, 127, 130, 133, 144 ppm.

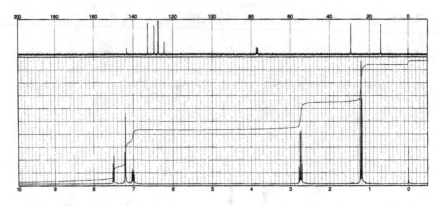

Figure 44.

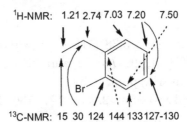

32. Study the NMR spectrum of 3-bromochlorobenzene (C_6H_4BrCl) (Figure 45).

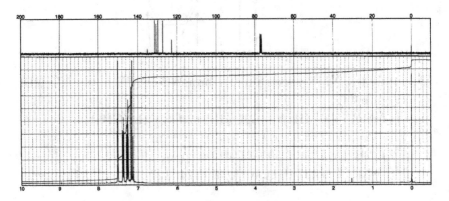

Figure 45.

^{1}H-NMR (300 MHz): δ 7.15 (1H, t, $J = 6$ Hz, Ph), 7.25 (1H, $J = 5$ and 3 Hz, dd, Ph), 7.37 (1H, dd, $J = 5$ and 4 Hz, Ph), 7.51 (1H, s, Ph) ppm.

^{13}C-NMR (75.43 MHz): δ 123, 127, 130, 131, 132, 135 ppm.

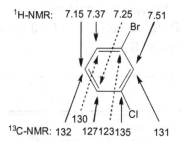

33. Study the NMR spectrum of 2-iodoanisole (C_7H_7BrO) (Figure 46).

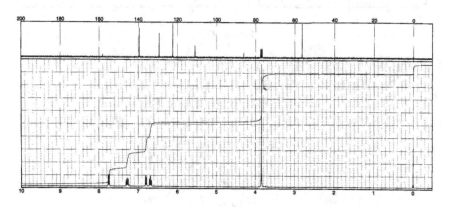

Figure 46.

^{1}H-NMR (300 MHz): δ 3.84 (3H, s, OCH_3), 6.69 (1H, t, $J = 5$ Hz, Ph), 6.80 (1H, d, $J = 4$ Hz, Ph), 7.28 (1H, td, $J = 5$ and 4 Hz, Ph), 7.75 (1H, d, $J = 4$ Hz, Ph) ppm.

^{13}C-NMR (75.43 MHz): δ 56, 86, 111, 123, 130, 139, 158 ppm.

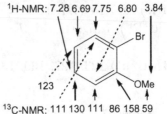

¹H-NMR: 7.28 6.69 7.75 6.80   3.84

¹³C-NMR: 111 130 111 86 158 59

34. Study the NMR spectrum of 3-chloro-5-methoxyphenol ($C_7H_7Cl_2O$) (Figure 47).

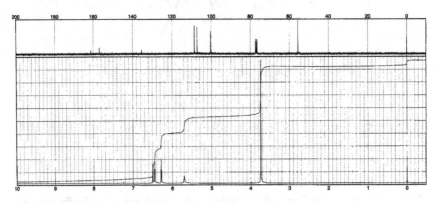

Figure 47.

¹H-NMR (300 MHz): δ 3.75 (3H, s, OCH₃),
5.70 (1H, br s, –OH), 6.29 (1H, t, $J = 5$ Hz, Ph),
6.45 (1H, t, $J = 4$ Hz, Ph), 6.50 (1H, s, Ph) ppm.
¹³C-NMR (75.43 MHz): δ 56, 100, 107, 109, 136, 157, 161 ppm.

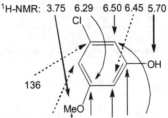

¹H-NMR: 3.75 6.29 6.50 6.45 5.70

¹³C-NMR: 109 56 161 100 157 106

35. Study the NMR spectrum of 4-butylaniline ($C_{10}H_{15}N$) (Figure 48).
¹H-NMR (300 MHz): δ 0.90 (3H, t, $J = 6$ Hz, CH₃),
1.32 (2H, h, $J = 6$ Hz, CH₂), 1.53 (2H, q, $J = 6$ Hz, CH₂),

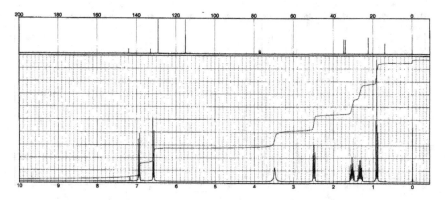

Figure 48.

2.48 (1H, t, $J = 6\,\text{Hz}$, CH_2), 3.48 (2H, br s, $-NH_2$),
6,59 (2H, d, $J = 8\,\text{Hz}$, Ph), 6.95 (2H, d, $J = 8\,\text{Hz}$, Ph) ppm.
^{13}C-NMR (75.43 MHz): δ 14, 22, 34, 35, 115, 129, 133, 144 ppm.

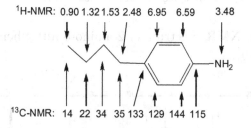

36. Study the NMR spectrum of 3-nitrostyrene ($C_8H_7O_2N$)
(Figure 49).
^{1}H-NMR (300 MHz): δ 5.42 (1H, d, $J = 8\,\text{Hz}$, $=CH_2$), 5.88 (1H,
d, $J = 12\,\text{Hz}$, $=CH_2$), 6.75 (1H, d, $J = 12$ and $8\,\text{Hz}$, $=CH$), 7.46
(1H, d, $J = 8\,\text{Hz}$, Ph), 7.7 (1H, d, $J = 6\,\text{Hz}$, Ph), 8.07 (1H, dd,
$J = 8$ and $6\,\text{Hz}$, Ph), 8.21 (1H, s, Ph) ppm.
^{13}C-NMR (75.43 MHz): δ 117, 121, 123, 129, 132, 135, 139,
148 ppm.

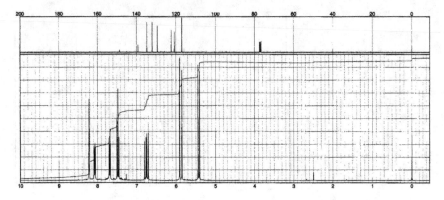

Figure 49.

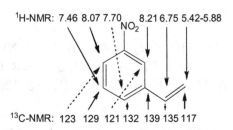

¹H-NMR: 7.46 8.07 7.70   8.21 6.75 5.42-5.88

¹³C-NMR: 123 129 121 132 139 135 117

37. Study the NMR spectrum of 2-amino-5 nitrophenol ($C_6H_6O_3N$) (Figure 50).

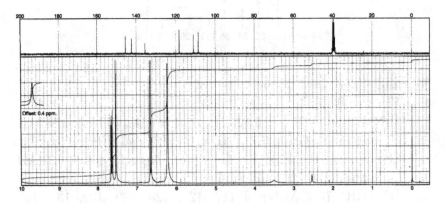

Figure 50.

^{1}H-NMR (300 MHz): δ 6.21 (2H, d, NH$_2$), 6.64 (1H, d, $J = 8$ Hz, Ph), 7.53 (1H, s, Ph), 7.64 (1H, dd, $J = 8$ and 6 Hz, Ph), 10.3 (1H, s, OH) ppm.
^{13}C-NMR (75.43 MHz): δ 109, 111, 118, 135, 142, 145 ppm.

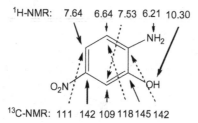

EXERCISES PROPOSED TO THE STUDENT

In all cases, the objective is to deliver the structure of the molecule that, likely, gives rise to the spectra depicted in each exercise.

Do not look at the solution before you try solving the exercise. Despite how tough the temptation may appear, be strong and resist it because, otherwise, you might be too confident on your interpretation capabilities. Just hide the solution with a piece of paper!

Compound (a) C$_6$H$_{14}$ (Figure 51)

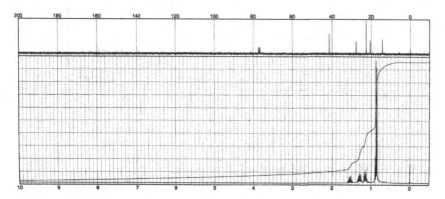

Figure 51.

Solution: 2-Methylpentane

Compound (b) C_7H_{16} (Figure 52)

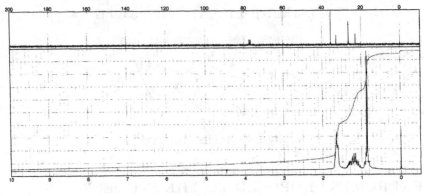

Figure 52.

Solution: 3-Methylhexane

Compound (c) C_7H_{14} (Figure 53)

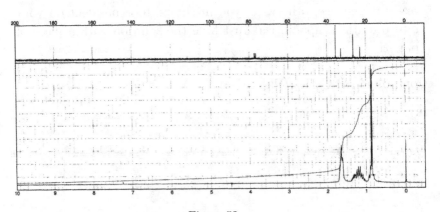

Figure 53.

Solution: Methylcyclohexane

Compound (d) C_4H_9Cl (Figure 54)

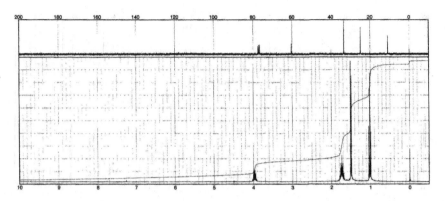

Figure 54.

Solution: 2-Chlorobutane

Compound (e) C_4H_9Br (Figure 55)

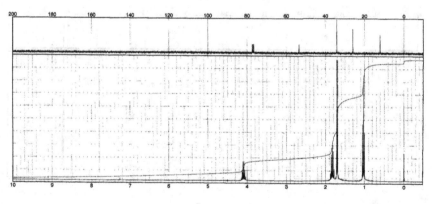

Figure 55.

Solution: 2-Bromobutane

Compound (f) $C_4H_{10}O$ (Figure 56)

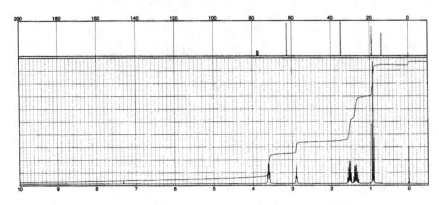

Figure 56.

Solution: 1-Butanol

Compound (g) $C_5H_{12}O$ (Figure 57)

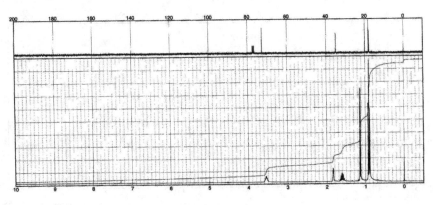

Figure 57.

Solution: 3-Methyl-2-butanol

Compound (h) $C_7H_{16}O$ (Figure 58)

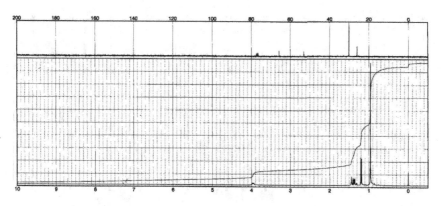

Figure 58.

Solution: 4,4-Dimethyl-2-pentanol

Compound (i) $C_2H_4O_2$ (Figure 59)

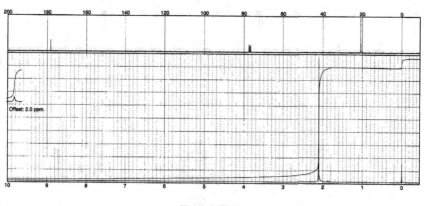

Figure 59.

Solution: Ethanoic acid and acetic acid

Compound (j) $C_9H_{22}O_2$ (Figure 60)

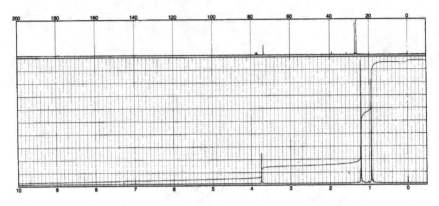

Figure 60.

Solution: Neopentyl pivalate

Compound (k) C_6H_{12} (Figure 61)

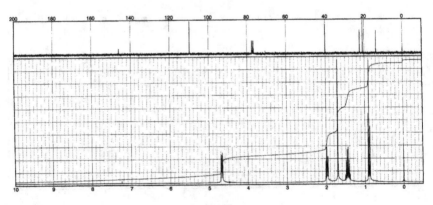

Figure 61.

Solution: 2-Methyl-1-pentene

Compound (l) C_6H_{12} (Figure 62)

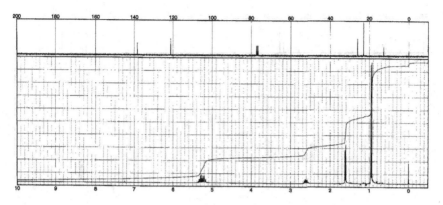

Figure 62.

Solution: *cis*-4-Methyl-2-pentene

Compound (m) C_6H_{10} (Figure 63)

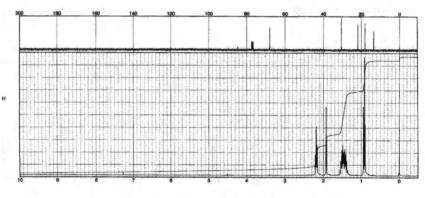

Figure 63.

Solution: 1-Hexyne

Compound (n) C_8H_{10} (Figure 64)

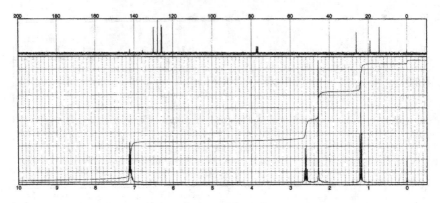

Figure 64.

Solution: 2-Ethyltoluene

Compound (o) $C_6H_4Br_2$ (Figure 65)

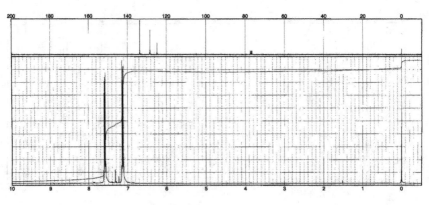

Figure 65.

Solution: 1,2-Dibromobenzene

Compound (p) $C_8H_{10}O$ (Figure 66)

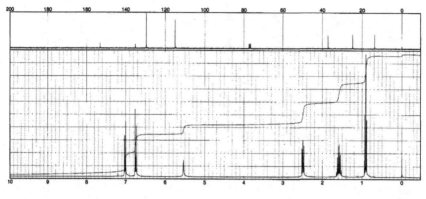

Figure 66.

Solution: 4-Propylbenzene

INDEX

Printed in the United States
By Bookmasters